Mathematical Learning and Cognition in Early Childhood

Katherine M. Robinson · Helena P. Osana ·
Donna Kotsopoulos
Editors

Mathematical Learning and Cognition in Early Childhood

Integrating Interdisciplinary Research into Practice

Springer

Editors
Katherine M. Robinson
Department of Psychology
University of Regina
Regina, SK, Canada

Helena P. Osana
Department of Education
Concordia University
Montreal, QC, Canada

Donna Kotsopoulos
Department of Mathematics
Wilfred Laurier University
Waterloo, ON, Canada

ISBN 978-3-030-12894-4 ISBN 978-3-030-12895-1 (eBook)
https://doi.org/10.1007/978-3-030-12895-1

Library of Congress Control Number: 2019934797

This Springer imprint is published by the registered company Springer Nature Switzerland AG
The registered company address is: Gewerbestrasse 11, 6330 Cham, Switzerland

Foreword

Over a decade ago, a group of researchers set out to identify the factors in early childhood development that are key predictors of school readiness (Duncan et al., 2007). By combining data from existing, large-scale, longitudinal data sets, they were able to estimate how strongly early math, early reading, early attention, and early social skills predicted later academic achievements. What came as a surprise to many at the time, and resulted in a significant public attention, was their finding that early math skills were not only a stronger predictor of later math skills, but also a robust predictor of children's later reading. Furthermore, Duncan et al.'s meta-analysis of all the reported relationships between early competencies and later skills revealed that overall, early math was the strongest predictor of later academic performance. These findings had a huge impact on the field because they brought into sharp focus the importance of early math skills and provoked greater attention to the study of how children learn mathematics from an early age. Furthermore, these results also lead to a greater level of interest in how to use evidence from research coming from a diversity of fields, including cognitive psychology, educational psychology, cognitive science, and cognitive neuroscience, to design evidence-based programs that foster mathematical skills and understanding in young children.

In the 11 years that have passed since Duncan et al.'s seminal finding, many researchers with different theoretical and methodological backgrounds have focused their efforts to better characterize the mathematical minds of young children. Doing so has led to the development of approaches for the assessment and characterization of children's early understanding of mathematics and the use of resulting knowledge to find ways to optimally foster children's mathematical skills and understanding to set them on a trajectory of learning and growth (Bailey et al., 2017).

Now is an optimal time to take stock of what fruits this period of research and application of research has brought to bear. We have additionally obtained a better understanding of the questions that remain unanswered, the novel avenues that have emerged for research and mathematics education, and the directions that should be the focus going forward. The present edited volume entitled "Early Mathematical Minds" does exactly this and more. The present volume is edited by three esteemed scholars of child development and mathematics education: Katherine M. Robinson,

Helena P. Osana, and Donna Kotsopoulos. These editors have brought together a well-regarded group of scholars who have contributed chapters that represent an accessible, rich, diverse, and interdisciplinary synthesis of what we currently know about the mathematical minds of young children. This volume is a must-read for those seeking a broad overview of recent advances in our understanding of what factors contribute to the successful development of young mathematical minds and how to best foster early math skills and understanding. The contributions are written in accessible language and thus are suitable for a multidisciplinary readership, ranging from educators and educational policy makers to undergraduate and graduate students as well as researchers studying the emergence of mathematical minds.

The contributions within this volume are reflective of the breadth and complexity of research on young children's mathematical minds. By addressing such topics as spatial thinking, computational thinking, the relationship between proportional reasoning and fractions, and spatial and mathematical language spoken in the home environment, the present volume sets itself apart from related books by going beyond a sole focus on factors that influence the development of mathematical minds. The collection offers perspectives on what constitutes effective ways of screening young children's mathematical skills and deeper understandings of how best to intervene in early development in diverse educational settings, including language immersion classrooms. Furthermore, the contributions cover important and widely debated subjects such as the role of gender in mathematics, the role played by manipulatives in early math education, and the potential of technology as a support for early math learning, both in the classroom and in the home environment. By integrating contributions that focus on the latest insights from empirical research into how children develop mathematical minds with explorations of how to best foster this development, the present volume successfully traverses the bridge between basic research on children's early development of mathematical skills, on the one hand, and understanding the application of that research to create both formal and informal learning environments to optimally support and engage young mathematical minds on the other.

Daniel Ansari
Department of Psychology and
Faculty of Education Western University
London, Ontario

References

Bailey, D., Duncan, G. J., Odgers, C. L., & Yu, W. (2017). Persistence and fadeout in the impacts of child and adolescent interventions. *Journal of Research on Educational Effectiveness*, *10*, 7–39. https://doi.org/10.1080/19345747.2016.1232459

Duncan, G. J., Dowsett, C. J., Claessens, A., Magnuson, K., Huston, A. C., Klebanov, P., … Japel, C. (2007). School readiness and later achievement. *Developmental Psychology*, *43*(6), 1428–1446. https://doi.org/2007-16709-012[pii]; https://doi.org/10.1037/0012-1649.43.6.1428

Contents

Editors and Contributors

About the Editors

Katherine M. Robinson, Ph.D. is Professor in the Department of Psychology at Campion College at the University of Regina, Saskatchewan. She received her B.A. from Bishop's University and M.Sc. and Ph.D. from the University of Alberta. Her research focuses on cognitive development with special emphases on the development of conceptual and procedural knowledge in arithmetic in both children and adults.

Helena P. Osana, Ph.D. is Professor in the Department of Education and Research Chair in Mathematical Cognition and Instruction at Concordia University, Montreal, Quebec. She received her B.Sc. from McGill University, M.A. from the University of British Columbia, and Ph.D. from the University of Wisconsin–Madison. She conducts research on the instructional factors that impact children's mathematical thinking. Her recent work centers on the impacts of external knowledge representations, including manipulatives, on children's learning and transfer in elementary mathematics. She is Member of the Center for Research in Human Development at Concordia University.

Donna Kotsopoulos, Ph.D. is Professor in the Faculty of Arts and Social Science at Huron University (Western University), London, Canada. She received her B.A. and B.Ed. at York University, and her M.Ed. and Ph.D. at the University of Western Ontario. Her research focuses on early mathematical cognition and learning, and organizational behavior. Her research has resulted in social innovation with the LittleCounters® program—a community and early child care center, play-based early numeracy program. She is Director of the Mathematical Brains Laboratory, and she is Co-Director of the Mathematics Knowledge Network (Ministry of Education, Ontario).

Contributors

Sabrina Shajeen Alam Department of Educational and Counselling Psychology, McGill University, Montreal, QC, Canada

Corey Brady Department of Teaching and Learning, Vanderbilt University, Nashville, TN, USA

Chad Buckland Department of Psychology, Western University, London, ON, Canada

Bert De Smedt Parenting and Special Education Research Unit, KU Leuven, Leuven, Belgium

Kim Desmarais Department of Psychology, Concordia University, Montreal, QC, Canada

Adam K. Dubé Department of Educational and Counselling Psychology, McGill University, Montreal, QC, Canada

Cheryll L. Fitzpatrick Department of Psychology, Memorial University of Newfoundland, St. John's, NL, Canada

Lisa Floyd Faculty of Education, Western University, London, ON, Canada

Darcy Hallett Department of Psychology, Memorial University of Newfoundland, St. John's, NL, Canada

Sarah Hodgins Department of Psychology, Wilfrid Laurier University, Waterloo, ON, Canada

Gulsah Kacmaz Department of Educational and Counselling Psychology, McGill University, Montreal, QC, Canada

Donna Kotsopoulos Huron University (at Western University), London, ON, Canada

Anne Lafay Department of Education, Concordia University, Montreal, QC, Canada

Joanne Lee Department of Psychology, Wilfrid Laurier University, Waterloo, ON, Canada

Sandra M. Linder Early Childhood Mathematics Education, College of Education, Clemson University, Clemson, SC, USA

Samantha Makosz Faculty of Education, Wilfrid Laurier University, Waterloo, ON, Canada

José Manuel Martínez Michigan State University, East Lansing, MI, USA

Michael Moes Department of Psychology, King's University College at Western University, London, ON, Canada

Vivian Nelson Faculty of Science, Wilfrid Laurier University, Waterloo, ON, Canada

Helena P. Osana Department of Education, Concordia University, Montreal, QC, Canada

Marcie Penner Department of Psychology, King's University College at Western University, London, ON, Canada

Nicole Pitsolantis Department of Education, Concordia University, Montreal, QC, Canada

Katherine M. Robinson Department of Psychology, Campion College at the University of Regina, Regina, SK, Canada

Serife Sevinc Department of Mathematics and Science Education, Middle East Technical University, Ankara, Turkey

Joke Torbeyns Centre for Instructional Psychology and Technology, KU Leuven, Leuven, Belgium

Lieven Verschaffel Centre for Instructional Psychology and Technology, KU Leuven, Leuven, Belgium

Run Wen Department of Educational and Counselling Psychology, McGill University, Montreal, QC, Canada

Nore Wijns Centre for Instructional Psychology and Technology, KU Leuven, Leuven, Belgium

Eileen Wood Department of Psychology, Wilfrid Laurier University, Waterloo, ON, Canada

Chu Xu Department of Educational and Counselling Psychology, McGill University, Montreal, QC, Canada

Joanna Zambrzycka Department of Applied Psychology and Human Development, Ontario Institute for Studies in Education, University of Toronto, Toronto, ON, Canada

Part I
Infancy and Preschool

Early Mathematical Minds: Interdisciplinary Perspectives on Early Mathematical Learning and Cognition

Donna Kotsopoulos, Katherine M. Robinson and Helena P. Osana

Supporting the mathematical development and learning of children is complex and involves multiple stakeholder groups including parents, caregivers, early childhood educators, teachers, researchers, and policymakers. Adding to the complexity is the reality that diverse research disciplines inform conversations about children's mathematical understanding and learning. Research informing mathematical cognition and learning stems from numerous disciplines and different methodological and theoretical traditions, including education, psychology, educational psychology, cognitive science, mathematics, and neuroscience. These diverse research traditions are often constructed for different audiences, for different purposes, and independently of one another, often resulting in siloed research.

It is our view that research, methods, and theories from different disciplines can complement each other to advance children's mathematics learning. An interdisciplinary approach may result in the creation and validation of approaches that are best suited to support the learning of mathematics, perhaps more so than research typically aligned with traditional psychological methods and theories (Popescu, 2014). Popescu (2014) proposed that it becomes difficult to capture the phenomena under investigation when researchers do not collaborate or communicate fully. It could be argued that educational psychology rests at the intersection between psychology (including its broad range of subfields) and education, and as such, can serve as a vehicle for such interdisciplinary collaboration.

There is also the enduring divide between research and practice (Farley-Ripple, May, Karpyn, & Tilley, 2018; Penuel, Allen, Coburn, & Farrell, 2015). Farley-Ripple

D. Kotsopoulos (✉)
Huron University (at Western University), London, ON, Canada
e-mail: dkotsopo@uwo.ca

K. M. Robinson
Department of Psychology, Campion College at the University of Regina, Regina, SK, Canada

H. P. Osana
Department of Education, Concordia University, Montreal, QC, Canada

K. M. Robinson et al. (eds.), *Mathematical Learning and Cognition in Early Childhood*,
https://doi.org/10.1007/978-3-030-12895-1_1

and colleagues (2018) described the problem as a "bidirectional problem in which characteristics of both the research and practice communities must be understood and addressed to strengthen ties between research and practice in education" (p. 235). While numerous hypotheses have been proposed to explain the lack of uptake of research evidence in the practices of parents, early childhood educators, and teachers (Lysenko, Abrami, Bernard, Dagenais, & Janosz, 2014), little consideration has been given to the role the disciplinary tenets within silos have had on the chasm between research and practice.

Educational psychology has evolved over the last 50 years. The original tenets of educational psychology were based on cognitive psychology (Derry, 1992; Mayer, 1993). Although much of the field still has its roots there, the field itself has become more situated in nature and directed toward the lives and experiences of individuals in school settings (e.g., discourse processes in mathematics classrooms, teachers' day-to-day practices, students' interpretations of mathematical representations). The degree of application may differ from one study to another, but many educational psychologists conduct research in schools through close collaboration and consultation with practitioners, such as school personnel.

Given the concerns about research uptake by school practitioners, mere collaboration may be insufficient. Investigations of problems of practice that emerge *from* classrooms are less common in educational psychology. Yet, such a shift may facilitate the bidirectional flow that Farley-Ripple and colleagues (2018) described because it would require that the experiences and expertise of partners, such as researchers and school personnel, mutually inform the questions that are asked and the way studies are developed. In this way, educators may claim more ownership of the solutions that emerge from their experiences and practices.

The main objective of this book is to explore early mathematical learning and cognition from interdisciplinary perspectives. The book aims to make a holistic contribution to understanding the conditions under which children sharpen and extend their mathematical thinking in a variety of settings. We are particularly proud of the contributions to this edited volume. They reflect interdisciplinary perspectives that are, in our view, appropriate springboards for conversations about bidirectionality in educational research. The studies intentionally borrow from different disciplines, but also reside in both research and practice, with clear implications for practice—whether that is in home settings, preschool settings or schools.

The three sections of the book are organized around developmental periods in cognition: (1) infancy and preschool; (2) the beginning of formal schooling; and (3) elementary education. Our first section, infancy, and preschool, begins with this chapter and than a chapter by Makosz, Zambryzcka, and Kotsopoulos, who explore what they define as the "girl crisis." Early childhood origins of gender differences and future mathematics learning and participation are explored. The authors draw from the extant literature, including their own. Evidence is explored to consider whether different patterns of participation are motivated by cognitive, behavioral, attitudinal, or socialized differences. Of considerable interest are their reflections on the role of males, including fathers, in what we understand about gender differences and the learning of mathematics. Their analysis suggests a lack of conclusiveness about gender gaps and a fade effect showing that these gender differences diminish as children progress through the school years.

In Chap. 3, Lee, Hodgins, and Wood address spatial learning, technology, and the role of parental talk. With technology-enhanced toys being widely available to parents, this contribution is timely. The authors introduce the important differences in spatial talk during technology-based play versus play with traditional manipulatives and discuss ways to capitalize on the affordances offered by each to enhance children's spatial learning.

In Chap. 4, Linder describes a five-week take-home "mathematics bags" intervention designed to support and increase mathematics play interactions between parents and preschool-aged children. The bags focused on various mathematics strands and were designed to encourage mathematical inquiry in the home. Linder found high levels of engagement and interest for both children and parents. Moreover, the intervention demonstrated to other practitioners the efficacy and feasibility of such an initiative.

Our second section, addressing the beginning of formal schooling, starts with a contribution from Penner, Buckland, and Moes, who explore early identification and interventions for kindergarten children at risk of mathematical difficulty. Using research from longitudinal studies, these authors identified cognitive predictors of numeracy skills and then identified evidence-based early screening tools for teachers and researchers. This interdisciplinary work illustrates how such interventions can happen in classrooms, with the ultimate goal of improving the long-term outcomes of students.

In Chap. 6, Kotsopoulos, Floyd, Nelson, and Makosz examine the differences between mathematical thinking and computational thinking. As the authors point out, many young children begin school as significant users of technology and as such, computational thinking is a topic of great interest among educators and innovators alike. To explore differences between mathematical and computational thinking, kindergarten children's free play was examined. Instances of free play that were viewed as computational thinking were captured by teachers using digital devices and then subsequently analyzed collaboratively by the teachers and the researchers. Considerable overlaps between mathematical and computational thinking were discovered, and teachers also enhanced their understanding of the distinctions between both types of thinking.

In Chap. 7, Osana and Pitsolantis explore the meaningful use of manipulatives in kindergarten. These authors investigate the instructional conditions that support the development of children's dual representation of manipulatives and the moderating effects of prior numeracy knowledge. An important finding was that children with higher prior knowledge were more successful at transferring their learning between different types of tasks and demonstrated superior performance on an application task.

In Chap. 8, Sevinc and Brady share classroom-based research that involved a three-phased instructional cycle: (a) narrative introduction, (b) model development, and (c) model sharing. The aim was to explore the extent to which activities elicited model representations of length and area. This chapter illustrates how young learners are capable of developing models representing length and area through the use of story, inquiry, and collaborative activity conducted in whole-class and small-group settings.

In Chap. 9, Wijns, Torbeyns, De Smedt, and Verschaffel investigate patterning in preschool and kindergarten settings. Patterning is proposed to have an important role in children's mathematical development. The authors share recent research, compare different definitions and operationalizations of patterning, and elaborate on the association between mathematical patterning abilities and other domain-specific and domain-general cognitive abilities. Finally, interventions aimed at stimulating patterning abilities in young children are explored.

The final section of the book focuses on the elementary school years and begins with Chap. 10. Robinson explores the importance of arithmetic concepts and how they fit with children's knowledge of arithmetic facts and arithmetic problem-solving procedures. A review of how concepts are assessed by researchers is presented as well as current research on the development of six specific arithmetic concepts: identity, negation, commutativity, inversion, associativity, and equivalence. Robinson explores several ways in which teachers and parents can increase children's understanding of arithmetic and promote the use of that knowledge to improve their mathematical skills.

In Chap. 11, Martínez articulates a theoretical framework based on situated cognition and one from sociolinguistics about second language education (i.e., communicative language teaching) to understand mathematics language integration. Using classroom examples, Martinez illustrates how pedagogical practices consistent with a situated perspective on mathematics education provide opportunities to engage with the second language and how pedagogical practices consistent with a communicative perspective on second language education provide opportunities to engage in mathematical activity.

In Chap. 12, Desmarais, Osana, and Lafay present an interesting chapter related to children who have learning difficulties or intellectual disabilities. The authors engaged in a classroom-based intervention called schema-based instruction (SBI; Jitendra & Star, 2011). SBI uses visual representations to teach students the mathematical structure of word problems. The chapter outlines the literature, the intervention, and the results. Their analysis of the ways in which disability intersects with mathematical instruction may account for the students' performance after the intervention.

In Chap. 13, Dubé, Xu, Kacmaz, Alam, and Ren also consider technology; more specifically, their chapter is about the role of tablets as an elementary mathematics education tool for both parents and teachers. As the authors point out, there is little consensus on whether or not tablets are effective tools for teaching mathematics, and studies seem to provide contradictory explanations about their effects on children's learning. This chapter is a systematic literature review of the tablet literature up to grade 5, published between 2012 and 2017.

In our final chapter, Fizpatrick and Hallett provide a comprehensive review of the literature examining children's early understanding of proportional reasoning and division, and how these early conceptions contribute to children's later understanding of fractions. Although early literature on proportional reasoning suggested that only adolescents have a true understanding of proportional reasoning, more recent research suggests that very young children, if asked appropriately, do demonstrate a basic or intuitive understanding of proportional structures.

In addition to each chapter reflecting interdisciplinary perspectives on early mathematical learning, each chapter also articulates applications to practice—be it in the home, early learning center, or school. A commitment to articulating application to practice is a significant contribution of this collective work. This commitment is in line with the tenets of this proposed new focus on bidirectionality in educational research. We anticipate the book will be of interest to developmental psychologists, neuroscientists, mathematics teachers, mathematics education researchers, and early childhood researchers and practitioners.

The book would be an ideal text for an introductory course in early mathematical cognition in a variety of disciplines, including psychology, education, educational psychology, educational neuroscience, child development, and cognitive development, particularly given the range of developmental periods, mathematical domains, methodological approaches, and contexts of application that are represented. Our sincere gratitude to our colleague authors who contributed so thoughtfully to this work.

References

Derry, S. J. (1992). Beyond symbolic processing: Expanding horizons for educational psychology. *Journal of Educational Psychology, 84,* 413–418.

Farley-Ripple, E., May, H., Karpyn, A., & Tilley, K. (2018). Rethinking connections between research and practice in education: A conceptual framework. *Educational Researcher, 47*(4), 235–245.

Jitendra, A. K., & Star, J. R. (2011). Meeting the needs of students with learning disabilities in inclusive mathematics classrooms: The role of schema-based instruction on mathematical problem-solving. *Theory Into Practice*, *50*(1), 12–19. https://doi.org/10.1080/00405841.2011.534912.

Lysenko, L., Abrami, P., Bernard, R., Dagenais, C., & Janosz, M. (2014). Educational research in educational practice: Predictors of use. *Canadian Journal of Education, 37*(2), 1–26.

Mayer, R. E. (1993). Educational psychology—past and future: Comment on Derry (1992). *Journal of Educational Psychology*, *85*(3), 551–553.

Penuel, W. R., Allen, A. R., Coburn, C. E., & Farrell, C. (2015). Conceptualizing research-practice partnerships as joint work at boundaries. *Journal of Education for Students Placed at Risk, 20*(1–2), 182–197.

Popescu, G. (2014). From psychological-disciplinary knowledge to a transdisciplinary one. *Procedia—Social and Behavioral Sciences, 128,* 438–441.

The "Girl Crisis": The Relationship Between Early Gender Differences and Future Mathematical Learning and Participation

Samantha Makosz, Joanna Zambrzycka and Donna Kotsopoulos

There is a dominant view in society that early participation in the areas of science, technology, engineering, and mathematics (STEM) is an important gateway to a successful future—both professionally and financially. Evidence from a variety of studies supports this perspective (Change the Equation, 2017; Conference Board of Canada, 2014). By the end of high school and within most STEM-based disciplines, a gender difference in terms of participation is evident. More young men than women choose to study in STEM-based disciplines in a post-secondary setting at the end of high school (You, 2013). Further, women are less likely to choose a STEM program in post-secondary, regardless of mathematical achievement in high school (Hango, 2013a), though by the end of high school, males' mathematics scores are higher than females' (Hango, 2013b). In advanced graduate education in STEM disciplines, men are overrepresented in most fields of study (Gillen & Tanenbaum, 2014). Indeed, by the end of high school, one might argue that a "girl crisis" emerges in STEM education that has serious implications for women in terms of future career prospects and economic prosperity.

Our focus in this chapter is to explore the early childhood origins of this trajectory, drawing from the extant of the literature and also referring to our own recent research. We explore the evidence to consider whether different patterns of participation are motivated by cognitive, behavioral, attitudinal, or socialized differences. Specifically, we will examine gender differences and similarities, reasons for gender differences (perception and beliefs, parent and teacher influences, gender stereotypes, and interest and motivation), and the lack of representation of males and fathers in research.

S. Makosz
Faculty of Education, Wilfrid Laurier University, Waterloo, ON, Canada

J. Zambrzycka
Department of Applied Psychology and Human Development, Ontario Institute for Studies in Education, University of Toronto, Toronto, ON, Canada

D. Kotsopoulos (✉)
Huron University (at Western University), London, ON, Canada
e-mail: dkotsopo@uwo.ca

K. M. Robinson et al. (eds.), *Mathematical Learning and Cognition in Early Childhood*,
https://doi.org/10.1007/978-3-030-12895-1_2

In addition, we will discuss the lack of conclusiveness in some of the gender gap research.

Gender Differences and Similarities

Children's mathematical knowledge begins to develop much sooner than the start of formal schooling (Resnick, 1989). Infants as young as six months are sensitive to numerosities (Starr, Libertus, & Brannon, 2013; Xu & Arriaga, 2007; Xu & Spelke, 2000; Xu, Spelke, & Goddard, 2005), but studies that have focused on infants typically do not take into account the role that gender may play. One study by Starkey (1992) looked at gender and infants by examining addition and subtraction concepts in 18- to 42-month-year-olds, but found no gender differences. A key issue when attempting to detect gender differences in infants is the sample size across studies. For instance, Spelke (2005) explains that most studies on infants do not report gender differences because of the lack of effects and that infant studies have not been incorporated in more powerful analyses such as meta-analyses.

The research on preschool- and kindergarten-aged children presents mixed findings on gender differences. Ginsburg and Russell (1981) found that the only gender differences for four- to five-year olds were in addition and subtraction tasks, with girls performing better, but the study consisted of a small sample of children. However, studies that have incorporated larger samples with this age group have not found gender or cultural differences on various numerical tasks, such as counting or arithmetic (Lummis & Stevenson, 1990; Song & Ginsburg, 1987). From these findings, there is little evidence to support gender differences in emergent numeracy abilities from infancy to four years of age.

To accurately detect gender differences, sample size needs to be considered. Studies that have examined mathematics achievement from the Early Childhood Longitudinal Study, Kindergarten Class of 1998–1999 (ECLS-K), which is a nationally representative sample (approximately 21,400 kindergarteners) in the USA, found that overall, there are no significant gender differences in mathematics at the start of kindergarten (Penner & Paret, 2008; Robinson & Theule Lubienski, 2011). Conversely, Penner and Paret (2008) did find gender differences at the start of kindergarten in favor of males among those at the top distribution of the 95th percentile, while a female advantage was present at the lower end of the distribution (1–40th percentile). Notably, when examining the top scores for kindergarteners' mathematics achievement, only 15% of females were in the top 1%. By eighth grade, 37% of females were in the top 1% (Robinson & Theule Lubienski, 2011). Although an increase is promising, this underrepresentation may account for the gender disparity of females in mathematics-related careers, given that females still only represent a third of the top achievers by eighth grade.

Further data from the ECLS-K demonstrated that parental education has a mediating effect on the male advantage found in kindergarten mathematics. The sons of parents with higher education levels had the greatest advantage; however, males from lower parental education levels still sustained an advantage when compared to females. Females at the bottom of the distribution had similar results regardless of parental education levels. By the time students progressed to the third grade, there was no longer a female advantage for those in the lower end of the distribution, and the male advantage had spread throughout the distribution (Penner & Paret, 2008).

Husain and Milimet (2009) coined the term "boy crisis," which makes reference to the fact that boys are lagging behind girls across multiple academic fields, except for mathematics. Husain and Milimet used the data from the ECLS-K and found that, by the start of kindergarten, males are marginally outperforming females, but by the end of kindergarten, this gender gap doubles. Likewise, the gap continues to double until the end of third grade. It is important to note that this early male advantage was predominantly related to white males, and the same results were not found with African–American or Hispanic children.

While it appears that boys are outperforming girls at the top of distribution, the findings are inconsistent when assessing participants' complex mathematical problem solving versus less complex mathematical tasks (e.g., computation). A meta-analysis conducted by Hyde, Fennema, and Lamon (1990) examined 100 studies and demonstrated that children as young as five years old do not show gender differences on more complex mathematical problems, but that males outperformed females in high school on similar measures. The meta-analysis did find that by grade two, girls have better mathematical computation and problem solving skills than males, but by the time students reach high school, males have better problem-solving skills (Hyde et al., 1990). In another study, Pargulski and Reynolds (2017) examined mean and variance differences on mathematical problem solving and numerical operations for over 2000 participants between the ages of 4–19. The authors found a significant male advantage for those categorized as high performers on problem solving, but they did not examine at what age these gender differences emerge. Pargulski and Reynolds did not find a gender difference for numerical calculations. Overall, the emerging gender gap in complex mathematical problem solving is troublesome, given the need for complex skills in order to enter STEM occupations.

Other studies have not necessarily found differences between boys' and girls' mathematical abilities, but have demonstrated differences in growth of mathematical abilities. For instance, Aunola, Leskinen, Lerkkanen, and Nurmi (2004) found no differences from preschool to the second grade using a sample of almost 200 Finnish students, but males showed a faster increase in performance compared to females and had more variability in their performance. For those who were ranked with high ability in mathematics in kindergarten, their gender was able to predict their performance in the second grade. Counting ability was also a predictor of mathematical performance. This finding highlights that girls may benefit from more exposure to and practice with counting, particularly prior to the start of formal schooling.

The most recent meta-analysis of over seven million American students from grades two to eleven did not find any gender differences on standardized American mathematics assessments (Hyde, Lindberg, Linn, Ellis, & Williams, 2008). Standard school assessments have been criticized for assessing lower-level mathematical skills; thus, consequently, the authors also examined data from the National Assessment of Educational Progress (NAEP), which includes more complex mathematical problems. The findings showed that by high school, females demonstrated similar performance to males. In comparison with the meta-analysis (Hyde et al., 1990), the NAEP data appear to show that gender differences in complex problem solving are disappearing.

Given that the research suggests that gender differences diminish with age, it may be that students' motivational levels explain why females are less likely to pursue mathematics in their future education and careers (Hyde, 2014). A review of various meta-analyses proposed a gender similarity hypothesis in mathematics, whereby there are more gender similarities than differences (Hyde, 2014). One area of mathematics that Hyde (2014) suggested has moderate gender differences is in 3D mental rotation, found in children as early as four years old, with a male advantage (Levine, Huttenlocher, Taylor, & Langrock, 1999).

Other researchers have hypothesized that gender differences are disappearing because of changes to education policies, such as No Child Left Behind (Cimpian, Lubienski, Timmer, Makowski, & Miller, 2016). In response, Penner and Paret (2008) analyzed the ECLS-K: 2011 dataset and found that gender results were remarkably similar to the 1999 dataset. Females are still less than one-third represented at the top of the distribution as early as in the spring of kindergarten, thus suggesting that these differences emerge before the start of kindergarten and may indeed be more influenced by environmental factors.

A common limitation among the meta-analyses conducted is that the studies included often lack samples that include children prior to the start of formal schooling. Robinson, Abbott, Berninger, and Busse (1996) prescreened children during preschool in order to examine children with high mathematical abilities. They found that on standardized mathematical tests, boys scored higher than girls on most of the quantitative measures. Nevertheless, the gender results found for this study were not based on the whole sample of students, and thus, there is a lack of generalizability to young students of various mathematical abilities.

Reasons for Gender Differences

Researchers have attempted to tease apart the underlying causes for observed gender differences in mathematics, arguing they are more complex than the nature versus nurture debate. In fact, there are a variety of factors that interact with one another such as psychology, biology, and socialization, along with environmental factors, attitudes, and beliefs (Halpern, Wai, & Saw, 2005; Jacobs, Lanza, Osgood, Eccles, & Wigfield, 2002; Wood & Eagly, 2002). Other reasons that have been proposed to explain why gender differences in mathematics emerge in the first few years of formal schooling include perception and beliefs, parent and teacher influences, lack of representation of males and fathers in research, gender stereotypes, and interest and motivation (e.g., Dickhauser & Meyer, 2006; Eccles, Wigfield, Harold, & Blumenfeld, 1993; Fredricks & Eccles, 2002; Gunderson, Ramirez, Levine, & Beilock, 2012; Jacobs &

Eccles, 1992; Neuburger, Jansen, Heil, & Quaiser-Pohl, 2012). These factors will be further discussed throughout the chapter.

Perception and Beliefs

Some researchers have found that males and females start school with the perception that they have strong abilities in mathematics (Bouffard, Marcoux, Vezeau, & Bordeleau, 2003; Eccles et al., 1993), though this perception changes as children continue through elementary school. In a meta-analysis across a variety of ages, females typically report negative attitudes toward mathematics (Hyde et al., 1990). By the first grade, males have reported higher perceptions of their mathematical abilities compared to females (Eccles et al., 1993).

Furthermore, Dickhauser and Meyer (2006) reported that males and females between the ages of eight to nine years have different perspectives concerning their personal attributions to mathematical ability. In this study, girls were more likely to attribute failure in mathematics to their low ability; conversely, they were less likely to attribute mathematical success to high ability. Additionally, girls did not consider their actual mathematical performance (i.e., grades) when concluding their ability attributions, even though their performance was usually positive. Boys did incorporate their actual mathematical performance when deciding their ability attributions. This implies that children at the elementary level have developed their own perceptions of their mathematical ability and why they are or are not good at mathematics, and that girls appear to have low confidence in their mathematical ability.

Parent and Teacher Influences

Parents have their own perceptions regarding their children's attributes. This is important because if females at a young age are less interested in mathematics and perceive that their parents do not value their competence in mathematics, they are less likely to pursue mathematics in the future (Jacobs, Davis-Kean, Bleeker, Eccles, & Malanchuk, 2005).

Research has found that mothers from the USA, Korea, Japan, and Taiwan tend to believe that their sons are better at mathematics; consequently, mothers have higher expectations for their sons in comparison with their daughters (Lummis & Stevenson, 1990). Parents are also more likely to attribute their eight- to nine-year-old daughters' success in mathematics to their effort and their sons' success to talent (Yee & Eccles, 1988). This perspective underestimates daughters' abilities, which in turn leads young girls to potentially underestimate their own ability. Girls may not be given the same confidence as boys. Interestingly, fathers have higher standards for boys who already have low mathematical abilities than for girls (Yee & Eccles, 1988). This perspective appears to be true among teachers as well, where males' failures in mathematics are attributed to lack of effort, but for females, their failure is attributed to lack of ability (Fennema, Peterson, Carpenter, & Lubinski, 1990; Tiedemann, 2000).

Although one study that examined parents' perceptions of their children from kindergarten to the third grade found no gender differences in their beliefs, parents believed mathematics was more important for their sons than for their daughters (Eccles, Jacobs, & Harold, 1990). Many of these studies are almost three decades old, and efforts have since been made to highlight the need for females to partake in mathematics (Change the Equation, 2017).

There is ample research that supports teachers as socializing agents in children's mathematical abilities and beliefs (Beilock, Gunderson, Ramirez, & Levine, 2010; Else-Quest, Hyde, & Linn, 2010; Gunderson et al., 2012; Upadyaya & Eccles, 2014). Therefore, it is vital for teachers to be aware of their own mathematics anxieties and beliefs, particularly because of the influence it may have on their students. For example, one study examined first-grade teachers who exhibited mathematics anxiety and found that the teachers' female students performed more poorly in mathematics than males, which was mediated by the students' own ability beliefs; there was no influence of anxiety on first-grade males (Beilock et al., 2010).

Lack of Representation of Males and Fathers

Interestingly to note, studies that include teachers and parents tend to include more female teachers and mothers. Beilock et al. (2010) analyzed only female teachers' mathematics anxiety, with a rationale that over 90% of teachers in the USA are female. Many studies examining only gender differences do not report which caregiver consented to participate, but those that include caregivers tend to only include mothers, with the justification that mothers typically respond to participate and spend the most time with their children.

In addition, Lummis and Stevenson (1990) interviewed only mothers of kindergarten children with the reasoning that it would be too difficult to find the time to interview fathers. Jacobs and Eccles' (1992) study was from a larger study (Michigan Study of Adolescent Life Transitions), where both mothers and fathers were asked to participate, but the authors used mothers as the sample given that more mothers returned the survey questionnaires. No comparison numbers were provided in the study, though the authors stated that similar results were found with fathers, which were not presented in the paper.

Another large dataset is the Childhood and Beyond study, which started in 1983 and includes both cross-sectional and longitudinal information on children from kindergarten to grade three (Jacobs et al., 2005). Both fathers and mothers responded to various survey questions about their beliefs and interests and how often they engage in mathematical activities with their child. Jacobs and colleagues (2005) used the Childhood and Beyond dataset to track children longitudinally from kindergarten to grade three and to track parent involvement and parents' perception in mathematics, along with other influencing factors.

Although the Childhood and Beyond dataset includes survey reports from fathers and mothers, mothers were used for the majority of the analyses because more mothers completed the surveys. For example, when examining what sorts of toys parents buy for their child, they asked mothers instead of fathers, reasoning that mothers are more likely to do the shopping. The mothers reported buying more mathematical toys for their sons compared to daughters. This implies that boys may have more opportunity and access to mathematical play experiences. Yet, mothers and fathers were more likely to report being involved with mathematical activities with their daughters, possibly because they believed their daughters needed more guidance in the area (Jacobs et al., 2005).

Simpkins, Fredricks, and Eccles (2015a) used the Childhood and Beyond dataset most recently and reported a total of 987 children, with 723 mothers and 541 fathers. In their study, they focused on the larger sample with mothers because of higher statistical power. They also included a section of fathers, arguing that it was necessary because of the dearth of research with participating fathers or data being indirectly collected about fathers. Fathers' behaviors predicted their child mathematical abilities starting in grade two, but mothers' did not, which may be explained by the stereotype of who is "better at" and who values mathematics. Simpkins, Fredricks, and Eccles (2015b) explained that they "could not directly test for mother and father differences and that the samples of mothers and fathers are drawn from overlapping, but not equivalent families" (p. 135), thus emphasizing the challenge of collecting data from fathers even over a longitudinal study spanning more than 12 years. More recent meta-analyses are needed to understand the influence fathers have on their children, given that fathers spend more time with their children when the mother is employed outside the home (Sandberg & Hofferth, 2001).

In our own research, we considered the same sorts of questions about mothers and fathers and their engagement with their young children. Eighteen mother–child dyads between the ages of two to five years old ($M_{\text{age}} = 39.39$ months; SD = 15.38; 10 boys) and 18 matched sample father–child dyads also between the ages of two to five years old ($M_{\text{age}} = 39.72$ months; SD = 15.07; 10 boys) participated in the study. Parents completed a demographic questionnaire, a mental rotation task (MRT; Vandenberg & Kuse, 1978), and two activity surveys adapted from Dearing and colleagues (2012) exploring spatial (e.g., building with blocks, puzzles) and mathematical activities (e.g., sing counting songs) in the home. Children's nonverbal quantitative reasoning was measured by the Stanford-Binet Intelligence Scales for Early Childhood, Fifth Edition (SB5: Roid, 2003). The Nonverbal Quantitative Reasoning subtest contains 18 items that require the child to answer questions based on quantity size, nonverbal mental addition, number recognition, estimation, three-dimensional block counting, and the relative magnitude of numbers.

Mothers were found to be more likely to report teaching their sons mathematics compared to their daughters. There were no significant differences in what fathers reported about their interactions with their daughters relative to their sons. A possible explanation for this may be that mothers typically spend more time with their children at home, thus are better able to report who they teach more often.

Correlational analyses were conducted separately for mothers and fathers to determine any associations between their child's quantitative reasoning and a variety of cognitive and social factors. The first set of correlational analyses were conducted with the mother–child dyads and explored the relationship between children's nonverbal quantitative reasoning to the following factors: child's age, gender, mother's education level, mother's mental rotation scores, at home teaching activities, and the frequency of spatial activities, including overall average frequency and the frequency of each spatial activity. The correlational analyses for the mother–child dyads found child's gender to be significantly correlated with their quantitative reasoning ability ($r = 0.52$, $p = 0.02$), indicating that girls were more likely to outperform the boys. For the father–child dyads, the correlational analyses revealed that child's age ($r = 0.55$, $p = 0.01$) and the frequency with which they engaged in building with construction toys ($r = 0.56$, $p = 0.01$) were significantly correlated with their quantitative reasoning scores. This indicates that as children aged their quantitative reasoning scores improved and children who played more often, with construction toys had higher quantitative reasoning scores. Our own results, therefore, support the notion that gendered engagement in the home by mothers, but not by fathers, is evident before formal schooling.

Gender Stereotypes

Gender stereotypes emerge early (Gelman, Taylor, & Nguyen, 2004). In conversation, subtle messages (i.e., reference to categories of gender, labeling of gender, and contrasting males vs. females) about gender by mothers can have an influence on their toddlers' gender beliefs (Gelman et al., 2004). In turn, the numeracy performance of children as young as five years old is influenced by such stereotypes (Ambady, Shih, Kim, & Pittinsky, 2001). Specifically, five-year-old Asian–American girls performed worse on a numeracy task when their gender identity was activated.

In kindergarten, females are also susceptible to stereotype threat, given that they are more likely to perform worse on a mathematical task when their gender stereotype is activated—mothers who view mathematics as a field that is male dominated are more likely to have daughters who perform worse on mathematical tasks (Tomasetto, Romana Alparone, & Cadinu, 2011). Additional research by Tomasetto, Mirisola, Galdi, and Cadinu (2015), who studied 253 six-year-olds (131 girls and 122 boys), and both their mothers and fathers, found that daughters' math self-perception was predicted by their mother's math stereotypes. However, both mothers and fathers did not differ in their math–gender stereotypes according to the gender of their child, though fathers were more likely to endorse math as a male-dominated field. Moreover, there was an association between fathers' evaluations of their child's ability and children's self-perception of ability. This was significant even after controlling for the effect of mothers. These findings shed light on the important influence of fathers, who are often underrepresented in the research.

More evidence examining children in grades one, two, and four highlighted the development of gender stereotypes (Freedman-Doan et al., 2000). Both males and females provided gender-stereotyped answers when asked what task they were the worst at. The majority of females stated science and computers, while males stated reading. In relation to mathematics, females stated less often that they were good at mathematics compared to males. Nonetheless, both genders believed they could improve on the task that they believed to be the worst at, but by the fourth grade, most students believed that they could not improve. Their reasoning for not being able to improve was attributed to lack of ability. This finding verifies that as children of both genders get older, they presume they are less capable of improving in their worst qualities (Freedman-Doan et al., 2000). Further, it suggests that early on, students are more confident in their abilities, but as they go through formal schooling, their self-efficacy decreases.

Interest and Motivation

Children's mathematical interest at the start of formal schooling is vital, as the higher the mathematical interest displayed by the child, the more likely teachers attribute their success to effort and ability, which then results in an increase in children's interest in mathematics (Upadyaya, Viljaranta, Lerkkanen, Poikkeus, & Nurmi, 2012). In one study, children's mathematical interest was tracked from kindergarten to grade six and teachers' perception of students' perceived performance, and effort was the most consistent factor in students' motivation throughout the years (Upadyaya & Eccles, 2014); there were no differences between genders with regards to children's interest and teacher beliefs, however. Nevertheless, both genders were equally sensitive to teacher feedback.

Early motivation is also imperative, as those who display higher mathematical motivation at the start of kindergarten perform higher on an arithmetic assessment at the end of the school year, without showing any gender differences (Viljaranta, Lerkkanen, Poikkeus, Aunola, & Nurmi, 2009). This is consistent with other studies that have not found gender differences in mathematical motivation at the start of formal schooling (Jacobs et al., 2002). Unfortunately, by the third grade, girls' motivation tends to decrease, but boys' motivation remains stable (Bouffard et al., 2003). This may explain why boys with high mathematical abilities in kindergarten have significantly increased in their abilities by the third grade, whereas girls have not (Husain & Millimet, 2009).

Children need to engage in mathematical play prior to the start of formal schooling, so they enter kindergarten with an interest and motivation in mathematics. Gender differences in mathematical motivation do not exist in preschool (Viljaranta et al., 2009), which implies that prior to the start of formal schooling, both genders are equally motivated in mathematics.

Conclusion

Overall, the literature that has examined gender differences in mathematical abilities prior to formal schooling suggests that they may be more socialized than innate. The most substantial studies examining gender differences begin in kindergarten and only find a male advantage at the top of the distribution. The most recent meta-analysis (Hyde, 2014) found that gender differences are disappearing: "nonsignificant gender difference, that is, a gender similarity, is as interesting and important as a gender difference" (Hyde, 2014, p. 393).

Awareness of gender similarity in mathematical ability is an important and critical mind shift that is necessary. Early childhood experiences at home are crucial, and more research is needed to examine how fathers spend time with their children. Awareness of gender similarity in mathematical ability is also an important mind shift for schools. One implication of this review is that parents, caregivers, and educators need to be informed of the importance of creating equal mathematical opportunities for both boys and girls, not only at an early age, but throughout childhood. Although most parents and teachers would readily agree that mathematical learning is important, the extent to which the adults recognize their own biases in their interactions with children may be limited.

The subversive nature of systemic gender bias in mathematical engagement, which results in under participation of women in post-secondary STEM disciplines, regardless of ability, consequently impacts women's future career choices, financial security, and health access, particularly in the USA, where health care is less accessible for low-income earners and their futures as women are more likely to be as single parents (Andersen & Newman, 2005; US Census Bureau, 2016). This demonstrates precisely how complex and yet utterly necessary it is to frame the situation as both a "girl crisis" and one of gender similarity.

References

Ambady, N., Shih, M., Kim, A., & Pittinsky, T. L. (2001). Stereotype susceptibility in children: Effects of identity activation on quantitative performance. *Psychological Science, 12*(5), 385–390.

Andersen, R., & Newman, J. F. (2005). Societal and individual determinants of medical care utilization in the United States. *The Milbank Quarterly, 83*(4), 1–28. https://doi.org/10.1111/j.1468-0009.2005.00428.x.

Aunola, K., Leskinen, E., Lerkkanen, M. -K., & Nurmi, J. -E. (2004). Developmental dynamics of math performance from preschool to grade 2. *Journal of Educational Psychology, 96,* 699–713. https://doi.org/10.1037/0022-0663.96.4.699.

Beilock, S. L., Gunderson, E. A., Ramirez, G., & Levine, S. C. (2010). Female teachers' math anxiety affects girls' math achievement. *PNAS Proceedings of the National Academy of Sciences of the United States of America, 107,* 1860–1863. https://doi.org/10.1073/pnas.0910967107.

Bouffard, T., Marcoux, M. -F., Vezeau, C., & Bordeleau, L. (2003). Changes in self-perceptions of competence and intrinsic motivation among elementary schoolchildren. *British Journal of Educational Psychology, 73*(2), 171–186. https://doi.org/10.1348/00070990360626921.

Change the Equation. (2017). *STEM help wanted: Demand for science, technology, engineering and mathematics weathers the storm*. Retrieved from http://changetheequation.org/sites/default/files/CTEq_VitalSigns_Supply%20%282%29.pdf.

Cimpian, J. R., Lubienski, S. T., Timmer, J. D., Makowski, M. B., & Miller, E. K. (2016). Have gender gaps in math closed? Achievement, teacher perceptions, and learning behaviors across two ECLS-K cohorts. *AERA Open, 2*(4), 1–19. https://doi.org/10.1177/2332858416673617.

Conference Board of Canada. (2014). *Graduates in science, math, computer science, and engineering*. Retrieved from http://www.conferenceboard.ca/hcp/provincial/education/sciencegrads.aspx.

Dearing, E., Casey, B. M., Ganley, C. M., Tillinger, M., Laski, E., & Montecillo, C. (2012). Young girls' arithmetic and spatial skills: The distal and proximal roles of family socioeconomics and home learning experiences. *Early Childhood Research Quarterly, 27*(3), 458–470.

Dickhauser, D., & Meyer, W. U. (2006). Gender differences in young children's math ability attributions. *Psychology Science, 48*(1), 3–16.

Eccles, J. S., Jacobs, J. E., & Harold, R. D. (1990). Gender role stereotypes, expectancy effects, and parents' socialization of gender differences. *Journal of Social Issues, 46,* 183–201. https://doi.org/10.1111/j.1540-4560.1990.tb01929.x.

Eccles, J. S., Wigfield, A., Harold, R. D., & Blumenfeld, P. (1993). Age and gender differences in children's self- and task perceptions during elementary school. *Child Development, 64,* 830–847. https://doi.org/10.2307/1131221.

Else-Quest, N. M., Hyde, J. S., & Linn, M. C. (2010). Cross-national patterns of gender differences in mathematics: A meta-analysis. *Psychological Bulletin, 136*(1), 103–127. https://doi.org/10.1037/a0018053.

Fennema, E., Peterson, P. L., Carpenter, T. P., & Lubinski, C. A. (1990). Teachers' attributions and beliefs about girls, boys and mathematics. *Educational Studies in Mathematics, 21*, 55–69.

Fredricks, J. A., & Eccles, J. S. (2002). Children's competence and value beliefs from childhood through adolescence: Growth trajectories in two male-sex-typed domains. *Developmental Psychology, 38*(4), 519–533.

Freedman-Doan, C., Wigfield, A., Eccles, J. S., Blumenfeld, P., Arbreton, A., & Harold, R. D. (2000). What am I best at? Grade and gender differences in children's beliefs about ability improvement. *Journal of Applied Developmental Psychology, 21*(4), 379–402. https://doi.org/10.1016/S0193-3973(00)00046-0.

Gelman, S. A., Taylor, M. G., & Nguyen, S. P. (2004). Mother-child conversations about gender: Understanding the acquisition of essentialist beliefs. *Monographs of the Society for Research in Child Development, 69*(1), i+v+vii+1–142. https://doi.org/10.1111/j.1540-5834.2004.06901001.x.

Gillen, A., & Tanenbaum, C. (2014). *Exploring gender imbalance among STEM doctoral degree recipients*. Washington, DC: American Institutes for Research.

Ginsburg, H. P., & Russell, R. L. (1981). Social class and racial influences on early mathematical thinking. *Monographs of the Society for Research in Child Development, 46*(6), 1–69. https://doi.org/10.2307/1165946.

Gunderson, E. A., Ramirez, G., Levine, S. C., & Beilock, S. L. (2012). The role of parents and teachers in the development of gender-related math attitudes. *Sex Roles, 66*(3–4), 153–166. https://doi.org/10.1007/s11199-011-9996-2.

Halpern, D. F., Wai, J., & Saw, A. (2005). A psychobiosocial model: Why females are sometimes greater than and sometimes less than males in math achievement. In A. M. Gallagher & J. C. Kaufman (Eds.), *Gender differences in mathematics: An integrative psychological approach* (pp. 48–72, Chap. xvi, 351 p.). New York, NY: Cambridge University Press.

Hango, D. (2013a). *Ability in mathematics and science at age 15 and program choice in university: Differences by gender*. Ottawa, Ontario: Tourism and Centre for Education Statistics Division, Ministry of Industry.

Hango, D. (2013b). *Gender differences in science, technology, engineering, mathematics and computer science (STEM) programs at university*. Ottawa, Ontario: Ministry of Industry.

Husain, M., & Millimet, D. L. (2009). The mythical 'boy crisis'? *Economics of Education Review, 28*(1), 38–48. https://doi.org/10.1016/j.econedurev.2007.11.002.
Hyde, J. S. (2014). Gender similarities and differences. *Annual Reviews of Psychology, 65,* 373–398. https://doi.org/10.1146/annurev-psych-010213-115057.
Hyde, J. S., Fennema, E., & Lamon, S. J. (1990). Gender differences in mathematics performance: A meta-analysis. *Psychological Bulletin, 107*(2), 139–155. https://doi.org/10.1037/0033-2909.107.2.139.
Hyde, J. S., Lindberg, S. M., Linn, M. C., Ellis, A. B., & Williams, C. C. (2008). Gender similarities characterize math performance. *Science, 321*(5888), 494–495.
Jacobs, J. E., & Eccles, J. S. (1992). The impact of mothers' gender-role stereotypical beliefs on mothers' and children's ability and perceptions. *Journal of Personality and Social Psychology, 63*(6), 932–944.
Jacobs, J. E., Lanza, S., Osgood, D. W., Eccles, J. S., & Wigfield, A. (2002). Changes in children's self-competence and values: Gender and domain differences across grades one though twelve. *Child Development, 73,* 509–527. https://doi.org/10.1111/1467-8624.00421.
Jacobs, J. E., Davis-Kean, P., Bleeker, M., Eccles, J. S., & Malanchuk, O. (2005). "I can, but I don't want to": The impact of parents, interests, and activities on gender differences in math. In A. M. Gallagher & J. C. Kaufman (Eds.), *Gender differences in mathematics: An integrative psychological approach* (pp. 246–263). New York, NY: Cambridge University Press.
Levine, S. C., Huttenlocher, J., Taylor, A., & Langrock, A. (1999). Early sex differences in spatial skill. *Developmental Psychology, 35*(4), 940–949. https://doi.org/10.1037//0012-1649.35.4.940.
Lummis, M., & Stevenson, H. W. (1990). Gender differences in beliefs and achievements. *Developmental Psychology, 26*(2), 254–263.
Neuburger, S., Jansen, P., Heil, M., & Quaiser-Pohl, C. (2012). A threat in the classroom: Gender stereotype activation and mental-rotation performance in elementary-school children. *Zeitschrift für Psychologie, 220*(2), 61–69. https://doi.org/10.1027/2151-2604/a000097.
Pargulski, J. R., & Reynolds, M. R. (2017). Sex differences in achievement: Distributions matter. *Personality and Individual Differences, 104,* 272–278. https://doi.org/10.1016/j.paid.2016.08.016.
Penner, A. M., & Paret, M. (2008). Gender differences in mathematics achievement: Exploring the early grades and the extremes. *Social Science Research, 37*(1), 239–253. https://doi.org/10.1016/j.ssresearch.2007.06.012.
Resnick, L. B. (1989). Developing mathematical knowledge. *American Psychologist, 44*(2), 162–169. https://doi.org/10.1037/0003-066X.44.2.162.
Robinson, J. P., & Theule Lubienski, S. (2011). The development of gender achievement gaps in mathematics and reading during elementary and middle school: Examining direct cognitive assessments and teacher ratings. *American Educational Research Journal, 48*(2), 268–302. https://doi.org/10.3102/0002831210372249.
Robinson, N. M., Abbott, R. D., Berninger, V. W., & Busse, J. (1996). Structure of abilities in math-precocious young children: Gender similarities and differences. *Journal of Educational Psychology, 88*(2), 341–352. https://doi.org/10.1037/0022-0663.88.2.341.
Roid, G. (2003). *Stanford-Binet intelligence scales* (5th ed.). Itasca, IL: Riverside Publishing.
Sandberg, J. F., & Hofferth, S. L. (2001). Changes in children's time with parents: United States, 1981–1997. *Demography, 38*(3), 423–436. https://doi.org/10.1353/dem.2001.0031.
Simpkins, S. D., Fredricks, J. A., & Eccles, J. S. (2015a). The role of parents in the ontogeny of achievement-related motivation and behavioral choices: II. Methods. *Monographs of the Society for Research in Child Development, 80*(2), 23–43. https://doi.org/10.1111/mono.12158.
Simpkins, S. D., Fredricks, J. A., & Eccles, J. S. (2015b). The role of parents in the ontogeny of achievement-related motivation and behavioral choices: VIII. Discussion. *Monographs of the Society for Research in Child Development, 80*(2), 119–137. https://doi.org/10.1111/mono.12164.
Song, M. -J., & Ginsburg, H. P. (1987). The development of informal and formal mathematical thinking in Korean and U.S. children. *Child Development, 58,* 1286–1296.

Spelke, E. S. (2005). Sex differences in Intrinsic aptitude for mathematics and science? A critical review. *American Psychologist, 60,* 950–958. https://doi.org/10.1037/0003-066X.60.9.950.

Starkey, P. (1992). The early development of numerical reasoning. *Cognition, 43*(2), 93–126.

Starr, A., Libertus, M. E., & Brannon, E. M. (2013). Infants show ratio-dependent number discrimination regardless of set size. *Infancy, 18,* 927–941. https://doi.org/10.1111/infa.12008.

Tiedemann, J. (2000). Gender-related beliefs of teachers in elementary school mathematics. *Educational Studies in Mathematics, 41*, 191–207.

Tomasetto, C., Romana Alparone, F., & Cadinu, M. (2011). Girls' math performance under stereotype threat: The moderating role of mothers' gender stereotypes. *Developmental Psychology, 47*(4), 943–949. https://doi.org/10.1037/a0024047.

Tomasetto, C., Mirisola, A., Galdi, S., & Cadinu, M. (2015). Parents' math-gender stereotypes, children's self-perception of ability, and children's appraisal of parents' evaluations in 6-year-olds. *Contemporary Educational Psychology, 42,* 186–198. https://doi.org/10.1016/j.cedpsych.2015.06.007.

Upadyaya, K., & Eccles, J. S. (2014). How do teachers' beliefs predict children's interest in math from kindergarten to sixth grade? *Merrill-Palmer Quarterly, 60*(4), 403–430. https://doi.org/10.13110/merrpalmquar1982.60.4.0403.

Upadyaya, K., Viljaranta, J., Lerkkanen, M. -K., Poikkeus, A. -M., & Nurmi, J. -E. (2012). Cross lagged relations between kindergarten teachers' causal attributions, and children's interest value and performance in mathematics. *Social Psychology of Education, 15*(2), 181–206. https://doi.org/10.1007/s11218-011-9171-1.

U.S. Census Bureau. (2016). *Current population survey: America's families and livingarrangements*. Retrieved from https://www.census.gov/topics/families.html.

Vandenberg, S. G., & Kuse, A. R. (1978). Mental rotations. A group test of three-dimensional spatial visualization. *Perceptual and Motor Skills, 47,* 599–604.

Viljaranta, J., Lerkkanen, M. -K., Poikkeus, A. -M., Aunola, K., & Nurmi, J. -E. (2009). Cross-lagged relations between task motivation and performance in arithmetic and literacy in kindergarten. *Learning and Instruction, 19*(4), 335–344. https://doi.org/10.1016/j.learninstruc.2008.06.011.

Wood, W., & Eagly, A. H. (2002). A cross-cultural analysis of the behavior of women and men: Implications for the origins of sex differences. *Psychological Bulletin, 128*(5), 699–727. https://doi.org/10.1037/0033-2909.128.5.699.

Xu, F., & Arriaga, R. I. (2007). Number discrimination in 10-month-old infants. *British Journal of Developmental Psychology, 25*(1), 103–108. https://doi.org/10.1348/026151005X90704.

Xu, F., & Spelke, E. (2000). Large number discrimination in 6-month-old infants. *Cognition, 74,* B1–B11.

Xu, F., Spelke, E. S., & Goddard, S. (2005). Number sense in human infants. *Developmental Science, 8*(1), 88–101. https://doi.org/10.1111/j.1467-7687.2005.00395.x.

Yee, D. K., & Eccles, J. S. (1988). Parent perceptions and attributions for children's math achievement. *Sex Roles, 19*(5), 317.

You, S. (2013). Gender and ethnic differences in precollege mathematics coursework related to science, technology, engineering, and mathematics (STEM) pathways. *School Effectiveness and School Improvement, 24*(1), 64–86.

Spatial Learning and Play with Technology: How Parental Spatial Talk Differs Across Contexts

Joanne Lee, Sarah Hodgins and Eileen Wood

Importance of Foundational Spatial Skills

Consider San Francisco's Golden Gate Bridge, Toronto's CN Tower, or the Eiffel Tower in Paris. Each of these amazing feats of human engineering required spatial knowledge as a foundation for success. Spatial visualization, spatial perception, spatial orientation, and mental rotation serve as the foundation for a multitude of everyday skills that adults and children perform. For example, we use simple spatial skills to identify and describe two-dimensional (2D) and three-dimensional (3D) shapes, the varying sizes, and orientations of these shapes and to represent and model objects in the environment. In addition, simple spatial skills allow us to gauge spatial relations between objects as well as between objects and ourselves. Using these basic spatial skills, we are able to organize wardrobes, stack our groceries in grocery bags, and avoid diving into shallow pools. We use more complex spatial skills to navigate from one location to another such as when we plan a trip or select a good hiding spot when playing hide-and-seek. Complex spatial skills are also evident when we apply geometric concepts in real-world contexts including computer modeling. Overall, spatial knowledge and skills direct and coordinate many of our informal and formal experiences. They allow us to understand spatial properties and spatial relations, and these, in turn, allow us to navigate the world, as well as to design the tools and structures that define some of our most amazing feats (e.g., National Governors Association Center for Best Practices, Council of Chief State School Officers, 2010; National Research Council, 2006; Newcombe, Uttal, & Sauter, 2013).

We communicate much of our spatial knowledge through visual and verbal representations. Language is a particularly important avenue of communication as spatial words occur frequently in everyday exchanges between parents and their children starting early in life. For example, spatial words are used to describe the *features*

J. Lee (✉) · S. Hodgins · E. Wood
Department of Psychology, Wilfrid Laurier University, Waterloo, ON, Canada
e-mail: jlee@wlu.ca

K. M. Robinson et al. (eds.), *Mathematical Learning and Cognition in Early Childhood*,
https://doi.org/10.1007/978-3-030-12895-1_3

of an object, the *rotation* of an object, or the spatial *relations* between objects (e.g., Casasola & Bhagwat, 2007; Casasola, Bhagwat, & Burke, 2009). Research suggests that, consistent with most complex cognitive skills, earlier acquisition of simple spatial concepts leads to better performance in later academic and practical contexts (e.g., Casey, Nuttall, Pezaris, & Benbow, 1995; Gunderson, Ramirez, Beilock, & Levine, 2012; Shea, Lubinski, & Benbow, 2001; Verdine et al., 2014). Specifically, foundational spatial skills are predictive of subsequent mathematics attainment in elementary and even high school (e.g., Duncan et al., 2007; Magnuson, Duncan, Lee, & Metzger, 2016; Mix et al., 2017; Watts, Duncan, Clements, & Sarama, 2018; Wolfgang, Stannard, & Jones, 2003). Moreover, Science, Technology, Engineering and Mathematics (STEM) careers typically are pursued by high school students who have high spatial competence compared to those with lower spatial competence (Shea et al., 2001; Wai, Lubinski, & Benbow, 2009). Thus, attention has been directed toward adult–child contexts, especially those involving parents and early childhood educators, where the earliest influences on spatial skills and knowledge can occur (e.g., McClure et al., 2017; Newcombe, 2010). In addition, technology, especially software programs designed for infants, toddlers, and preschoolers, has recently become a focus for research as these software programs are often explicitly involved in providing instruction important for spatial development.

The following chapter will review current and unfolding research that examines how spatial talk is provided to young children in traditional and technology-based contexts, the factors that might influence the production of spatial language in these contexts, and the caveats regarding technology use of applications (apps) on mobile devices. Exploring language related to spatial concepts in these early interactions provides a foundation for understanding children's subsequent mathematics competence.

Why Is Spatial Language Important to Early Spatial Development?

Language is a symbolic system used to embody and express our thoughts, feelings, and conceptual understanding. Spatial language serves as a representational tool to communicate mathematical thinking (e.g., Kuhn, 2000). It can be used to direct children's attention and facilitate encoding and understanding of spatial concepts (Gentner, 2003; Gentner & Lowenstein, 2002; Gentner, Özyürek, Gürcanli, & Goldin-Meadow, 2013). A body of research supports the early introduction of spatial language as a means for facilitating the acquisition of spatial knowledge and skills. For example, Casasola and Bhagwat (2007) provided 18-month-old children with verbal labels to explain "support relations" between objects. "Support relations" involves knowing whether there is appropriate support available to provide sufficient balance to maintain a structure when stacking objects vertically or horizontally. As such, spatial knowledge would be necessary to ensure a tower or bridge built from blocks does not collapse because of insufficient support. Having labels that clearly defined relations between objects (e.g., two things placed *on top* of each other vs. one thing *in/inside* another) facilitated learning even among very young learners. Among slightly older infants (21- and 22-month-olds), understanding of tight-fit support relations was also enhanced when children were provided with language that allowed them to discriminate between an object that was *tightly in* versus *loosely in* the other (Casasola et al., 2009). Gains based on provision of linguistic labels are also evident for older preschool-aged children. For example, hearing relational spatial words such as *on top*, *under*, and *to the left* was strongly related to the enhanced understanding of spatial locations between objects (Dessalegn & Landau, 2008; Loewenstein & Gentner, 2005; Plumert & Nichols-Whitehead, 1996). In addition, hearing locational spatial sentences such as "The toy is hidden by the frog" (where the frog is the closest landmark to the toy location) versus non-locational spatial sentences such as "I'm hiding the toy here" helped 4-year-olds perform significantly better at recalling the toy location than the non-locational sentences group (Miller, Patterson, & Simmering, 2016). These laboratory-based studies identify the importance of early spatial language exposure.

Predictive effects of spatial language on gains in spatial understanding among young children have subsequently been demonstrated in naturalistic contexts. For example, Pruden, Levine, and Huttenlocher (2011) found that parents who provided descriptive words to identify shapes, size (e.g., long, small), and properties of 2D and 3D objects (e.g., edge, corner) during everyday activities promoted spatial word production and competence in their children. Similarly, Foster and Hund (2012) demonstrated that four- and five-year-old children showed a more proficient understanding and use of spatial relational words (e.g., *between* and *in the middle*) when these words were used by parents during daily interactions. Longitudinal benefits have also been observed. Specifically, Pruden and colleagues (2011) found that 4.5-year-old children of parents who used more spatial language when their children were younger (i.e., 14 and 46 months of age) produced more spatial language (up to 525 spatial words over the nine 90-min home visits), than their peers whose parents did not (5 spatial words over the same period). Gains in spatial word production and competence at 54 months were evident even after controlling for other non-spatial talks. Thus, early exposure to parental spatial language clearly resulted in gains in children's spatial language and competence; more importantly, these gains persisted over time. These same advantages have been demonstrated for children with learning challenges. Specifically, Landau, Spelke, and Gleitman (1984) found that hearing spatial words facilitated the acquisition of spatial concepts, such as object features and locations for children with visual impairments. In sum, the research evidence from laboratory and naturalistic contexts consistently indicates that hearing more spatial words, especially during ongoing natural interactions, enables young children to acquire more spatial concepts. Having exposure to language regarding spatial concepts allows these young children to form abstract representations of different spatial relations early in development (Casasola, 2008; Levinson, Kita, Haun, & Rasch, 2002; Munnich, Landau, & Dosher, 2001).

Supporting Foundational Spatial Development Through Play

Play is a naturally occurring context that provides opportunities for spatial language exposure to and learning of spatial concepts. For example, children can increase their knowledge about shapes and dimensions by exploring objects on their own or through observation or interaction with a parent (Ginsburg, 2006; Seo & Ginsburg, 2004). Children's innate curiosity and exploration of the world around them provides the foundation for parents to support their children's development (Piaget, 1932; Vygotsky, 1978). Indeed, a wide body of literature supports parent–child interactions as a means for promoting learning of mathematical concepts (e.g., Foster & Hund, 2012; Jirout & Newcombe, 2015; Levine, Suriyakham, Rowe, Huttenlocher, & Gunderson, 2010; Levine, Ratliff, Huttenlocher, & Cannon, 2012; Piaget, 1932; Pruden et al., 2011). Parents and caregivers can facilitate children's discoveries and cognitive development by providing rich and diverse play opportunities (Bruner, 1972; Lancy, 2007; Vygotsky, 1978). Play offers an informal and interactive context for children to form conceptual representations based on their experiences interacting with objects and people around them (e.g., Ginsburg, 2006; Hirsh-Pasek, Michnick Golinkoff, Berk, & Singer, 2009; National Association for the Education of Young Children (NAEYC), 2009). For example, conceptual representations of objects' affordances are formed by infants discovering the properties of objects, exploring and manipulating them, and acting on them to produce positive or negative effects in their environment (Bourgeois, Khawar, Neal, & Lockman, 2005; Fontelle, Kahrs, Neal, Newton, & Lockman, 2007; Gibson, 1988). Through play, parents and other adults have opportunities to facilitate children's acquisition of concepts by guiding or showing them how things work. According to Vygotsky (1978), cognitive development takes place in what he described as "the zone of proximal development," which defines the distance between what children can do independently and what they can achieve when a knowledgeable adult scaffolds them. Recent research demonstrates that scaffolding can improve shape knowledge for typical, atypical, and non-valid exemplars of triangles, rectangles, pentagons, and hexagons. Four- and five-year-old children who explored these concepts with an adult's guidance performed better compared to those who played on their own, without adult's guidance (Fisher, Hirsh-Pasek, Golinkoff, & Newcombe, 2013).

Opportunities for scaffolding to be effective depend on the abilities of the child and the play context. Some play contexts may not be as conducive for scaffolding as others. As noted above, early exposure to spatial language enhances early spatial development (e.g., Casasola & Bhagwat, 2007; Casasola et al., 2009; Casey, Andrews, Schindler, Kersh, Samper, & Copley, 2008; Ferrara, Hirsh-Pasek, Newcombe, Golinkoff, & Lam, 2011; Pruden et al., 2011). However, not all parent–child engagement contexts equally elicit production of spatial words. For example, more spatial language is generated when parents and their young children are engaged with blocks and puzzles than when they are engaged in non-spatial activities such as reading and drawing with their young children (Hermer-Vazquez, Moffet, & Munkholm, 2001; Ferrara et al., 2011; Levine et al., 2012; Pruden et al., 2011). In

Ferrara et al., (2011), parents used 90% more spatial language (average proportion of spatial talk relative to the total of words produced) with their preschoolers during free/unstructured block play compared to parents playing with their preschoolers during free/unstructured play involving other toys such as puppets, dolls, pretend food, and kitchen utensils (average proportion of spatial talk relative to the total of words produced). Differences in parental production of spatial language such as these suggest that some forms of play are more conducive to naturally exploring spatial features.

Engaging in block play, for example, provides opportunities to use spatial language related to location (e.g., "We will put this triangular block on top of the tower"), spatial dimension (e.g., "Mine block is taller than yours"), and spatial features (e.g., "A square block has four sides"). As more sophisticated block structures are created, spatial relational language (e.g., between, beside, on top) is introduced to help children form spatial relationships between objects in the block structures. The nature of puzzle play, on the other hand, elicits use of spatial words such as flat, straight, and upside down to describe spatial features and locations of the shape puzzle pieces (Levine et al., 2012). However, both block and puzzle play require spatial reasoning skills and, thus, encourage language to support this reasoning. For example, parent–child dyads must determine the exact location that fits each puzzle piece or block by examining and discussing the spatial characteristics of the different puzzle pieces or blocks (e.g., Caldera et al., 1999). Engaging in these discussions, and utilizing relevant spatial language to do so, provides naturally occurring opportunities to facilitate learning of spatially relevant concepts and language.

The production of spatial language, even within spatially relevant play contexts, may be impacted as a result of instructional approaches (e.g., free play, structured play) and individual characteristics of parents and children (e.g., including level of interest and knowledge base). For example, the amount of parental spatial talk can be influenced by the instructional information available through the toys themselves. Specifically, less spatial language (i.e., 8.2%) was generated by parents when they and their child were engaged with "talking" electronic shape sorters that provide labels for shapes as well as other sounds than the parent–child dyads who played with traditional 3D shape sorters that did not "talk" (i.e., 15.8%; Zosh et al., 2015). Different instructional approaches adopted by parents also impact learning opportunities. For example, parents engaged in free block play with their preschool children produced 68% less spatial talk (average proportion of spatial talk relative to the total of words produced) than parents engaged in guided block play where they were asked to follow step-by-step instructions to construct a structure (Ferrara et al., 2011). These empirical findings identify features within parent–child play contexts in which richer spatial language is most likely elicited when engaged with traditional three-dimensional (3D) spatial toys.

In addition, parental effect and anxiety toward mathematics can influence parents' engagement in mathematical activities and talk (e.g., Blevins-Knabe, Austin, Musun, Eddy, & Jones, 2000; Maloney, Ramirez, Gunderson, Levine, & Beilock, 2015b). For example, parents with positive effect toward mathematical activities, as measured by their enjoyment level in the study, were found to engage in more mathematical activities with their 4- to 6-year-olds than those who did not (Blevins-Knabe et al., 2000). Hence, the opportunities for children to hear mathematical talk could be reduced if their parents were less likely to engage in mathematical activities (e.g., Levine et al., 2012) because of their negative effect toward mathematics. Recent research has revealed that mathematical anxiety among parents of first and second graders was associated with lower mathematical performance for children who received frequent help with their school work compared to peers whose parents did not provide frequent help (Maloney et al., 2015a, 2015b). It is possible, then, that parents who experience math anxiety may express more frustration and negativity while interacting with their child as suggested by Maloney and colleagues (2015a, 2015b). Consequently, we hypothesize that parents with heightened math anxiety may use fewer and more rigid instructional approaches in mathematical contexts than other parents, thereby negatively impacting children's performance. Although Maloney and colleagues (2015a, 2015b) did not directly examine the nature of parental math talk, these differences in instructional approaches due to heightened math anxiety, in turn, can negatively impact parent–child engagement and talk about mathematical concepts during homework assistance (Maloney et al., 2015a, 2015b). Hence, there is a renewed call by researchers to offer effective intervention or support programs such as providing tip sheets or video models to help parents to identify and engage in best practices during interactions with their children at home (e.g., Blevins-Knabe & Austin, 2016; Maloney, Converse, Gibbs, Levine, & Beilock, 2015a; McClure et al., 2017; Robinson, 2014).

Touch-Screen Technology and Early Spatial Development

Today, many parent–child interactions occur in the context of technology (e.g., Eagle, 2012; Flynn & Richert, 2015; Kabali et al., 2015; Rideout, 2014). With decreasing costs, increasing mobility and a continuously growing array of software applications, mobile, virtual, two-dimensional (2D) interactive devices (e.g., smartphones and tablets) are becoming an increasingly important part of playtime for very young children (e.g., Common Sense Media, 2017; Kabali et al., 2015). Among the plethora of software programs available online are a multitude of programs targeting early math learning (Baccaglini-Frank & Maracci, 2015; Ginsburg, 2017; Moyer-Packenham et al., 2015). Affordances inherent in newer interactive touch-screen devices (e.g., immediate feedback, multimedia, and anywhere anytime learning) may differ from affordances traditionally associated with tangible 3D toys (Geist, 2014; Guernsey & Levine, 2015; National Association for the Education of Young Children (NAEYC) & Fred Rogers Centre for Early Learning and Children's Media, 2012). Mobile touch-screen devices are lightweight and afford easy user interaction through swiping and tapping gestures (e.g., Kucirkova, 2014; Neumann & Neumann, 2014). Most software employs a "game-like format" which increases attention to and engagement with the content (e.g., Abdul Jabbar & Felicia, 2015; Gee, 2008; Vogel et al., 2006). How adult–child interactions translate from traditional contexts to mobile technology learning contexts is an emerging area of research, especially with respect to young children acquiring spatial concepts.

Considerable literature in early childhood mathematics education supports the need for children to explore and mathematize key concepts through application to many varied novel contexts (e.g., Newcombe, 2010; Uttal, 2000). This would require exposure to spatial concepts represented physically in the real-world and abstract representations such as visual images, maps, and models. For example, understanding the concept of a cylinder can involve exposure to physical examples found in children's everyday lives such as tinned goods, water bottles, and cookie jars. Similarly, opportunities to manipulate physical representations can occur by providing children with various cylindrical shapes in block building, puzzle solving, or other contexts. The concrete nature of these types of physical manipulatives aligns with children's cognitive capabilities. However, because they are static and concrete, it may be necessary for children to be exposed to multiple types of physical manipulatives before they can extract the key spatial concepts.

Virtual 2D manipulatives, on the other hand, allow learners to dynamically manipulate on-screen objects through simple swiping and tapping tools that are not possible with physical 3D manipulatives (Moyer, Niezgoda, & Stanley, 2005; Yerushalmy, 2005). For example, virtual 2D manipulatives afford young children opportunities to explore the concepts of composition and decomposition (e.g., taking apart a hexagon to make two pentagons and vice versa; Clements & Sarama, 2007; Moyer-Packenham & Westernskow, 2013). Understanding of the salient features, especially abstract ones, such as those specific to triangles, may be easier to communicate through dynamic media. For example, the visual prototype of a triangle for young children between four and six years old is an isosceles triangle. Identifying triangles (other than an isosceles triangle) is difficult for most young children (Aslan & Aktas-Arnas, 2007; Clarke, 2004; Clements, Swaminathan, Hannibal, & Sarama, 1999; Yin, 2003). Software representations can make salient features of different triangles more readily apparent to young children by systematically resizing defining attributes such as aspect ratios of width and height as well as skewness (or lack of symmetry) as a child expands or shrinks the image of a triangle or series of triangles on screen. As such, an extensive and "manipulatable" representation of various types of triangles—isosceles, equilateral, right-angled—would be made available quickly and seamlessly. Such dynamic and multiple variations of triangles are not easily achieved through single integrated presentations using physical manipulatives. Thus, the flexibility of dynamic presentations offers the potential for parents and children to explore and discuss a greater range of spatial concepts in any given learning session. Dynamic representations of spatial features and relationships in a virtual medium, therefore, provide diverse and varied opportunities to enhance children's acquisition of abstract spatial concepts (e.g., Sarama & Clements, 2016).

Touch-Screen Technology and Parental Use of Spatial Words

To date, research on spatial input has mainly focused on spatial play with three-dimensional (3D) objects such as tangible blocks and jigsaw puzzles, that children can physically touch, hold, rotate, and manipulate in their hands (e.g., Levine et al., 2012; Needham, 2009; Pruden et al., 2011). Initial examination of 2D touchpad devices (e.g., tablets and smartphones) suggests that use of these devices impacts parent–child interactions involving spatial language.

In a recent study, we examined the nature of spatial input—quantity and types of spatial words—produced by parents when engaged with their preschoolers in two separate 30-min play sessions at their home: one with 3D tangible blocks and puzzles, and the other using an iPad® featuring block and puzzle apps (Ho, Lee, Wood, Kassies, & Heinbuck, 2017). Overall, 6.2% of the parental talk during the sessions reflected spatial talk—words describing spatial properties and features, shapes, spatial dimensions, orientations and transformations, location and directions, deictics as well as continuous amount—in the 3D blocks and puzzles play context compared to 5.8% spatial talk in the 2D iPad® play context. These overall percentages did not differ statistically. However, developmental differences were evident in the amount of spatial talk generated. Specifically, when using the 2D iPad® with their older preschoolers, parents used fewer spatial words. Furthermore, although there was no overall difference in the quantity of spatial talk across the two play contexts, there were differences in the *types* of spatial talk that were elicited through each context. More words associated with spatial dimensions (e.g., big, tall, small), location, directions, and continuous amount (e.g., same, match, piece) were used by parents in the 3D play contexts than in the 2D play contexts. In contrast, more words related to orientations and transformations (e.g., turn, spin, rotate) as well as deictics (e.g., here, there, where) were produced in the 2D play contexts in comparison with the 3D context.

Thus far, the major focus of existing research on technology use of electronic toys during early childhood has been on the amount of parental input (e.g., Sosa, 2016; Zosh et al., 2015) and this initial study by Ho and colleagues (2017) suggests that amount may not differ. However, language acquisition research has shown that it is not simply a matter of quantity over quality. Both quantity and quality (i.e., the diversity or variation of words) of parental input influence young children's subsequent vocabulary development (Huttenlocher, Waterfall, Vasilyeva, Vevea, & Hedges, 2010; Rowe, 2012). Our research suggests that 2D and 3D play contexts may complement the types of spatial talk generated by parents.

In a subsequent study using similar methodology, we again found differences in the diversity of parental spatial words used in both the 3D (an average of 31 different spatial words) and the 2D iPad® play contexts (an average of 25 different spatial words) (Lee, Hodgins, Douglas, & Wood, 2017). Specifically, in 3D play contexts, parents produced more diverse words related to spatial dimensions, shapes (e.g., shape, square, triangle), and continuous amount than in the 2D play contexts. In the 2D play contexts, parents used more diverse words associated with orientations and transformations than in the 3D play contexts. Aspects of the software programs may contribute to these differences. For example, in the 2D context, software demands required specific actions to be performed to complete the task. Specifically, one task required *turning* an upside-down isosceles triangle 45° counterclockwise to fit into the existing geometric shape in the game in order to progress to the next step in the app. Importantly, the findings from both studies contrasting 2D iPad® and 3D traditional toys demonstrate that these contexts elicit differences in the nature of parental spatial talk and lend support to introducing touch-screen technology into play contexts at home to complement the types of parental spatial input and engagement elicited in the traditional 3D play contexts.

Instructional Affordances and Caveats Regarding Pedagogical Content of Apps in Spatial Development

Although our two research studies, combined with earlier research, suggest that the introduction of technologies could enhance spatial talk and facilitate mathematical thinking, caution is required before fully integrating technology into home and even early childhood settings. In particular, the design and content of software needs to be more critically examined to ensure high pedagogy and developmental standards are present. This would ensure that parents are able to tap into the full potential of touch-screen technology as a learning tool (e.g., Carbonneau & Marley, 2015; Hirsh-Pasek et al., 2015).

Unfortunately, emerging findings suggest shortcomings in many apps. For example, in Larkin's (2016) evaluation of geometry apps designed for children between 5 and 12 years old, only 7 of the 53 apps received a rating of 6 or higher on their 10-point scale based on three criteria: content, pedagogy (e.g., ease of use without instruction), and facilitation of the learner's thought process. Likewise, only 4 of the 19 geometry apps targeted for children between 3 and 5 years old covered at least 4 out of the 5 spatial concepts in a taxonomy reflecting developmental progression of key concepts (Lee, Douglas, Wood, & Andrade, 2017). Additionally, when the apps were evaluated with respect to instructional affordances, such as providing immediate, accurate feedback and moving children to higher or lower levels of difficulty based on performance, all but two of the 19 apps received a rating of 2 or lower on a 5-point scale. This is problematic because such instructional affordances (e.g., automatic leveling and feedback) have been found to improve mathematical performance of 4- to 7-year-old students using a tablet in the classroom (Outhwaite, Gulliford, & Pitchford, 2017).

These findings underscore the need for caution in using software to complement traditional math play contexts. Although apps may be labeled and advertised as "educational," the educational value of the app needs to be carefully examined. As Hirsh-Pasek and colleagues (2015) explain, some apps simply "masquerade" as educational software, for example, by appearing to introduce important concepts such as shape recognition, but encouraging only rote memorization rather than the underlying knowledge of how geometry develops in terms of features and transformation of shapes. As seen in other cognitive domains such as reading (e.g., Parish-Morris, Mahajan, Hirsh-Pasek, Golinkoff, & Collins, 2013), instructional shortcomings in software design and delivery can negatively impact learning and the nature of parent–child interactions. For example, when parent–child dyads engaged in dialogic reading with e-books compared to traditional book contexts, the nature of the questions posed by parents differed across the two contexts. In addition, children had lower levels of comprehension of key story elements in the e-book condition than in the traditional text condition. Differences in parent–child interactions apparent in other domains, such as reading, may also arise in mathematical contexts. Software design could thus also influence spatial language production and learning.

At present, parents and other caregivers have few reliable resources that identify developmentally appropriate apps based on the formal evaluation of their educational content (Guernsey & Levine, 2015; Rideout, 2014). Although a handful of Web-based consumer concern groups (e.g., Common Sense Media, Moms with Apps, Best Apps for Kids) provide some evaluation of the educational content of the apps, these evaluations are not systematic or exhaustive. Drawing from recent research regarding early literacy and reading software (Grant et al., 2012; Wood et al., 2016), there is a critical need for (a) the construction of tools that would allow parents and child-care educators to evaluate apps themselves, as well as (b) research that evaluates the content, instructional affordances, and outcomes associated with children's math apps. These elements are necessary to determine what underlying mathematical skills and processes are facilitated and to ensure that technology can be used to provide value-added learning experiences and opportunities when parents and children engage them.

Emerging research reinforces the need to investigate the impact of software design on learning outcomes and, in particular, parent–child spatial talk. Interactivity in software design, although typically associated with high engagement, may need to be altered to meet the needs of very young learners (Choi & Kirkorian, 2016; Kirkorian, Choi, & Pempek, 2016). Toddlers and preschoolers, for example, have less developed executive functions such as the ability to focus attention and impulse control which can be supported through app design that allows these children to easily disengage from distracting interactive features and focus their attention on educational content. Recent studies (Choi & Kirkorian, 2016; Kirkorian et al., 2016) have demonstrated that programs requiring 2- and 3-year-olds to tap a localized spot within the screen rather than tapping anywhere on screen facilitated their learning while older children performed better when they could move forward by tapping anywhere on the screen. In another study, design affordances such as repetitive and interactive features that reduce cognitive task demands have been found to aid in learning of 4- to 7-year-olds, especially for children with poor memory abilities (Outhwaite et al., 2017). This emerging research reinforces the need to examine individual differences in order to better understand for whom and when design affordances will maximize learning. To date, little is known about how design affordances may change the way parents use spatial language with their young children at play. Hence, ongoing research needs to address the impact of various affordances within specific content areas, such as spatial skills, in order to optimize learning opportunities for children when parents engage them with technologies.

Conclusions

Individual differences in mathematics emerge before the age of four (e.g., Levine et al., 2010; Verdine et al., 2014), and these differences persist into formal schooling (e.g., Aunola, Leskinen, Lerkkanen, & Nurmi, 2004; Duncan et al., 2007; Jordan, Kaplan, Olah, & Locuniak, 2006). With respect to spatial-visual skills, foundational skills can be introduced very early in life and can be facilitated through scaffolding from parents. Today, developmentally appropriate activities have expanded to include technology-based platforms. Digital technologies provide the potential for unique learning opportunities by exposing young children to developmentally appropriate cognitive (including math) skills as well as enriched and spatially diverse parental talk through play (e.g., Kabali et al., 2015; NAEYC & the Fred Rogers Centre for Early Learning and Children's Media, 2012; Rideout, 2014). Both 2D and 3D forms of play offer opportunities for children to acquire different spatial representations ranging from sensory-concrete to abstract. The diversity in media also influences how parents scaffold and talk to their children during play, with early evidence suggesting that 2D and 3D representations yield different but complementary spatial talk from parents. Before advocating for the integration of 2D and 3D presentations as the standard for play, however, considerable development is required regarding our understanding of how affordances in software design impact children's learning. In addition, research needs to examine how different media contexts and different software design elements impact parental engagement and language production in math contexts. Although digital technologies are deemed as a potential significant learning tool, parents play an important role in early spatial development. Extending our current initial understandings of these two important influences on early development is an important goal for ongoing research.

References

Abdul Jabbar, A., & Felicia, P. (2015). Gameplay engagement and game-based learning: A systematic review. *Review of Educational Research, 85*(4), 740–779.

Aslan, D., & Aktas-Arnas, Y. (2007). Three- to six-year-old children's recognition of geometric shapes. *International Journal of Early Years Education, 15*(1), 83–104.

Aunola, K., Leskinen, E., Lerkkanen, M. -K., & Nurmi, J. -F. (2004). Developmental dynamics of math performance from preschool to grade 2. *Journal of Educational Psychology, 96*, 699–713.

Baccaglini-Frank, A., & Maracci, M. (2015). Multi-touch technology and preschoolers' development of number-sense. *Digital Experiences in Mathematics Education, 1*(1), 7–27.

Blevins-Knabe, B., & Austin, A. B. (2016). *Early childhood mathematics skill development in the home environment*. Cham, Switzerland: Springer International Publishing AG.
Blevins-Knabe, B., Austin, A. B., Musun, L., Eddy, A., & Jones, R. M. (2000). Family home care providers' and parents' beliefs and practices concerning mathematics with young children. *Early Child Development and Care, 165,* 41–58.
Bourgeois, K. S., Khawar, A. W., Neal, S. A., & Lockman, J. J. (2005). Infant manual exploration of objects, surfaces, and their interrelations. *Infancy, 8*(3), 233–253.
Bruner, J. (1972). Nature and uses of immaturity. *American Psychologist, 27,* 687–708.
Caldera, Y. M., Culp, A. M., O'Brien, M., Truglio, R. T., Alvarez, M., & Huston, A. C. (1999). Children's play preferences, construction play with blocks and visual-spatial skills. Are they related? *International Journal of Behavioral Development, 23,* 855–872.
Carbonneau, K. J., & Marley, S. C. (2015). Instructional guidance and realism of manipulatives influence preschool children's mathematics learning. *The Journal of Experimental Education, 83*(4), 495–513.
Casasola, M. (2008). The development of infants' spatial categories. *Current Direction in Psychological Science, 17*(1), 21–25.
Casasola, M., & Bhagwat, J. (2007). Does a novel word facilitate 18-month-olds' categorization of a spatial relation? *Child Development, 78,* 1818–1829.
Casasola, M., Bhagwat, J., & Burke, A. S. (2009). Learning to form a spatial category of tight-fit relations: How experience with a label can give a boost. *Developmental Psychology, 45,* 711–723.
Casey, B. M., Andrews, N., Schindler, H., Kersh, J. E., Samper, A., & Copley, J. (2008). The development of spatial skills through interventions involving block building activities. *Cognition and Instruction, 26*, 269–309.
Casey, B. M., Nuttall, R., Pezaris, E., & Benbow, C. P. (1995). The influence of spatial ability on gender differences in mathematics college entrance test scores across diverse samples. *Developmental Psychology, 31*(4), 697–705.
Choi, K., & Kirkorian, H. L. (2016). Touch or watch to learn? Toddlers' object retrieval using contingent and noncontingent video. *Psychological Science, 27,* 726–736.
Clarke, B. A. (2004). A shape is not defined by its shape: Developing young children's geometric understanding. *Journal of Australian Research in Early Childhood Education, 11*(2), 110–127.
Clements, D. H., & Sarama, J. (2007). Effects of a preschool mathematics curriculum: Summative research on the building blocks project. *Journal for Research in Mathematics Education, 38,* 136–163.
Clements, D. H., Swaminathan, S., Hannibal, M. A., & Sarama, J. (1999). Young children's concepts of shape. *Journal of Research in Mathematics Education, 30,* 192–212.
Common Sense Media. (2017). *The common sense consensus: Media use by kids age zero to eight.* San Francisco, CA: Common Sense Media. Retrieved from https://www.commonsensemedia.org/sites/default/files/uploads/research/csm_zerotoeight_fullreport_release_2.pdf.
Dessalegn, B., & Landau, B. (2008). The role of language in binding and maintaining feature conjunctions. *Psychological Science, 19*(2), 189–195.
Duncan, G. J., Dowsett, C., Claessens, A., Magnuson, K., Huston, A., Klebanov, P., et al. (2007). School readiness and later achievement. *Developmental Psychology, 43*(6), 1428–1446.
Eagle, S. (2012). Learning in the early years: Social interactions around picturebooks and digital technologies. *Computers & Education, 59,* 38–49.
Ferrara, K., Hirsh-Pasek, K., Newcombe, N. S., Golinkoff, R. M., & Lam, W. S. (2011). Blocktalk: Spatial language during block play. *Mind, Brain, and Education, 5*(3), 143–151.
Fisher, K. R., Hirsh-Pasek, K., Golinkoff, R. M., & Newcombe, N. (2013). Taking shape: Supporting preschoolers' acquisition of geometric knowledge through guided play. *Child Development, 84*(6), 1872–1878.
Flynn, R. M., & Richert, R. A. (2015). Parents support preschoolers' use of a novel interactive device. *Infant and Child Development, 24*(6), 624–642.
Fontelle, S. A., Kahrs, B. A., Neal, S. A., Newton, A. T., & Lockman, J. J. (2007). Infant manual exploration of composite substrates. *Journal of Experimental Child Psychology, 98,* 153–167.

Foster, E. K., & Hund, A. M. (2012). The impact of scaffolding and overhearing on young children's use of the spatial terms between and middle. *Journal of Child Language, 39*(2), 338–364.
Gee, J. (2008). Good videogames, the human mind, and good learning. In E. Wood & T. Willoughby (Eds.), *Children learning in a digital world* (pp. 40–63). Oxford, UK: Blackwell Publishing.
Geist, E. (2014). Using tablet computers with toddlers and young preschoolers. *Young Children, 69,* 58–63.
Gentner, D. (2003). Why we're so smart. In D. Gentner & S. Goldin-Meadow (Eds.), *Language in mind: Advances in the study of language and cognition* (pp. 195–235). Cambridge, MA: MIT Press.
Gentner, D., & Lowenstein, J. (2002). Relational language and relational thought. In E. Amsel & P. Brynes (Eds.), *Language, literacy, and cognitive development: The development and consequences of symbolic communication* (pp. 87–120). Mahwah, NJ: Erlbaum.
Gentner, D., Özyürek, A., Gürcanli, Ö., & Goldin-Meadow, S. (2013). Spatial language facilitates spatial cognition: Evidence from children who lack language input. *Cognition, 127,* 318–330.
Gibson, E. J. (1988). Exploratory behavior in the development of perceiving, acting, and the acquiring of knowledge. *Annual Review of Psychology, 39,* 1–41.
Ginsburg, H. P. (2006). Mathematical play and playful mathematics: A guide for early education. In D. Singer, R. M. Golinkoff, & K. Hirsh-Pasek (Eds.), *Play = learning: How play motivates and enhances children's cognitive and social-emotional growth* (pp. 145–165). New York, NY: Oxford University Press.
Ginsburg, H. P. (2017). Maya loves Tubby and you should too, if you want to understand and evaluate children's math learning in the digital age. *Teachers College Record, 119*(3), 1–26.
Grant, A., Wood, E., Gottardo, A., Evans, M., Phillips, L., & Savage, R. (2012). Assessing the content and quality of commercially available reading software programs: Do they have the fundamental structures to promote the development of early reading skills in children? *NHSA: Dialogue: A Research to Practice Journal for the Early Intervention Field, 15*(4), 319–342.
Guernsey, L., & Levine, M. H. (2015). *Tap, click, read: Growing readers in a world of screens*. San Francisco, CA: Wiley/Jossey-Bass.
Gunderson, E. A., Ramirez, G., Beilock, S. L., & Levine, S. C. (2012). The relation between spatial skill and early number knowledge: The role of the linear number line. *Developmental Psychology, 48*(5), 1229–1241.
Hermer-Vazquez, L., Moffet, A., & Munkholm, P. (2001). Language, space, and the development of cognitive flexibility in humans: The case of two spatial memory tasks. *Cognition, 79,* 263–299.
Hirsh-Pasek, K., Michnick Golinkoff, R., Berk, L. E., & Singer, D. G. (2009). *A mandate for playful learning in preschool.* New York, NY: Oxford University Press.
Hirsh-Pasek, K., Zosh, J. M., Golinkoff, R. M., Gray, J. H., Robb, M. B., & Kaufman, J. (2015). Putting education in "educational" apps: Lessons from science of learning. *Psychological Science in the Public Interest, 16*(1), 3–34.
Ho, A., Lee, J., Wood, E., Kassies, S., & Heinbuck, C. (2017). Tap, swipe & build: Parental spatial input during iPad® and toy play. *Infant and Child Development, 27*(1).https://doi.org/10.1002/icd.2061.
Huttenlocher, J., Waterfall, H., Vasilyeva, M., Vevea, J., & Hedges, L. V. (2010). Sources of variability in children's language growth. *Cognitive Psychology, 61*(4), 343–365.
Jirout, J. J., & Newcombe, N. (2015). Building blocks for developing spatial skills: Evidence from a large, representative U.S. sample. *Psychological Science, 26*(3), 302–308.
Jordan, N. C., Kaplan, D., Olah, L. N., & Locuniak, M. N. (2006). Number sense growth in kindergarten: A longitudinal investigation of children at risk for mathematics difficulties. *Child Development, 77*(1), 153–175.
Kabali, H. K., Irigoyen, M. M., Nunez-Davis, R., Budacki, J. G., Mohanty, S. H., Leister, K. P., et al. (2015). Exposure and use of mobile media devices by young children. *Pediatrics, 135*(6), 1044–1050.
Kirkorian, H. L., Choi, K., & Pempek, T. A. (2016). Toddlers' word learning from contingent and non-contingent video on touchscreens. *Child Development, 87,* 405–413.

Kucirkova, N. (2014). iPads in early education: Separating assumptions and evidence. *Frontiers in Psychology, 5*(715), 1–3.

Kuhn, D. (2000). Metacognitive development. *Current Directions in Psychological Science, 9*(5), 178–181.

Lancy, D. (2007). Accounting for variability in mother-child play. *American Anthropologist, 109,* 273–284.

Landau, B., Spelke, E., & Gleitman, H. (1984). Spatial knowledge in a young blind child. *Cognition, 16,* 225–260.

Larkin, K. (2016). Geometry and iPads in primary schools: Does their usefulness extend beyond tracing an oblong? In P. S. Moyer-Packenham (Ed.), *Mathematics education in the digital era: International perspectives on teaching and learning mathematics with virtual manipulatives* (pp. 247–274). AG, Switzerland; Springer International Publishing.

Lee, J., Douglas, E., Wood, E., & Andrade, S. (2017). *How educational are math apps for preschoolers?* Poster presented at the 2017 Society for the Research of Child Development biennial meeting, Austin, TX.

Lee, J., Hodgins, S., Douglas, E., & Wood, E. (2017, June). *Does parental spatial talk differ during iPad® and toy play?* Paper presented at the Canadian Society for Brain, Behaviour and Cognitive Science (CSBBCS) 27th annual meeting, Regina, SK.

Levine, S. C., Suriyakham, L. W., Rowe, M. L., Huttenlocher, J., & Gunderson, E. A. (2010). What counts in the development of young children's number knowledge? *Developmental Psychology, 46,* 1309–1319.

Levine, S. C., Ratliff, K. R., Huttenlocher, J., & Cannon, J. (2012). Early puzzle play: A predictor of preschoolers' spatial transformation skills. *Developmental Psychology, 48*(2), 530–542.

Levinson, S. C., Kita, S., Haun, D. B. M., & Rasch, B. H. (2002). Returning the tables: Language affects spatial reasoning. *Cognition, 84*(2), 155–188.

Loewenstein, J., & Gentner, D. (2005). Relational language and the development of relational mapping. *Cognitive Psychology, 50,* 315–353.

Magnuson, K., Duncan, G., Lee, K. T. H., & Metzger, M. (2016). Early school adjustment and educational attainment. *American Educational Research Journal, 53*(4), 1198–1228.

Maloney, E. A., Converse, B. A., Gibbs, C. R., Levine, S. C., & Beilock, S. L. (2015a). Jump-starting early childhood education at home: Early learning, parent motivation, and public policy. *Perspectives on Psychological Science, 10*(6), 727–732.

Maloney, E. A., Ramirez, G., Gunderson, E. A., Levine, S. C., & Beilock, S. L. (2015b). Intergenerational effects of parents' math anxiety on children's math achievement and anxiety. *Psychological Science, 26,* 1480–1488.

McClure, E. R., Guernsey, L., Clements, D. H., Bales, S. N., Nichols, J., Kendall-Taylor, N., et al. (2017). *STEM starts early: Grounding science, technology, engineering, and math education in early childhood.* New York, NY: The Joan Ganz Cooney Center at Sesame Workshop.

Miller, H. E., Patterson, R., & Simmering, V. R. (2016). Language supports young children's use of spatial relations to remember locations. *Cognition, 150,* 170–180.

Mix, K. S., Levine, S. C., Cheng, Y. -L., Young, C. J., Hambrick, D. Z., & Konstantopoulos, S. (2017). The latent structure of spatial skills and mathematics: A replication of the two-factor model. *Journal of Cognition and Development, 18*(4), 465–492.

Moyer, P. S., Niezgoda, D., & Stanley, J. (2005). Young children's use of virtual manipulatives and other forms of mathematical representations. In W. Masalski & P. C. Elliot (Eds.), *Technology-supported mathematics learning environments: 67th yearbook* (pp. 17–34). Reston, VA: National Council of Teachers of Mathematics.

Moyer-Packenham, P. S., & Westenskow, A. (2013). Effects of virtual manipulatives on student achievement and mathematics learning. *International Journal of Virtual and Personal Learning Environments, 4*(3), 35–50.

Moyer-Packenham, P. S., Shumway, J. F., Bullock, E., Tucker, S. I., Anderson-Pence, K. L., Westenskow, A., … Jordan, K. (2015). Young children's learning performance and efficiency when using virtual manipulative mathematics iPad apps. *Journal of Computers in Mathematics and Science Teaching, 34*(1), 41–69.

Munnich, E., Landau, B., & Dosher, B. A. (2001). Spatial language and spatial representation: A cross linguistic comparison. *Cognition, 81,* 171–207.

National Association for the Education of Young Children (NAEYC). (2009). *Position statement: Developmentally appropriate practice in early childhood programs serving children from birth through to age 8.* Washington, DC: NAEYC.

National Association for the Education of Young Children (NAEYC), & the Fred Rogers Center for Early Learning and Children's Media at Saint Vincent College. (2012). *Technology and interactive media as tools in early childhood programs serving children from birth through age 8.* Joint position statement. Washington, DC: NAEYC.

National Governors Association Center for Best Practices, Council of Chief State School Officers. (2010). *Common core state standards for mathematics*. Retrieved from http://www.corestandards.org/Math/.

National Research Council. (2006). *Learning to think spatially: GIS as a support system in the K-12 curriculum.* Washington, DC: National Academies Press.

Needham, A. (2009). Learning in infants' object perception, object-directed action, and tool use. In A. Woodward & A. Needham (Eds.), *Learning and the infant mind* (pp. 208–226). New York, NY: Oxford University Press.

Neumann, M. M., & Neumann, D. L. (2014). Touch screen tablets and emerging literacy. *Early Childhood Education Journal, 42,* 231–239.

Newcombe, N. S. (2010). Picture this: Increasing math and science learning by improving spatial thinking. *American Educator*, 29–43.

Newcombe, N. S., Uttal, D. H., & Sauter, M. (2013). Spatial development. In P. Zelazo (Ed.), *Oxford handbook of developmental psychology, Vol. 1: Body and mind* (pp. 564–590). New York, NY: Oxford University Press.

Outhwaite, L. A., Gulliford, A., & Pitchford, N. J. (2017). Closing the gap: Efficacy of a tablet intervention to support the development of early mathematical skills in UK primary school children. *Computers & Education, 108,* 43–58.

Parish-Morris, J., Mahajan, N., Hirsh-Pasek, K., Golinkoff, R. M., & Collins, M. (2013). Once upon a time: Parent–child dialogue and storybook reading in the electronic era. *Mind Brain and Education, 7,* 200–211.

Piaget, J. (1932). *Play, dreams, and imitation in childhood.* New York, NY: Norton.

Plumert, J. M., & Nichols-Whitehead, P. (1996). Parental scaffolding of young children's spatial communication. *Developmental Psychology, 32*(3), 523–532.

Pruden, S. M., Levine, S. C., & Huttenlocher, J. (2011). Children's spatial thinking: Does talk about the spatial world matter? *Developmental Science, 14*(6), 1417–1430.

Rideout, V. J. (2014). *Learning at home: Families' educational media use in America.* A report of the families and media project. New York, NY: The Joan Ganz Cooney Center at Sesame Workshop.

Robinson, K. (2014). *The broken compass*. Cambridge, MA: Harvard University Press.

Rowe, M. L. (2012). A longitudinal investigation of the role of quantity and quality of child-directed speech in vocabulary development. *Child Development, 83*(5), 1762–1774.

Sarama, J., & Clements, D. H. (2016). Physical and virtual manipulatives: What is "concrete"? In P. S. Moyer-Packenham (Ed.), *International perspectives on teaching and learning mathematics with virtual manipulatives. Mathematics education in the digital era 7* (pp. 71–93). AG, Switzerland: Springer International Publishing.

Seo, K.-H., & Ginsburg, H. P. (2004). What is developmentally appropriate in early childhood mathematics education? Lessons from new research. In D. H. Clements, J. Sarama, & A. - M. DiBiase (Eds.), *Engaging young children in mathematics: Standards for early childhood mathematics education* (pp. 91–104). Hillsdale, NJ: Erlbaum.

Shea, D. L., Lubinski, D., & Benbow, C. (2001). Importance of assessing spatial ability in intellectually talented young adolescents: A 20-year longitudinal study. *Journal of Educational Psychology, 93*(3), 604–614.

Sosa, A. V. (2016). Association of the type of toy used during play with the quantity and quality of parent-infant communication. *JAMA Pediatrics, 170*(2), 132–137.

Uttal, D. H. (2000). Seeing the big picture: Map use and the development of spatial cognition. *Developmental Science, 3*(3), 247–264.

Verdine, B. N., Golinkoff, R. M., Hirsh-Pasek, K., Newcombe, N. S., Filipowicz, A. T., & Chang, A. (2014). Deconstructing building blocks: Preschoolers' spatial assembly performance relates to early mathematical skills. *Child Development, 85*(3), 1062–1076.

Vogel, J. J., Vogel, D. S., Cannon-Bowers, J., Bowers, C. A., Muse, K., & Wright, M. (2006). Computer gaming and interactive simulations for learning: A meta-analysis. *Journal of Educational Computing Research, 34*(3), 229–243.

Vygotsky, L. S. (1978). *Mind in society: The development of higher mental processes* (M. Cole, V. John-Steiner, S. Scribner, & E. Souberman, Eds. and Trans.). Cambridge, MA: Harvard University Press.

Wai, J., Lubinski, D., & Benbow, C. P. (2009). Spatial ability for STEM domains: Aligning over 50 years of cumulative psychological knowledge solidifies its importance. *Journal of Educational Psychology, 101*(4), 817–835.

Watts, T. W., Duncan, G. J., Clements, D. H., & Sarama, J. (2018). What is the long-run impact of learning mathematics during preschool? *Child Development, 89*(2), 539–555.

Wolfgang, C. H., Stannard, L. L., & Jones, I. (2003). Advanced constructional play with LEGOs among preschoolers as a predictor of later school achievement in mathematics. *Early Child Development and Care, 5,* 467–475.

Wood, E., Petkovski, M., De Pasquale, D., Gottardo, A., Evans, M. A., & Savage, R. (2016). Parent scaffolding of young children when engaged with mobile technology. *Frontiers in Psychology, 7,* 690. https://doi.org/10.3389/fpsyg.2016.00690.

Yerushalmy, M. (2005). Functions of interactive visual representations in interactive mathematical textbooks. *International Journal of Computers for Mathematical Learning, 10,* 217–249.

Yin, S. H. (2003). Young children's concept of shape: Van Hiele visualization level of geometric thinking. *The Mathematics Educator, 7*(2), 71–85.

Zosh, J. M., Verdine, B. N., Filipowicz, A., Golinkoff, R. M., Hirsh-Pasek, K., & Newcombe, N. S. (2015). Talking shape: Parental language with electronic versus traditional shape sorters. *Mind, Brain, and Education, 9*(3), 136–144.

Supporting Mathematics Play in Home Environments: A Feasibility Examination of a Take-Home Bag Intervention

Sandra M. Linder

Family engagement is a critical component of early childhood education. A body of research supports the notion that partnerships between high-quality childcare opportunities and high-quality family engagement can best support young children as they progress through the infant/toddler and preschool years into kindergarten (Ginsburg, 2007; Linder, Ramey, & Zambak, 2013; McNeal, 2015; Wu & Qi, 2006). While it is clear that family engagement is important, it is unclear what this engagement could look like across some content areas.

Early literacy development is a good example. The value of high-quality home literacy environments has been established as a predictor of success as children transition from preschool to more formal education settings, such as kindergarten, and as they progress through early childhood and elementary grade levels (Burgess, Hecht, & Lonigan, 2002; High, 2008). These high-quality home literacy environments include materials that support literacy development (e.g., accessible library of children's books), but also include parent literacy interactions with children where parents model the value of reading by engaging in the act of reading alongside of their children (Gottfried, Schlackman, Gottfried, & Boutin-Martinez, 2015). Family literacy bags have been used in the past to increase family engagement and provide families with concrete ideas they could use to support children's literacy development (Barbour, 1998; Brand, Marchand, Lilly, & Child, 2014; Crawford & Zygouris-Coe, 2006; Dever & Burts, 2002).

There is also an established body of research that demonstrates children's mathematical growth when engaging with mathematical tasks in home environments (Claessens & Engel, 2013; LeFevre, Skwarchuk, Smith-Chant, Fast, Kamawar, & Bisanz, 2009; Ramani & Siegler, 2008; Skwarchuk, Sowinski, & LeFevre, 2014). Research supports the notion that parent and child interactions in relation to mathematics can support mathematical growth in young children (Anders et al., 2012;

S. M. Linder (✉)
Early Childhood Mathematics Education, College of Education, Clemson University, 407F Tillman Hall, Clemson, SC 29634-0705, USA
e-mail: sandram@g.clemson.edu

K. M. Robinson et al. (eds.), *Mathematical Learning and Cognition in Early Childhood*,
https://doi.org/10.1007/978-3-030-12895-1_4

Huntsinger, Jose, & Luo, 2016; LeFevre, Polyzoi, Skwarchuk, Fast, & Sowinski, 2010). For example, Anders and colleagues (2012) examined mathematics play through numeracy-related activities in home environments and found strong associations between early numeracy play and school success (Anders et al., 2012). Mathematics play can take many forms. Direct mathematics activities such as board games (Ramani & Siegler, 2008) or direct mathematics questioning (how many items do you see?) are one type of mathematics play. Other types include more informal scenarios such as cooking with a parent, reading, and counting items in a story, building structures with blocks or Legos, puzzle-making, and so forth (LeFevre et al., 2010). These studies suggest that home mathematics environments could promote young children's cognitive growth in mathematics.

Despite this research, many parents or caregivers (herein called parents) are simply unsure of how to support early mathematical development (Huntsinger et al., 2016). Examples of how parents can support mathematical thinking are necessary and may be helpful to ensure that high-quality family engagement in mathematical play occurs in home environments. The tenets of high-quality home literacy environments can potentially be translated to the mathematics domain to increase mathematical play in home environments. These strategies should be play-based, grounded in exploration, engaging enough that parents and children are motivated to complete them continuously over time, and non-intrusive enough that parents and children can explore with them in whatever way makes sense to their contexts and schedules.

This chapter examines the feasibility of using play-based mathematics bags to support mathematical play of preschoolers in home environments. While "take-home bag" type products for school-aged children related to mathematics or STEM are flooding the marketplace, empirical evidence to examine the impact of these products on young children's cognitive growth lags.

Some authors have examined mathematics bag interventions for parents of school-aged children and found that the practices outlined within these bags (primarily numeracy focused) supported increased mathematical engagement in home environments (Goos & Jolly, 2004; Kokoski & Patton, 1997; Muir, 2009, 2012). This chapter builds on research related to school-aged children by examining the feasibility of a take-home mathematics bag intervention over a five-week period for families of preschool age children. Two questions guided our work: (1) How and to what extent did families interact with the materials in each mathematics bag? (2) What did families learn about their children's mathematical knowledge as a result of each mathematics bag?

The interdisciplinary nature of this chapter draws from the worlds of psychology and mathematics. Specifically, Bronfenbrenner's Ecological Systems Theory of Development (Bronfenbrenner, 1986) guides the development of the mathematics bags used in this study. By capitalizing on the worlds that exist for each child within that child's individual microsystems (e.g. home and school), we can better support the connections for the child within their mesosystem or where that child's microsystems interact. The mesosystem represents the connections that children can make between home and school, between parent and teacher, and between the mathematical content explored with their peers in school and how it can relate to their world outside of school (Bronfenbrenner, 1986). It is these connections that support cognitive growth across domains (Downer & Pianta, 2006).

Preschool—Home Connections

This project started as part of a larger study focused on providing early mathematics professional development for childcare teachers in southeastern United States. The project, *Building Environments for Early Mathematics Success* (Project BEEMS) examined the role of the teacher in creating effective classroom environments for early mathematical thinking by making explicit connections between the child, the mathematical content and processes, and the classroom space (Linder & Simpson, 2016; Simpson & Linder, 2016). In the second year of Project BEEMS, teacher participants in four-year-old classrooms discussed strategies for connecting their students' home lives to mathematical interactions within the classroom space. As part of these discussions, the project team and BEEMS participants began to explore possibilities for increasing mathematical interactions in home environments. From these discussions, we determined that, in addition to being play-based and mathematics focused, the intervention designed to increase family mathematical interactions must have the following characteristics:

- **The burden of time and cost on families must be small**. The majority of families at these childcare centers had non-traditional schedules, in that they worked more than one job or worked a second or a third shift in the evenings. These families had little time to spend attending workshops or meetings, even if the meetings were scheduled within the home setting, which has been successful for other early childhood family interventions (Olds, 2006; Olds et al., 2013; Wagner, Spiker, & Linn, 2002). Many of the families involved in this project were from single-parent families and did not have childcare support in the evenings. The strategy should not require parents to attend workshops or meetings, or spend time away from the home. Rather, the strategy must allow parents to have flexibility for when and how often they choose to engage in the task with their children.
- **The burden of time and cost on teachers must be small**. To ensure sustainability of the intervention over time and to set the stage for scale-up possibilities, the family mathematics strategy needed to be cost effective, in that the materials could be shared across families and most of the materials used were non-consumable and did not need to be replaced on a regular basis. In addition, teachers shared the same time constraints as parents in terms of having to implement this intervention on top of their already scheduled duties. To ensure that the strategy could be continued beyond the scope of the project, we had to make it self-servicing, meaning that families could implement it without a high level of external support from teachers.

Mathematics Bags

With these characteristics in mind, the BEEMS project team and teacher participants decided on take-home bags as a potential strategy for increasing family mathematical interactions through play. The project team developed the bags and then took teacher participants through the tasks during BEEMS PD sessions to ensure that they could support parents during the implementation of the intervention if necessary (although parents did not end up needing any support to use the mathematics bags). These bags were self-servicing, in that they contained all of the materials necessary to explore a mathematical content area through play. There were some bags that needed to be refurbished when parents returned them, but the items needed were common to childcare settings, such as crayons, stickers, and construction paper. The bags also contained a task sheet that provided families with directions and suggested questioning strategies for engaging with the materials. The mathematics bags allowed for flexibility. Parents had the ability to choose when to play with their children through the mathematics bags over the course of a week. Some played with each bag one time, some played with the bags every day of the week.

Each bag focused on a mathematical content area (early number, geometry, measurement, data analysis, algebra/patterning). Tasks were grounded in mathematical process skills including problem solving, reasoning and proof, representation, communication, and making connections (NCTM, 2014). We utilized the 5E learning cycle approach (Bybee, 2015) to structure the framework within each bag. This approach is commonly used in inquiry instruction to provide a framework for teachers to support inquiry practices in classroom spaces. The 5E cycle begins with an Engage phase where a topic or phenomenon is introduced and children often make connections to what they already know about the topic. The cycle then continues with the Explore phase where children engage in open-ended tasks designed to promote inquiry processes such as problem-solving, asking questions, and thinking critically. Following the Explore phase, children have an opportunity to reflect on the task through the Explain phase. Finally, the Elaborate phase provides an opportunity to children to extend what they learned through a new task. The fifth E stands for Evaluate, which takes place throughout the learning cycle (Bybee, 2015). While the learning cycle is often used in classroom settings, it was chosen for the mathematics bags to provide a framework for parents who are using these inquiry practices for what could be the first time at home.

Each bag contained a literature selection that connected to the mathematical content area of focus (Engage). Parents were asked to read and discuss the literature selection with their child. Following this read aloud, task sheets provided potential ways of exploring the materials that fostered mathematical play between parents and children and opportunities to reflect on what they did together (Explore and Explain). Task sheets also included ideas for extension and for making mathematical connections within the home environment (Elaborate). The use of reflection sheets within each bag acted as the evaluation portion of the 5E framework.

The contents of the data analysis math bag are included in Fig. 1 as an example of the types of materials that were included across all five bags. The data analysis bag began with a read aloud of "Do Like a Duck Does" by Judy Hindley and Ivan Bates. This book allowed parents and children to discuss the similarities and differences between the ducks and the fox in the story and the task sheet provided open-ended questions for parents to ask children as they discuss during and following the story. After the read aloud, families were encouraged to explore sorting with a variety of buttons. The goal of these sorting tasks was for families to discuss potential rules for sorting rather than parents telling their children to sort by a prescribed rule. Following the sorting tasks, parents were encouraged to implement sorting tasks throughout the day in their home environment (e.g., sorting toys or books, and sorting dishes.). Table 1 provides an outline of the bag topic, read aloud, short task description, and method of elaboration for each of the five bags in the study.

In total, ten families participated in the project. These ten families all had four-year-old children at two of the childcare centers involved in Project BEEMS. Each family was recommended by their child's classroom teacher and, when contacted, was willing to participate in the process. Although all family members, including siblings, were typically involved in the mathematics play occurring at home through the math bags, only female parents volunteered to engage in the pre-/post-data collection processes. The highest education level obtained by the parent in the ten participating families was as follows: two were high school graduates, three had attended some college coursework, four had completed as associates degree and one had completed a bachelor's degree.

Participating families received a weekly mathematics bag on a Friday and had the entire week to explore its contents. Families could engage with each mathematics bag in a single play session or spread the tasks out over a weeklong period. Families were encouraged to play with the materials multiple times throughout the week, but were only required to engage once with the materials in order to remain part of the project. On each subsequent Friday, families returned their mathematics bag to the

Fig. 1 Contents of data analysis math bag

Table 1 Outline of mathematics bags using the 5E learning cycle

	Engage	Explore	Explain	Elaborate
Early numeracy	Read Aloud: "How Many Bugs in a Box" (Carter, 1987)	Describing relative size of cups and predicting how many vegetables can fit in each cup. Composing and decomposing sets for each cup	Comparing predictions to results and identifying strategies for counting and describing a set	Counting items around the house throughout the week. Describing strategies for counting and describing sets of objects
Geometry	Read Aloud: "Color Zoo" (Ehlert, 1989)	Using pattern cards and pattern blocks to create animal designs with shapes, removing cards to create additional designs without support	Describing designs and attributes of shapes within designs, exploring how shapes can be comprised of other shapes within the design	Engaging in puzzle play to explore transformations, playing I Spy to identify shapes around the house
Measurement	"Who Sank the Boat" (Allen, 1996)	Using a balance and materials commonly found in classrooms/homes (paper clips, cotton balls, q-tips, rubber balls, marbles, etc.) to explore heavy and light	Comparing predictions about heavy and light to results from exploration. Describing attributes of objects in the bag in terms of weight	Examining heavy and light with water instead of solid objects and making connections to sinking and floating during bath time or outdoor play
Algebra	"Max Found Two Sticks" (Pickney, 1994)	Replicating, describing, and creating sound patterns using rhythm sticks	Discussing the rhythm patterns and comparing the patterns to repeating patterns made with counting animals	Replicating, describing, and creating patterns throughout the week using stickers and crayons/paper
Data Analysis	"Do Like a Duck Does" (Hindley & Bates, 2002)	Finding multiple ways to sort a set of fabric buttons	Discuss the various rules they chose for sorting, discuss the sorts and resulting groups	Sorting items around the house such as laundry, dishes, and toys

childcare center by putting it in their child's backpack. In addition to learning about the tasks during BEEMS PD sessions so they were able to help parents if necessary, teachers supported the intervention by refurbishing and switching the mathematics bags on Friday afternoon to ensure that each participating family had a new bag to explore the following week. Teachers also made phone calls, texts, or e-mails to participating families on Thursday evenings to remind them to send the mathematics bags on Friday mornings.

Engaging in Math Play Parent Feedback Form
Content Area: Data Analysis

Parent's name:

Child's name:

Date:

How many times did you and/or child complete the task? ____________

This task was appropriate for my child Agree Disagree

We enjoyed completing the task Agree Disagree

The directions were clear Agree Disagree

This was the first time my child and I did a task like this Agree Disagree

What math understandings (or misunderstandings) did your child show when participating in the task?

What questions or conversations did you add to the task as you were doing it with your child?

How did you continue this task throughout the week (perhaps with different materials)?

Do you have any other comments or questions about the activity?

Fig. 2 Parent feedback form

Included in each math bag was a parent feedback form (see Fig. 2) that asked for self-reported data on bag usage, clarity of directions, and level of enjoyment. The feedback form also contained open-ended questions regarding how families interacted with the math bag throughout the week and what mathematical understandings children showed while engaging in mathematical play. In addition to this weekly feedback form, parents completed semi-structured interviews following the five-week intervention to determine their overall reaction to the intervention.

What Did Parents Say and Do?

Parents had an overwhelmingly positive reaction to the mathematics bag intervention as a strategy for increasing mathematical play in home environments. All families completed the five-week intervention period and completed all the tasks within all five mathematics bags. As far as the number of times families engaged with each mathematics bag on a weekly basis, all families had a high level of engagement with every bag (see Fig. 3). Mean level of engagement for all of the bags ranged from 3.3 to 3.88 times per week, indicating that on average, families played with the content of the mathematics bags multiple times per week even though they were only asked to play at least one time per week.

Table 2 shows participant responses to closed-ended statements about each bag. Overall, parents felt that the bags were appropriate for the age level of the child and that the directions were clear. In addition, parents enjoyed completing the tasks in each bag with their children. In one case, a parent reported that they disliked the early number bag but added a note to the feedback form that it was because her child had difficulty recognizing numbers. Interestingly, many parents felt like this intervention

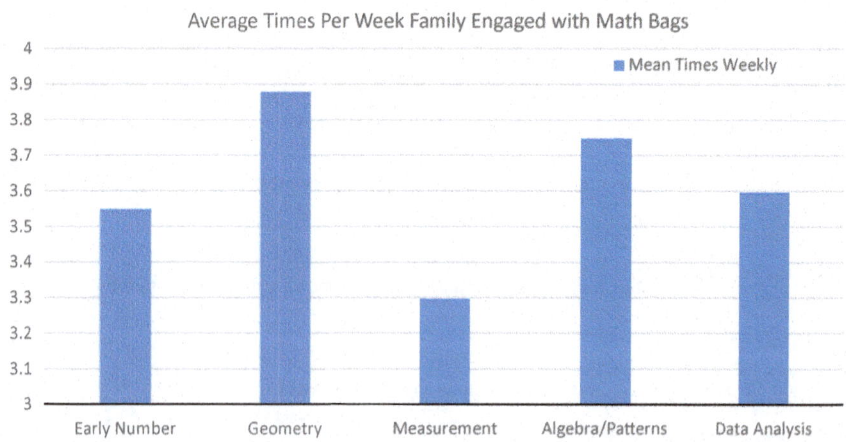

Fig. 3 Weekly interactions with each math bag

Table 2 Parent reactions to weekly math bags

		Early number (%)	Geometry (%)	Measurement (%)	Algebra (%)	Data analysis (%)
This task was appropriate for my child	Agree	100	100	90	90	100
	Disagree	0	0	10	10	100
We enjoyed completing the task	Agree	90	100	100	100	100
	Disagree	10	100	100	100	100
The directions were clear	Agree	100	100	100	100	100
	Disagree	0	0	0	0	0
This was the first time my child and I did a task like this	Agree	60	50	80	70	60
	Disagree	40	50	20	30	40

was the first time they experienced these types of tasks with their child. These results show a high level of motivation (by completing each bag multiple times) and positive reactions to the content of each of the five bags.

Content Specific Engagement and Parent Feedback

Early Number. Parents recognized that their children had the ability to count and label sets with a total amount. Two parents described their child's ability to use one-to-one correspondence when counting sets. However, while the task in the early number mathematics bag focused on composing and comparing sets, eight parents described asking their child to add or subtract the sets. For example, the task called for children to fill cups with a given amount of vegetable manipulatives (composing sets) and then to separate the set into two cups (decompose) in as many ways as they were able to find. Instead, parents would describe asking their children to take away certain amounts of the vegetable manipulatives from the total set (subtraction). Overall, parents felt their children did very well with the content of this mathematics bag. One parent, Melissa (pseudonyms used throughout), made this comment: "My daughter made me proud when she predicted a certain amount of veggies in the cup and realized she had too many. She quickly fixed her mistake."

Geometry. Parents described their child's ability to count the number of shapes or identify the names of the shapes. Some even described asking their children to add using the shapes even though the task did not call for parents and children to discuss number-related concepts. Oftentimes parents reported including questions that related to sorting the shapes by colors. There were very few mentions of attributes of shapes or higher-level analysis within geometric levels of thinking such as examining shapes in terms of their attributes. However, two parents did describe how their child worked to make bigger shapes using the smaller shapes. These two families also described additional questions that they asked their children while playing with the bags, such as "What smaller shapes do you see in the bigger picture?" and "What shapes do you see around the house?"

Measurement. Many of the parents recognized their child's understanding of balancing items on a scale and the concepts of heavy versus light. They also described their strategies for making connections around the house with measurement by exploring heavy and light with water or using different materials than the ones provided with the balance that was included in this mathematics bag. Only one parent (Bonnie) focused on the notion of counting when describing how they interacted with the materials, "We counted the paper clips, and the cotton balls, and feathers and put them into the balance to see what side would weigh more." While her approach was reasonable, she did not mention any discussion of the attributes of these items and why one object might be heavier than another object.

Algebra. Parents reported a high level of enjoyment from their children when creating or replicating sound patterns as part of this mathematics bag. However, even though patterning was the main focus of the read aloud and tasks in the bag, again parents focused any additional questions on counting or number concepts such as adding sets. While parents described creating patterns with stickers and representing sound patterns using sticks, when asked to describe what their child learned, responses often centered on early number.

Data. The majority of parents did not make any changes to the suggested tasks provided in the data analysis bag. One parent described having her children act out the parts of the book but did not make changes to the sorting suggestions. All of the parents described their child's process for counting and comparing each set of buttons, which was appropriate for the task outlined in the bag. However, along with counting and comparing, children were encouraged to engage in sorting sets with a variety of rules. Parents often described suggesting the rules that children should use to sort (color, design). Parents described extensions they made around the home such as sorting dishes or clothing by how they were similar and different.

In general, parents did not describe any negative reactions to the materials, tasks, or suggested questions within any of the bags. They all described ways that they added to the content of the bags, such as adding additional questions or extensions to focus on household connections, but no parents described intent to alter the mathematics tasks suggested in each bag. However, based on reported feedback, parents at times changed the *intent* of a mathematics bag task to be more traditionally oriented, such as directing a child toward a particular rule for sorting rather than discussing potential rules and deciding on one together. In addition, all but one parent reported children's increased understandings related to early number, even when early number concepts were not the focus of a particular mathematics bag. Only the early number and data analysis bags had explicit connections to number concepts. The measurement, geometry, and algebra/patterning bags had tasks that did not include explicit number concept connections. This result may indicate that parents perceive mathematics for young children as primarily related to counting or early number concepts. Additional interview data gathered as part of this project asked parents to define mathematics for young children. Parent responses to this line of questioning heavily focused on number concepts (counting, adding, subtracting). In terms of bag implementation, if parents perceive early mathematics as primarily related to early number concepts, they may be more likely to promote these concepts over all others. The impact of prioritizing these concepts warrants further investigation.

Analysis of the semi-structured interviews following the five-week implementation yielded three themes: (1) It was fun; (2) I was surprised by my child; and (3) I need to do more. These themes are described below.

It was fun. Every participant described the five-week intervention as something that was fun. While most of these families had overwhelming schedules with multiple jobs, sibling needs, and additional commitments, they all found time to play with their children on multiple occasions throughout each week. They described the mathematics bags as enjoyable and found that having set tasks to do with their child helped their feelings of insecurity at not knowing how to best encourage their child's mathematical growth. Further, they recognized that their children were having fun and found enjoyment in watching children explore the materials. Tonya, a parent of twin boys, commented, "The kids really enjoyed doing these activities. I feel the activities are a good learning experience for the kids." Even parents who had self-reported negative feelings or low levels of confidence related to mathematics saw this intervention as enjoyable and something that they would have continued past the five-week period and would like to repeat in the future. For example, one parent, Wanda, reflected, "We did it every day because she isn't going to let me forget, she'll be like, mama my bag, get my bag. I'm like ok-let me get your bag. She liked that though-she'd be on me like mama-my bag, I'd say ok come on let's get up to the table and she'd run to the table, she enjoyed that. And I didn't think she would but she found a lot of stuff interesting. You know some people don't but she really does."

Surprised by my child. Parents reported being surprised at what their children were capable of doing. Melinda, parent of four-year old Mason, stated, "I liked that he was learning. Because I hadn't really thought about putting shapes on top of pictures and how many different triangles does it take to make a pentagon I never really thought about asking him that and he actually wanted to count after asking the questions I could see his mind working so I really like that. I like to see him amazed when he learns something new." However, they also felt overwhelmed by the amount of mathematical content explored within each bag and it took some time to get used to the notion that their children were capable of more than what they expected. For example, Wanda stated,

> When I saw the first bag, I said oh my god, what are we going to have to do in this bag! But as I took it out and read the instructions and everything, and me and her were sitting at the table, everything we went through, the colors, the sticks, she loved the drum sticks like how the little boy had the drum…she loved the little boat thing, I mean cause she understood how to balance it because we were switching things out and she was showing me. And I said do you understand how to balance it and she said yes and I asked what else does it look like because she watches a lot of stuff, so she was like, it looks like a big boat mama. We had fun with those bags, it was a lot of stuff and I didn't realize we were going to do a lot of different stuff with them.

I need to do more. Parents reported that the bags helped them to understand some of the ways they could support their children throughout the day to encouraging mathematical growth. Even though parents often described children's understanding in terms of early number, three parents also began to expand their horizons of how mathematics can be connected to other content domains. One of these three parents, Miranda, commented, "I didn't think about math being associated with reading other than just numbers on a paper." The mathematics bags also pushed parents to reflect on the types of interactions they had with their children both during the suggested tasks and also during a typical day. Parents described the need to ask their children more questions and talk about mathematics throughout the day. Bonnie commented, "It was a lot of questions and I think the questions was what made him think, cause I ask questions but I don't ask them in the way the sheet was asking … Because a lot of those questions I never thought to ask him, but now it is like ok I need to start asking him a little bit more questions to open up his mind."

Intentional Connections

The math bags made explicit connections between content areas to support mathematical play between parents and children. Every bag first included a literature selection that connected to the mathematical content in the bag. On average, each mathematics bag took 45 min to an hour to complete in its entirety and all parents described their children wanted to do the entire bag each time they began to play. They also described completing the tasks in each bag multiple times throughout the week. Finally, the mathematics bags supported physical development by providing materials that would also enhance other skills such as fine motor development (e.g., stickers, crayons, animal manipulatives) and gross motor development (e.g. rhythm sticks and the possibility that they might make patterns with movement).

The overall feasibility of using mathematics bags as a way to encourage family mathematical play interactions was extremely high for the parents in this study. Not only did they go above and beyond the minimum requirements for the study, but every family voiced a willingness to continue with the intervention for a longer duration (two even stated that the five-week duration was too short and requested additional bags). Families saw the mathematics bags as fun, easy to follow, and appropriate for their child. In addition, they identified the flexibility to use the bags whenever they could find time within their schedule as a positive attribute to the intervention. Further, families did not describe the need for additional support from a teacher to complete the tasks within each bag. Each participant felt confident enough in their own mathematics understanding and the mathematics understanding of their child to engage with the tasks in each bag.

While families felt that the self-servicing aspect of the bags was a positive attribute, continued exploration of additional support could be worthwhile. Although parents reported completing each task with fidelity, there were no observations of these play periods in the home environments, although pre- and post-play periods did show a significant increase in mathematical interactions between parents and children as reported in a separate paper. Further, descriptions of how they engaged with each bag from feedback forms and interviews indicate a heavy emphasis on early number at the expense of other early childhood mathematical concepts. These content shifts suggest that parents may need increased support in understanding the depth of mathematical content that young children can explore.

A central goal of this intervention was for it to remain non-intrusive on participant or teacher time. Continued examination of this intervention could include an alternative strategy for increased support such as online forums or a reference sheet of frequently asked questions or common misconceptions included in each bag. Further, it would be worthwhile to explore additional mathematics bags as part of this intervention as participant fatigue did not occur as part of this study.

Conclusion

Mathematics bags seemed to be a fun and engaging intervention for increasing mathematical connections between home and school environments and for increasing the amount and quality of mathematical interactions between parents and their children. By capitalizing on the worlds that exist for each child within that child's individual microsystems (in this case, home and school), we can better support the connections for the child within their mesosystem, or where that child's microsystems interact. The mesosystem represents the connections that children can make between home and school, between parent and teacher, and between the mathematical content explored with their peers in school and how it can relate to their world outside of school (Bronfenbrenner, 1986). The mathematics bags supported these connections through inquiry-oriented tasks that allowed parents and children to interact through mathematical play.

These mathematics bags can be easily created with items taken directly from early childhood classrooms or with items that are commonly found in home environments. Teachers can use the learning cycle framework to develop bags that allow parents to engage their children with a read aloud, explore together with a mathematical task that connects to the story, and then elaborate on this task by extending it to their home environment. Throughout this cycle, parents should be encouraged to act as a facilitator, asking open-ended questions, playing with their child throughout, and allowing their child to make decisions within the task. This approach can be used within any mathematical content area.

This chapter acts as a small piece of a larger discussion of how to support high-quality home mathematics environments and what implications these environments can have on young children's mathematics achievement and dispositions toward mathematics. Future research should explore this type of non-intrusive mathematics intervention with an increased number of families from varying backgrounds to determine feasibility across multiple variables. Concurrently, this research should explore the impact of this type of intervention on the amount and quality of parent and child mathematical interactions. Finally, future research on mathematics bag interventions should explore the predictive value of high-quality home mathematics environments on children's success as they transition to kindergarten and progress toward increasingly difficult mathematical content over time.

References

Allen, P. (1996). *Who sank the boat?*. London, England: Puffin Books.

Anders, Y., Rossbach, H., Weinart, S., Ebert, S., Kuger, S., Lehrl, S., et al. (2012). Home and preschool learning environments and their relations to the development of early numeracy skills. *Early Childhood Research Quarterly, 27*(2), 231–244.

Barbour, A. C. (1998). Home literacy bags promote family involvement. *Childhood Education, 75*(2), 71–75.

Burgess, S. R., Hecht, S. A., & Lonigan, C. J. (2002). Relations of the home literacy environment (HLE) to the development of reading-related abilities: A one-year longitudinal study. *Reading Research Quarterly, 37*(4), 408–426.

Brand, S. T., Marchand, J., Lilly, E., & Child, M. (2014). Home-school literacy bags for twenty-first century preschoolers. *Early Childhood Education Journal, 42*(3), 163–170.

Bronfenbrenner, U. (1986). Ecology of the family as a context for human development: Research perspectives. *Developmental Psychology, 22*(6), 723–742.

Bybee, R. (2015). *The BSCS 5E instructional model: Creating teachable moments*. Arlington, VA: NSTA Press.

Carter, D. A. (1987). *How many bugs in a box?*. New York, NY: Little Simon.

Claessens, A., & Engel, M. (2013). How important is where you start? Early mathematics knowledge and later school success. *Teachers College Record, 115*(6), n6.

Crawford, P. A., & Zygourias-Coe, V. (2006). All in the family: Connecting home and school with family literacy. *Early Childhood Education Journal, 33*, 261–267.

Dever, M., & Burts, D. (2002). An evaluation of family literacy bags as a vehicle for parent involvement. *Early Child Development and Care, 174*(4), 359–370.

Downer, J. T., & Pianta, R. C. (2006). Academic and cognitive functioning in first grade: Associations with earlier home and childcare predictors and with concurrent home and classroom experiences. *School Psychology Review, 35*(1), 11.

Ehlert, L. (1989). *Color zoo*. New York, NY: HarperCollins Publishers.

Ginsburg, K. R. (2007). The importance of play in promoting healthy child development and maintaining strong parent-child bonds. *Pediatrics, 119*(1), 182–191.

Goos, M., & Jolly, L. (2004). Building partnerships with families and communities to support children's numeracy learning. *MERGA 27*.

Gottfried, A. W., Schlackman, J., Gottfried, A. E., & Boutin-Martinez, A. S. (2015). Parental provision of early literacy environment as related to reading and educational outcomes across the academic lifespan. *Parenting, 15*(1), 24–38.
High, P. C. (2008). School readiness. *Pediatrics, 121*(4), e1008–e1015.
Hindley, J., & Bates, I. (2002). *Do like a duck does*. Cambridge, MA: Candlewick Press.
Huntsinger, C. S., Jose, P. E., & Luo, Z. (2016). Parental facilitation of early mathematics and reading skills and knowledge through encouragement of home-based activities. *Early Childhood Research Quarterly, 37,* 1–15.
Kokoski, T. M., & Patton, M. M. (1997). Beyond homework: Science and mathematics backpacks. *Dimensions of Early Childhood, 25*(2), 11–16.
LeFevre, J. A., Polyzoi, E., Skwarchuk, S.-L., Fast, L., & Sowinski, C. (2010). Do home numeracy and literacy practices of Greek and Canadian parents predict the numeracy skills of kindergarten children? *International Journal of Early Years Education, 18,* 55–70.
LeFevre, J.-A., Skwarchuk, S.-L., Smith-Chant, B. L., Fast, L., Kamawar, D., & Bisanz, J. (2009). Home numeracy experiences and children's math performance in the early school years. *Canadian Journal of Behavioral Science, 41,* 55–66.
Linder, S. M., Ramey, M. D., & Zambak, S. (2013). Predictors of success for school readiness: A selective review of the literature. *Early Childhood Research and Practice, 15*(1), n1.
Linder, S. M., & Simpson, A. (2016). *Examining the Effects of mathematics professional development in child care settings*. Paper presented at the annual meeting of the American Educational Research Association. Washington, D.C.
McNeal, R. B., Jr. (2015). Parent involvement and student performance: The influence of school context. *Educational Research for Policy and Practice, 14*(2), 153–167.
Muir, T. (2009). At home with numeracy: Empowering parents to be active participants in their child's numeracy development. *MERGA, 1,* 395–402.
Muir, T. (2012). Numeracy at home: Involving parents in mathematics education. *International Journal for Mathematics Teaching and Learning, 25,* 1–13.
National Council of Teachers of Mathematics (NCTM). (2014). *Principles to action: Ensuring mathematical success for all*. Reston, VA: NCTM.
Olds, D. L. (2006). The nurse–family partnership: An evidence-based preventive intervention. *Infant Mental Health Journal, 27*(1), 5–25.
Olds, D., Donelan-McCall, N., O'Brien, R., MacMillan, H., Jack, S., Jenkins, T., et al. (2013). Improving the nurse–family partnership in community practice. *Pediatrics, 132,* 110–117.
Pickney, B. (1994). *Max found two sticks*. Reading Rainbow Book. New York, NY: Aladdin Books.
Ramani, G. B., & Siegler, R. S. (2008). Promoting broad and stable improvements in low-income children's numerical knowledge through playing number board games. *Child Development, 79*(2), 375–394.
Simpson, A., & Linder, S. M. (2016). The indirect effect of children's gender on early childhood educators' mathematical talk. *Teaching and Teacher Education, 54,* 44–53.
Skwarchuk, S. L., Sowinski, C., & LeFevre, J. A. (2014). Formal and informal home learning activities in relation to children's early numeracy and literacy skills: The development of a home numeracy model. *Journal of Experimental Child Psychology, 121,* 63–84.
Wagner, M., Spiker, D., & Linn, M. I. (2002). The effectiveness of the parents as teachers program with low-income parents and children. *Topics in Early Childhood Special Education, 22*(2), 67–81.
Wu, F., & Qi, S. (2006). Longitudinal effects of parenting on children's academic achievement in African American families. *The Journal of Negro Education, 75,* 415–429.

Part II
The Beginnings of Formal Schooling

Early Identification of, and Interventions for, Kindergarten Students at Risk for Mathematics Difficulties

Marcie Penner, Chad Buckland and Michael Moes

Introduction

There are large individual differences in numeracy skill, even as early as kindergarten (Jordan, Kaplan, Ramineni, & Locuniak, 2009). On average, there is a seven-year span in ability within a single elementary classroom (Cockcroft, 1982). Of particular concern is the finding that children who enter kindergarten with poor numeracy skills do not catch up (Aunola, Leskinen, Lerkkanen, & Nurmi, 2004; Clarren, Martin, & Townes, 1993; Duncan et al., 2007; Jordan, Kaplan, Ramineni, & Locuniak, 2009; Morgan, Farkas, & Wu, 2009; Shalev, Manor, & Gross-Tsur, 2005; Watts, Duncan, Siegler, & Davis-Kean, 2014), likely due to the lack of early identification and intervention tools (Mazzocco, 2005). One in ten children will persistently struggle to gain numeracy skills (Mazzocco & Myers, 2003). This is problematic, as basic numeracy skills are important not only for later mathematics achievement, but also for general academic performance and student retention (Duncan et al., 2007; Romano, Babchishin, Pagani, & Kohen, 2010; Parsons & Bynner, 2005). Indeed, school-entry numeracy skills were found to be the best predictor of later academic achievement (Duncan et al., 2007; Romano, Babchishin, Pagani, & Kohen, 2010). Numeracy skills are also important for life outcomes, including employment opportunities, obtaining and retaining employment, promotion opportunities, owning a home, income, quality of health care, and mental health (Bynner & Parsons, 1997; Parsons & Bynner,

M. Penner (✉)
Department of Psychology, King's University College at Western University,
266 Epworth Avenue, London, ON N6A 2M3, Canada
e-mail: mpennerw@uwo.ca

C. Buckland
Department of Psychology, Western University, London, ON, Canada

M. Moes
Department of Psychology, King's University College at Western University,
London, ON, Canada

K. M. Robinson et al. (eds.), *Mathematical Learning and Cognition in Early Childhood*,
https://doi.org/10.1007/978-3-030-12895-1_5

1997, 2005; Ritchie & Bates, 2013). Thus, research focused on developing tools to assess and predict children's numeracy skills in kindergarten, or earlier, is critical. This chapter will focus on (1) early cognitive predictors of numeracy skill, (2) diagnostic tools for early identification of at-risk students, and (3) early interventions for students identified as at risk for mathematics difficulties.

Predictors of Numeracy

In order to identify which children are at risk for mathematics difficulties, we need early, long-term (i.e., longitudinal) predictors of numeracy. In the past decade, applied research across disciplines, including cognitive science, developmental psychology, education, and developmental neuroscience, has focused on identifying these predictor skills (Cirino, 2011; Hornung, Schiltz, Brunner, Martin, 2014; Krajewski & Schneider, 2009; LeFevre et al., 2010; Martin, Cirino, Sharp, & Barnes, 2014; Mazzocco & Thompson, 2005; Sowinski et al., 2015; Xenadou-Dervou, Molenaar, Ansari, van der Schoot, & van Lieshout, 2016). LeFevre et al. (2010) provide a validated model, *Pathways to Mathematics*, which summarizes early cognitive precursors (i.e., preschool/kindergarten) to later numeracy skill. Three distinct types of cognitive skills predict numeracy outcomes concurrently and longitudinally: quantitative, working memory[1], and linguistic skills (LeFevre et al., 2010; Sowinski et al., 2015). The Pathways to Mathematics model (LeFevre et al., 2010) has been further validated with a wider range of tasks for each skill (Cirino, 2011; Hornung et al., 2014; Sowinski et al., 2015). Thus, we use it as a framework for investigating early predictors here.

Quantitative Skills

Quantitative skills include subitizing (i.e., quickly determining the number of items in a small set without counting), counting, number sequencing, number comparison (i.e., determining which set has more/less), non-symbolic arithmetic tasks (e.g., adding/subtracting with manipulatives), and estimation. These tasks may involve *symbolic* representations of number (e.g., Arabic digits, spoken or written number words) or *non-symbolic* representations of number (e.g., objects or pictures of objects). These early numeracy skills of quantifying, labeling, comparing, and manipulating sets provide the foundation for later mathematical achievement (Jordan & Dyson, 2016).

[1] LeFevre et al. (2010) focused specifically on spatial attention (or visuospatial working memory), but research has demonstrated a developmental trend from a contribution for visuospatial working memory to verbal working memory in mathematics (Krajewski & Schneider, 2009; McKenzie, Bull & Gray, 2003; Rasmussen & Bisanz, 2005), so in the current chapter we expand this predictor to working memory more generally.

Findings. To measure quantitative skills, LeFevre et al. (2010) used a subitizing task and a non-symbolic arithmetic task. For the subitizing task, children were shown small sets of dots on a computer screen and asked to state "how many" as quickly as they could without making too many mistakes. If children are able to subitize, their time to respond will be similar across sets of 1–3 items, which is considered the subitizing range for this age group (Trick, Enns, & Brodeur, 1996). If they are instead *counting* the objects, their time to respond should increase with each additional item to enumerate. For the non-symbolic arithmetic task, both the experimenter and the child had a toy barn and a set of animals. The experimenter lined up a set of animals, saying "Farmer Smith is going to put some animals in the barn. See?" and placed them in the barn. Once in the barn, the animals were not visible to the child. For some trials, animals were added to (i.e., addition trials) or taken out of (i.e., subtraction trials) the barn. The child was asked to "show me how many animals are in the barn now, using your own mat and animals." LeFevre et al. (2010) found that these quantitative skills in preschool and kindergarten predicted mathematics outcomes two years later, including performance on a standardized calculation test (Calculation subtest; Woodcock & Johnson, 1989) and standardized numeracy test (KeyMath Test-Revised—Numeration Subtest; Connolly, 2000), after controlling for gender, working memory, and linguistic skills. These findings are supported by those of other researchers; counting and estimation skills in kindergarten each predicted arithmetic performance in Grade 1 (Bartelet, Vaessen, Blomert, & Ansari, 2014).

Number comparison tasks are a common measure of quantitative skill (for a review see De Smedt, Noël, Gilmore, & Ansari, 2013), with numerical comparison proposed as the key foundational capacity for numeracy (Gersten & Chard, 1999; Vanbinst, Ansari, Ghesquière, & De Smedt, 2016). As shown in Fig. 1, number comparison tasks may involve symbolic stimuli or non-symbolic stimuli. Performance on symbolic number comparison tasks (e.g., with Arabic digits) in kindergarten predicted performance on multiple mathematics outcome measures longitudinally in Grade 1, including tests of mathematical achievement (De Smedt, Verschaffel, & Ghesquière, 2009), arithmetic fluency (Martin et al., 2014), computation (Martin et al., 2014), and word problems (Martin et al., 2014). Performance on symbolic number comparison tasks in kindergarten also predicted performance on multiple measures of mathematics longitudinally in Grade 2, including arithmetic and word problems (Desoete, Ceulemans, De Weerdt, & Pieters, 2010), mathematical fluency, and mathematical reasoning (Toll, Viersen, Kroesbergen, & Van Luit, 2015).

Non-symbolic number comparison also predicts numeracy outcomes. Performance on non-symbolic number comparison tasks (e.g., comparing one set of dots to another) in kindergarten predicted arithmetic performance in Grade 1 (Desoete et al., 2010) and mathematical fluency in Grade 2 (Toll et al., 2015). When considered together (Bartelet et al., 2014; Desoete et al., 2010; Toll et al., 2015), symbolic number comparison demonstrates more predictive power than non-symbolic ($r = 0.30$ and $r = 0.24$, respectively; Schneider et al., 2017 meta-analysis). However, non-symbolic comparison may be particularly useful when working with younger

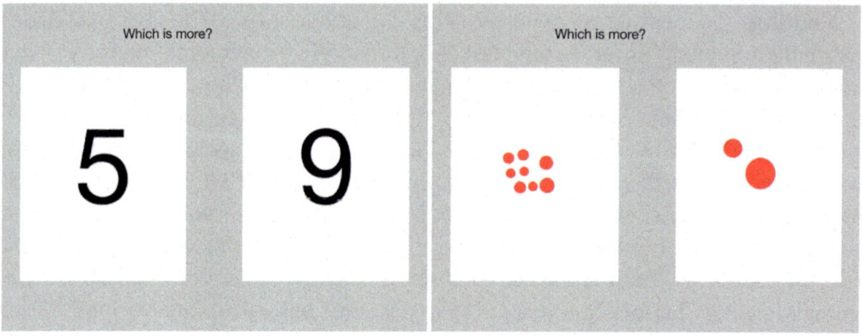

Fig. 1 Symbolic and non-symbolic number comparison tasks. Children respond by touching the larger number (Hume & Hume, 2014b)

children[2], who have not yet learned to link Arabic digits with the number of items in a set (Nosworthy, Bugden, Archibald, Evans, & Ansari, 2013).

Working Memory

Working memory is a cognitive system responsible for the active maintenance and temporary storage of task-relevant information (Baddeley, 1992; Miyake & Shaw, 1999) and is often measured using span tasks to determine how many items can be held in working memory. Visuospatial working memory is responsible for the maintenance and storage of visual and/or spatial information (Baddeley, 1992)—for example, mentally counting the number of windows in your childhood home. To measure visuospatial working memory, one common spatial span task, Corsi blocks, involves an experimenter pointing to an increasing sequence of locations with the child asked to copy the pattern, either in the same order or in reverse order (Orsini et al., 1987). Verbal working memory is responsible for the maintenance and storage of verbal information (i.e., inner speech; Baddeley, 1992)—for example, mentally recalling a grocery list while in the store. To measure verbal working memory, in one common verbal span task, digit span, the experimenter reads an increasing list of numbers (e.g., 7 2 8 6) and the child is asked to repeat the list, either in the same order or reverse order (Orsini et al., 1987). Working memory predicts both mathematics and general academic performance (Peng et al., 2016; Purpura & Ganley, 2014; Raghubar & Barnes, 2017). In mathematics, working memory supports the performance of multiple steps in counting, arithmetic, and problem-solving, the ability to keep track

[2]Non-symbolic number comparison, however, should not be considered as a developmental precursor that facilitates symbolic comparison performance, as evidence suggests that the direction of this relation is actually the reverse—improvement in symbolic number comparison facilitates performance on non-symbolic comparison (Lyons, Bugden, Zheng, De Jesus, & Ansari, 2017).

of intermediate results, and the ability to visualize problems and solutions (LeFevre et al., 2010; Peng et al., 2016).

Findings. To measure visuospatial working memory, LeFevre et al. (2010) used a computerized spatial span task (shown in Fig. 2). In this task, the child viewed a frog jumping in sequence from one lily pad to another. Once the frog was done jumping, the child was asked to "copy the pattern" by pointing. LeFevre et al. (2010) found that visuospatial working memory skill in preschool and kindergarten predicted mathematics outcomes two years later, including performance on a standardized calculation test (Calculation subtest; Woodcock & Johnson, 1989) and standardized tests of numeracy, geometry, and measurement (KeyMath Test-Revised—Numeration, Geometry, and Measurement subtests; Connolly, 2000), after controlling for gender, quantitative, and linguistic skills. LeFevre et al.'s findings are supported and extended by those of other researchers. Visuospatial working memory skill in kindergarten predicted arithmetic and word problem performance in Grade 1 (Toll, Kroesbergen, & Van Luit, 2016; Simmons, Singleton, & Horne, 2008).

Lee and Bull (2015) found that kindergarteners' working memory (a composite of visuospatial and verbal working memory) predicted both performance and growth in mathematics from Grade 1 to 9. Martin et al. (2014) measured both visuospatial and verbal working memory in kindergarten and found that the two measures differentially predicted Grade 1 mathematics outcomes; visuospatial working memory predicted performance on word problems, whereas verbal working memory predicted performance on applied problems (i.e., verbally presented word problems). A meta-analysis showed that visuospatial working memory and verbal working memory contribute equally to skill in mathematics (Peng et al., 2016).

Fig. 2 Visuospatial working memory task. Children watch the frog move to a sequence of locations and then copy the sequence by touching the locations (Hume & Hume, 2014a)

Linguistic Skills

Early linguistic skills include phonological awareness (i.e., knowledge of the sound structure of language) and receptive vocabulary (i.e., words the child understands). Beyond predicting numeracy, phonological awareness is considered the key foundational capacity for literacy (Gersten & Chard, 1999; Vanbinst, Ansari, Ghesquière, & De Smedt, 2016). In mathematics, linguistic skills support the learning of mathematics vocabulary (e.g., number names and numerals, more than/less than/equal to) and rules of the number system (e.g., base-ten; LeFevre et al., 2010; Purpura & Ganley, 2014; Raghubar & Barnes, 2017).

Findings. To measure linguistic skills, LeFevre et al. (2010) used a receptive vocabulary task and a phonological awareness task. Receptive vocabulary was measured using the Peabody Picture Vocabulary Test (Revised; Dunn & Dunn, 1997), wherein children are shown a set of four pictures and asked to choose the picture that corresponds to a verbally presented word (e.g., cobweb). Phonological awareness was measured using the Elision subtest of the Comprehensive Test of Phonological Processing (Wagner et al., 1999), wherein children are asked to repeat words, but with a sound missing (e.g., "Say *bold*. Now say *bold* without saying /b/").

LeFevre et al. (2010) found that linguistic skills in preschool and kindergarten, including phonological awareness and receptive vocabulary, predicted mathematics outcomes two years later, including performance on a standardized calculation test (Calculation subtest; Woodcock & Johnson, 1989) and standardized tests of numeracy, geometry, and measurement (KeyMath Test-Revised—Numeration, Geometry, and Measurement subtests; Connolly, 2000), after controlling for gender, quantitative, and working memory skill. Consistent with the LeFevre et al. findings, kindergarten phonological awareness predicted Grade 1 word-problem performance (Martin et al., 2014; Simmons, Singleton, & Horne, 2008). Phonological awareness at school entry was found to be the strongest predictor of both mathematics grades and national mathematics test scores two years later (over and above the existing school-readiness measure, free lunch eligibility, special education need, and gender), resulting in a recommendation for schools to include phonological awareness as part of baseline assessments to identify children who may need intervention (Savage & Carless, 2004).

Summary

There is strong support for quantitative skills, working memory, and linguistic skills as kindergarten predictors of later numeracy, across studies, countries, and measures. The relations between working memory and linguistic skills in kindergarten and later mathematics skill are mediated (or explained) by quantitative skills (Cirino, 2011; Krajewski & Schneider, 2009; Hornung et al., 2014; Passolunghi & Lafranchi, 2012). Children vary considerably in their numeracy skills (Cockcroft, 1982; Jordan et al.,

2009). Quantitative skills, working memory, and linguistic skills in combination account for 44–79% of this variability in children's arithmetic, 53–61% in word problems, 48–64% in number system knowledge, and 36–84% in geometry, one or more years later (Cirino, 2011; Hornung et al., 2014; LeFevre et al., 2010; Martin et al., 2014). That said, the specific set of predictors for any given mathematics outcome depends on the cognitive demands of the mathematics task (LeFevre et al., 2010; Martin et al., 2014; Hornung et al., 2014). For example, quantitative skills were found to predict arithmetic and knowledge of the number system, but not geometry (LeFevre et al., 2010). The benefit of this robust set of early predictors of later mathematics achievement, however, is that performance on these measures can be used to identify which children are likely to struggle to gain numeracy skills.

Early Identification

The first step in helping children at risk for mathematics difficulties is to identify which children are actually at risk. Given that children who enter kindergarten with poor numeracy skills do not catch up, at least without appropriate intervention, identification of at-risk students should happen in kindergarten, or sooner. Here, we focus on early identification tools appropriate for kindergarten students.

What is a numeracy screener? A numeracy screener is simply a test designed to identify which individual students are currently struggling, or at risk of struggling, to gain numeracy skills. Screeners are used to select children to receive intervention and to gauge their progress.

Criteria for evaluating screeners. In our review of reported early identification numeracy screeners, we focused on finding evidence-based screeners for use in kindergarten that were: (1) a test consisting of specific questions and explicit scoring instructions; (2) distributable, in either paper or computerized form; (3) practical for school-based use; (4) reliable (i.e., consistent) and valid, including demonstration of predictive validity longitudinally across grades; (5) normed or provided a cutoff criterion for judgment of at-risk status; and (6) published in English.

A surprising number of early identification screeners reported in the literature did not explicitly provide test items, and instead listed constructs or skills to be tested (e.g., Mazzocco & Thompson, 2005), or did not provide scoring methods. We focused on screeners that could be administered efficiently in classrooms by teachers, either to individual students or groups of students. Some screeners have demonstrated *concurrent validity*, they predict children's numeracy skills at the same point in time, but have not yet demonstrated *predictive validity*, predicting children's numeracy skills longitudinally (e.g., Nosworthy et al., 2013), across grades (e.g., Lembke & Foegen, 2009). Given the goal of early identification of children who will persistently struggle to gain numeracy skills, predictive validity is essential. Finally, simply having a child's screening test score is not sufficient for making an informed decision about whether the child is at risk; grade-level norms or cut-points are key for identification. It should be noted that available screeners are at different stages of development, and some that do not currently meet all the criteria may do so in the near future. For example, Ansari and colleagues are currently evaluating the predictive validity of the Numeracy Screener (Nosworthy et al., 2013; D. Ansari, personal communication, June 15, 2017). A more exhaustive list of potential screening tools can be found in Fuchs et al. (2007).

Evidence-based screeners. The following early identification screeners met the above criteria: Number Sets Test (Geary, Hoard, Byed-Craven, Nugent, & Numtee, 2007) and Number Sense Screener (Jordan & Glutting, 2012; Jordan, Glutting, & Ramineni, 2008, 2010; Jordan, Glutting, Ramineni, & Watkins, 2010). Both these screeners focus on assessing early quantitative skills. Researchers developing early numeracy screeners have focused on quantitative predictors, as they are seen to reflect the core foundational skill/deficit (Gersten & Chard, 1999). It should be noted that domain-general screeners exist for working memory (Nadler & Archibald, 2014) and linguistic skills (Bridges & Catts, 2011), though the validity of these screeners for numeracy outcomes has not been thoroughly investigated. Thus, working memory and linguistic screeners should be used in concert with evidence-based quantitative screeners to identify children at risk for mathematics difficulties, rather than as stand-alone early numeracy screeners.

Number Sets Test

The Number Sets Test (Geary et al., 2007, 2009) is a timed pencil-and-paper test for use in kindergarten to Grade 3 that can be group administered and takes about 10 min to complete. Age-/grade-level norms have not yet been compiled, but cutoffs have been determined to identify at-risk children (Geary et al., 2009). The four test sheets contain combinations of two and three sets of the numbers 0–9 in symbolic (Arabic digits), non-symbolic (objects, either the same or different), and mixed formats (both symbolic and non-symbolic). See Fig. 3 for an example. Children are instructed to circle all combinations on the sheet "that can be put together to make" the target number (either 5 or 9) as quickly and accurately as possible (Geary et al., 2009, p. 3). Key performance measures include the number of combinations correctly identified as matching the target (i.e., hits) and the number of combinations incorrectly identified as matching the target (i.e., false alarms). These scores are used to derive a measure of *sensitivity*, the difference between hits and false alarms.

The Number Sets Test demonstrates good concurrent validity in kindergarten through Grade 3 (Geary et al., 2009), relating strongly to performance on a standardized test of mathematics achievement (Numerical Operations subtest of the Wechsler Individual Achievement Test-II-Abbreviated; Wechsler, 2001). Predictive validity of the Number Sets Test has been demonstrated for Grade 1 scores, which moderately predicted mathematics achievement in Grade 3 (Geary et al., 2009), and strongly predicted functional numeracy scores (i.e., arithmetic, fraction comparison, fraction calculation, and word problems geared to employment) in Grade 7 (Geary, Hoard, Nugent, & Bailey, 2013). Moreover, this relation held despite controlling for Grade 1

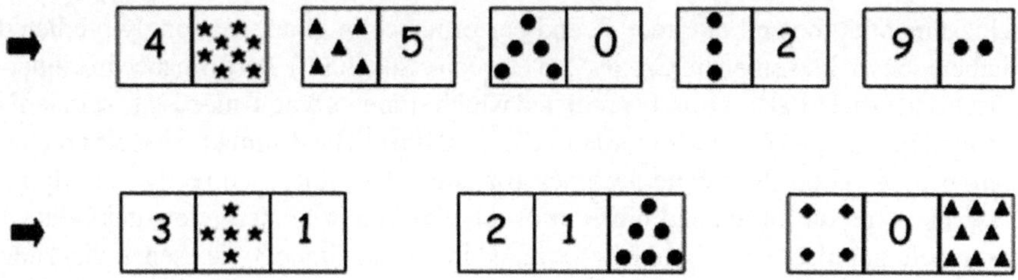

Fig. 3 Sample items from the Number Sets Test (Geary et al., 2009). Children are instructed to "Circle all the sets that add up to 9. Work as quickly as you can."

counting ability, reading ability, working memory, intelligence, and in-class attention. The Number Sets Test demonstrates good diagnostic accuracy (or clinical validity), correctly identifying two out of three children in Grade 1 that would be identified as having a mathematics learning disability in Grade 3 (i.e., sensitivity) and nine out of ten children that would not (i.e., specificity; Geary et al., 2009). Interested teachers and researchers can obtain the Number Sets Test by contacting the lead author, Dr. David Geary (D. Geary, personal communication, June 19, 2017).

Number Sense Screener

The Number Sense Screener (Jordan et al., 2008, 2010a, 2010b; Jordan & Glutting, 2012) is an untimed paper test for use in kindergarten and Grade 1 that is individually administered and takes about 15 min to complete. Grade-level norms have been compiled (Jordan & Glutting, 2012), and cutoffs have been determined to identify at-risk children (Jordan et al., 2010b). The most recent iteration of the screener (Jordan & Glutting, 2012) has 29 test items that cover six skills, including counting (e.g., enumerating five items, rote counting to 20), digit recognition (e.g., What number is this? 13), verbal symbolic number comparison (e.g., Which is bigger: 5 or 4?), nonverbal addition/subtraction (i.e., using manipulatives), verbal addition/subtraction facts (e.g., How much is 2 and 1?), and verbal story problems (e.g., Jill has 2 pennies. Jim gives her 1 more penny. How many pennies does Jill have now?). Raw accuracy scores are available for each skill as well as total raw scores, grade-standardized scores, and percentile ranks. Test–retest reliability, the consistency of individual results, in kindergarten was high (Jordan et al., 2010b).

The Number Sense Screener demonstrates good concurrent validity in Grade 1, relating strongly to performance on a standardized test of mathematics achievement (Applied Problems and Calculation subtests of the Woodcock–Johnson III; McGrew, Schrank, & Woodcock, 2007) within the same grade, but at different testing points (Jordan et al., 2009). Predictive validity of the Number Sense Screener has been demonstrated for kindergarten and Grade 1 scores. Performance on the Number Sense Screener at the beginning of kindergarten strongly predicted mathematics achievement at the end of Grade 3, and performance in Grade 1 strongly predicted mathematics achievement in Grades 2–3 (Jordan et al., 2009). Performance in kindergarten and Grade 1 also strongly predicted which students would meet state standards for mathematics in Grade 3 (Jordan et al., 2010b). The Number Sense Screener demonstrates good diagnostic accuracy (or clinical validity), correctly identifying between seven out of ten and nine out of ten children in kindergarten and Grade 1 that would not meet state standards for mathematics in Grade 3 (i.e., sensitivity) and between four out of ten and nine out of ten children that would meet state standards in Grade 3 (i.e., specificity; Jordan et al., 2010b). Interested teachers and researchers can purchase the published version of the Number Sense Screener (currently $46.95 CAD; Jordan & Glutting, 2012).

Summary

Validated early screeners exist to identify which children are at risk for mathematics difficulties. In addition to these quantitative tools, early screening for working memory and linguistic skills, which account for additional unique variance in children's numeracy, may be useful (Cirino, 2011; Hornung et al., 2014; LeFevre et al., 2010; Martin et al., 2014; Raghubar & Barnes, 2017). It is important to note that decisions on at-risk status must be re-evaluated at multiple time points within or across years (Mazzocco & Myers, 2003), as decisions based on a single data point may not reflect a persistent numeracy problem. It is also advisable to base decisions of at-risk status on the results of more than one screener, as different screeners will identify different groups of children as at risk (Mazzocco & Myers, 2003). Early identification is only the first step. Once children are identified as at risk for mathematics difficulties, evidence-based interventions are needed to help these children catch up to their typically developing peers. Although diagnostic tools may present tantalizing opportunities for the creation of intervention tools, training performance on screening measures is not necessarily an effective intervention that will impact broader numeracy skills (Ansari, De Smedt, & Grabner, 2012).

Early Intervention

Once children at risk for mathematics difficulties have been identified with appropriate screening tools, the next step is to intervene to change the developmental trajectory and improve the numeracy outcomes of these students. Without early intervention, students may experience persistent numeracy difficulties; more than 6 in 10 children identified as having mathematics difficulties throughout kindergarten continued to have difficulties in Grade 5 (Morgan, Farkas, & Wu, 2009). Here, we focus on early intervention tools appropriate for kindergarten students.

What Is a Numeracy Intervention?

A numeracy intervention is an educational program designed to address areas of need in a child who is struggling, or at risk of struggling, academically in mathematics. The goal of the intervention is long-term academic improvement in mathematics. Interventions are not one size fits all; the type and level of intervention should be tailored to a child's current level of numeracy performance (Clements & Sarama, 2011), rather than determined solely on the basis of their age or grade.

Criteria for Evaluating Interventions

In our review of reported early numeracy interventions, we focused on finding evidence-based interventions for use in kindergarten that were: (1) a program, consisting of a textbook, software and/or explicit instructional processes (Slavin, 2008); (2) distributable, in either paper or computerized form; (3) practical for school-based use; (4) intended for use with at-risk children, versus as a core curriculum; (5) published in English; and (6) demonstrated to be effective with use in kindergarten. In reviewing evidence of intervention efficacy, we identified interventions supported by studies that were shown to meet the following key criteria, adapted and expanded from Mononen, Aunio, Koponen, and Aro (2015), and Slavin (2008).

Studies must use children identified as at risk, based on low numeracy performance, and contain an explicit, quantitative criterion to define low performance. To determine if an intervention improves the mathematics performance of at-risk children, it must be evaluated using at-risk children. The results of a given intervention for typically developing children may vary considerably from those for at-risk children. Definitions of low-performing and at-risk children vary widely in the literature (from children performing below the 10th, 25th, or 50th percentile on a screening test, to teacher identification, to measures like family income in place of numeracy performance); the use of a criterion or cutoff explicitly specifies the group of children the intervention has been evaluated for.

Studies must include an at-risk comparison group. Without a comparison group of at-risk students, it is impossible to determine whether any effect of the intervention is actually due to the intervention and not just due to regular learning/development. The control group will ideally be an *active* control group (also called a seen control), a group that receives *another* intervention, not just regular classroom instruction. This is to control for *Hawthorne effects*—children improving based on receiving special attention (i.e., motivation bias) versus the specific intervention (Fawcett & Reid, 2009). A *passive* control group (also called an unseen control or business-as-usual control), a group that receives only normal classroom instruction, is, however, sufficient to demonstrate that any improvements are not due simply to regular learning/development. A typically developing control group may also be included, to determine whether at-risk children are closing the skills gap with their peers, but it cannot be substituted for an at-risk control group[3].

Group assignment (intervention/control) must be determined based on random assignment. Random assignment means that each student included in the study has an unbiased (or random) chance of being assigned to the intervention or the control group(s). Random assignment is crucial for being able to infer that the intervention caused any outcome differences between the groups. Without random assignment, outcomes may instead be due to preexisting systematic differences between the groups (e.g., the severity of numeracy difficulties, behavioral issues or English proficiency) rather than the intervention.

Studies must include a pretest, immediate posttest, and delayed posttest. A pretest, evaluating performance on the outcome measure prior to the intervention, is important to evaluate the equivalency of groups prior to intervention, to show growth (growth = posttest score − pretest score), and for use as a control variable in statistical tests (as children who started at a lower or higher skill level may benefit more from intervention). A posttest is when outcomes are evaluated at the end of an intervention. These posttests may be *immediate,* following immediately after the intervention, or *delayed*, following after some delay (e.g., 8 weeks after the intervention ended). Given that the goal of an intervention is long-term improvement in numeracy, it is important that an intervention demonstrates effects beyond the immediate time frame.

Numeracy outcome measures must be reliable, valid, and unbiased. A reliable measure is one that is consistent. Reliability is important for an outcome measure so that changes can be attributable to learning versus random error. Without a reliable and valid outcome measure, it is impossible to determine whether the intervention improves numeracy. In some studies, the outcome measure may be biased toward the intervention group; for example, the test measures precisely the same skills that the intervention teaches, which are not skills taught in normal mathematics instruction. Often, this criterion of an unbiased outcome is achieved through the use of a standardized mathematics test, though other measures can be reliable, valid, and unbiased.

Interventions must demonstrate numeracy gains, compared to an appropriate control group, that are both statistically significant and meet the Institute of Educational Sciences criterion for meaningful intervention effects. This criterion focuses on whether an intervention actually works—whether at-risk children using the intervention improve their numeracy skills more than those who do not. Intervention effects that are *statistically significant* (generally p values < 0.05) are unlikely to have occurred by chance. Effect sizes (e.g., Hedges' g, Cohen's d) allow you to determine how large an effect the intervention had on numeracy outcomes. Thus, effect sizes are useful, as they allow you to evaluate whether the effect of an intervention is meaningful. The Institute for Educational Science (2014) has established

[3]One reason is the possibility of regression to the mean. Children identified as at risk based on one time point may show improvement in later time points simply because they underperformed on that particular test relative to their actual ability.

a criterion of $g \geq 0.25$ for a meaningful intervention effect. It is important to consider both statistical significance and effect size, as interventions may demonstrate statistically significant results, but the results may be so small as to not be worth implementing, whereas other interventions may demonstrate large effects, but those results may be likely to have occurred by chance (i.e., not statistically significant).

Evidence-Based Interventions

Only one early numeracy intervention—Number Sense Interventions (Dyson et al., 2013; Jordan et al., 2012; Jordan & Dyson, 2014)—met the above criteria at immediate posttest and, in a single study, an adapted version additionally met the criteria at delayed posttest. Available interventions, however, are at different stages of development and some that do not currently meet all the criteria may do so in the near future. More exhaustive lists of potential intervention tools can be found in Chodura, Kuhn, and Holling (2015) or Mononen et al. (2015).

Other interventions reviewed either did not have an appropriate study to allow us to thoroughly evaluate intervention efficacy (e.g., Räsänen, Salminen, Wilson, Aunio, & Dehaene, 2009; Salminen, Koponen, Räsänen, 2015; Toll & Van Luit, 2013) or did not show significant and meaningful gains compared to a control group on an unbiased outcome measure (e.g., Fuchs, Fuchs, Hamlet, Powell, Capizzi, & Seethaler, 2006). Many studies identified students as at risk based on low socioeconomic status (SES; e.g., enrollment in a free/reduced lunch program) rather than low numeracy performance (e.g., Dyson et al., 2013; Jordan et al., 2012; Ramani & Siegler, 2008; Stacy, Cartwright, Arwood, Canfield, & Kloos, 2017). Although low SES children are more likely to demonstrate low numeracy performance than their higher SES peers (Jordan, Huttenlocher & Levine, 1992; Jordan, Kaplan, Olah Nabors, & Locuniak, 2006), selecting children to receive interventions based on SES creates a mixed group with low and higher numeracy skills. Intervention studies using SES to select children for participation show higher gains than those using actual numeracy performance (Jordan & Dyson, 2016), potentially because children with typical to high numeracy skills prior to intervention benefit more from the intervention.

Number Sense Interventions

Number Sense Interventions (Dyson et al., 2013; Jordan et al., 2012; Jordan & Dyson, 2014; for a review see Jordan & Dyson, 2016) is an instructor-lead intervention designed for small-group administration (recommended group size of four children per instructor). The intervention was designed for use in kindergarten to Grade 1 with children at risk for mathematics difficulties. The scripted guide (Jordan & Dyson, 2014), which currently costs $46.50 CAD, contains 24 half-hour lessons, including activities, printable worksheets, and games. The recommended intervention schedule is three times per week for a period of eight weeks. The intervention follows a developmental progression (Frye et al., 2013) to build quantitative skills related to number (e.g., subitizing, number recognition, mapping symbols to quantities, number sequencing, and the base-ten system), number relations (e.g., number comparison, before/after relations), and number operations (e.g., number combinations, non-symbolic, and symbolic arithmetic).

The efficacy of Number Sense Interventions has been evaluated in four randomized-control-trial studies (Dyson et al., 2013, Dyson et al., 2015; Hassinger-Das, Jordan, & Dyson, 2015; Jordan et al., 2012; for a review, see Jordan & Dyson, 2016). Two of these studies defined at-risk status based on low numeracy performance. Hassinger-Das et al. (2015) evaluated the efficacy of Number Sense Interventions in low numeracy children (defined as performance below the 25th percentile on the Number Sense Screener). Children were randomly assigned to three groups: the intervention group, an active control group (who received a mathematics vocabulary intervention), and a passive control group (receiving regular classroom instruction). All children completed a pretest, immediate posttest, and delayed posttest (eight weeks afterward). Numeracy outcome measures included the Number Sense Screener, and standardized tests of calculation skill and word-problem solution. Children who participated in the Number Sense Intervention performed significantly better than both the active and passive control groups on calculation skill at the immediate posttest and exceeded the effect size to demonstrate educational effectiveness (*g* of 0.58 and 0.59, respectively). No other immediate posttest gains, nor any delayed-posttest gains, were statistically significant. Thus, significant long-term effects of the intervention were not demonstrated in this study.

Dyson et al. (2015) also evaluated the efficacy of Number Sense Interventions in low numeracy children (defined as performance below the 25th percentile on the Number Sense Screener). Children were randomly assigned to three groups: the Number Sense Interventions group, a group that received Number Sense Interventions *plus number-fact practice* (five minutes of simple addition and subtraction flash cards per session), and an active control group (receiving the school's regular mathematics intervention for the same time period). A typically developing, passive control group was also included. All children completed a pretest, immediate posttest, and delayed posttest (six weeks afterward). Numeracy outcome measures included the Number Sense Screener, a standardized test of calculation skill (Woodcock–Johnson Calculation subtest), and a timed arithmetic task. Children who participated in the regular Number Sense Intervention performed significantly better than the active control group on timed arithmetic and calculation skill at the immediate posttest and exceeded the effect size to demonstrate educational effectiveness (*g* of 0.69 and 0.58, respectively). However, these gains did not remain significant in the delayed posttest. Children who participated in the Number Sense Intervention *plus number-fact practice* performed significantly better than the active control group on the Number Sense Screener, timed arithmetic, and calculation skill at the immediate posttest and exceeded the effect size to demonstrate educational effectiveness (*g* of 0.82, 0.78, and 0.60, respectively). Moreover, children who participated in the Number Sense Intervention plus number-fact practice performed significantly better than the active control group on the Number Sense Screener and timed arithmetic at the delayed posttest and demonstrated moderate educational effectiveness (*g* of 0.56 and 0.58, respectively). Comparing the gains made by the Number Sense Interventions plus number-fact practice group to those made by the typically developing control group, receiving normal classroom instruction, suggests that this combined intervention has the potential to close 44% of the achievement gap in number sense and 54% of the achievement gap in timed arithmetic (Dyson et al., 2015). Thus, significant long-term effects of the intervention were demonstrated in this study, but only when paired with five minutes of number-fact practice/session. Additional research from other laboratories is needed to independently replicate these findings, and further research is needed to provide stronger evidence of the lasting effectiveness of Number Sense Interventions and to tease apart the numeracy gains due to the intervention from those of number-fact practice alone.

Summary

Despite a proliferation of early numeracy interventions, both instructor-led and computer-based, only one (Number Sense Interventions; Jordan & Dyson, 2014) has currently been shown to improve the numeracy outcomes of at-risk kindergarten students when efficacy studies were evaluated using a rigorous set of criteria. Why be so rigorous? School-based interventions require educational resources, including instructional space, costs to purchase the intervention, purchase and maintain technology (in the case of computerized interventions), and to train and staff instructors. Struggling students typically miss some portion of regular class instruction time, which is instead devoted to the intervention. If the intervention does not improve student numeracy outcomes, these resources have been wasted along with an opportunity to have lasting effects on a child's academic and life success. Thus, it is important to thoroughly evaluate the effectiveness of interventions.

Although three distinct types of cognitive skills predict numeracy outcomes, quantitative, working memory, and linguistic skills (LeFevre et al., 2010; Sowinski et al., 2015), researchers developing early numeracy interventions have focused on building quantitative skills. This focus has been supported by research, as the only interventions to improve numeracy outcomes for children involve training quantitative skills, such as subitizing, counting, non-symbolic arithmetic, and number comparison (Raghubar & Barnes, 2017). It should be noted, however, that domain-general interventions exist for working memory (Agus, Mascia, Fastame, Melis, Pilloni, & Penna, 2015; Holmes, Gathercole, & Dunning, 2009; Kroesbergen, van't Noordende, & Kolkman, 2014) and linguistic skills (Hulme, Bowyer-Crane, Caroll, Duff, & Snowling, 2012; Bowyer-Crane et al., 2008), though these interventions have not convincingly been shown to improve numeracy outcomes (Dowker, 2016; Melby-Lervag & Hulme, Melby-Lervag and Hulme 2013; Melby-Lervag, Redick, & Hulme, 2016). Although combining training in quantitative skills with working memory and/or linguistic skills may lead to better student outcomes, this is likewise not yet supported by research (Kroesbergen et al., 2014; Raghubar & Barnes, 2017). Where then do working memory and linguistic skills come into play in early numeracy interventions? Working memory skills moderate the effect of numeracy interventions on numeracy outcomes. Specifically, children with better working memory skills benefit more from numeracy interventions (Toll & van Luit, 2013). In contrast, for linguistic skills, English-language learners (ELL) benefited more from the Number Sense Interventions than non-ELL students (Dyson et al., 2015). Thus, considering children's working memory and linguistic skills can help guide the selection of suitable interventions by capitalizing on an individual child's relative cognitive strengths and minimizing their weaknesses (Raghubar & Barnes, 2017).

Conclusion

This chapter outlined robust, early cognitive predictors of later numeracy skill (i.e., quantitative skills, working memory, and linguistic skills), indicated reliable and valid diagnostic screening tools for early identification of at-risk students, and provided and applied criteria to evaluate the efficacy of early numeracy interventions. The interdisciplinary body of applied numeracy research has grown substantially over the past decade. There was, nonetheless, a paucity of screening tools and interventions that met our stringent methodological criteria. Stringent criteria are important, however, lest we make poor educational choices for students. What educators need are evidence-based tools, screeners and interventions, derived from this newfound wealth of applied numeracy research. Moreover, we need a heightened level of methodological rigor in the design and interpretation of the effectiveness of these tools.

Given the strong, and pervasive link between basic numeracy skills and both later academic and life outcomes, early identification and intervention can make a meaningful difference in the lives of at-risk students, their families, and communities. Poor numeracy skills constrain the educational and employment opportunities available to individuals (Bynner & Parsons, 1997; Parsons & Bynner, 1997, 2005; Ritchie & Bates, 2013). These negative outcomes may become more pronounced in the current era where manufacturing and other skilled labor positions are disappearing, along with retail positions, while educational requirements for these jobs are increasing (Bynner & Parsons, 1997; Hicks & Devaraj, 2017). In addition to employment outcomes, poor numeracy skills are associated with poorer decision-making and health outcomes (Reyna, Nelson, Han, & Dieckmann, 2009), and higher levels of depression (Bynner & Parsons, 1997). Thus, interdisciplinary work moving from applied research on children's numeracy to *applications* of this knowledge—the development and rigorous testing of early screening and intervention tools—though challenging, is crucial.

Disclosures No conflicts of interest, financial or otherwise, are declared by the authors.

References

Agus, M., Mascia, M. L., Fastame, M. C., Melis, V., Pilloni, M. C., & Penna, M. P. (2015). The measurement of enhancement in mathematical abilities as a result of joint cognitive trainings in numerical and visual-spatial skills: A preliminary study. *Journal of Physics: Conference Series, 588*(1). https://doi.org/10.1088/1742-6596/588/1/012041.

Ansari, D., De Smedt, B., & Grabner, R. H. (2012). Neuroeducation—A critical overview of an emerging field. *Neuroethics, 5*(2), 105–117. https://doi.org/10.1007/s12152-011-9119-3.

Aunola, K., Leskinen, E., Lerkkanen, M., & Nurmi, J. (2004). Developmental dynamics of math performance from preschool to grade 2. *Journal of Educational Psychology, 96*(4), 699–713. https://doi.org/10.1073/0022-0663.96.4.699.

Baddeley, A. (1992). Working memory. *Science, 255*(5044), 556–559.

Bartelet, D., Vaessen, A., Blomert, L., & Ansari, D. (2014). What basic number processing measures in kindergarten explain unique variability in first-grade arithmetic proficiency? *Journal of Experimental Child Psychology, 117,* 12–28. https://doi.org/10.1016/j.jecp.2013.08.010.

Bowyer-Crane, C., Snowling, M. J., Duff, F. J., Fieldsend, E., Carroll, J. M., Miles, J. … Hulme, C. (2008). Improving early language and literacy skills: Differential effects of an oral language versus a phonology with reading intervention. *Journal of Child Psychology and Psychiatry, 49*(4), 422–432.

Bridges, M. S., & Catts, H. W. (2011). The use of a dynamic screening of phonological awareness to predict risk for reading disabilities in kindergarten children. *Journal of Learning Disabilities, 44*(4), 330–338. https://doi.org/10.1177/0022219411407863.

Bynner, J., & Parsons, S. (1997). *Does numeracy matter? evidence from the national child development study on the impact of poor numeracy on adult life*. London, England: The Basic Skills Agency.

Chodura, S., Juhn, J., & Holling, H. (2015). Interventions for children with mathematical difficulties: A meta-analysis. *Zeitschrift für Psychologie, 223*(2), 129–144. https://doi.org/10.1027/2151-2604/a000211.

Cirino, P. T. (2011). The interrelationships of mathematical precursors in kindergarten. *Journal of Experimental Child Psychology, 108*(4), 713–733. https://doi.org/10.1016/j.jecp.2010.11.004.

Clements, D. H., & Sarama, J. (2011). Early childhood mathematics intervention. *Science, 333*(6045), 968–970.

Clarren, S. B., Martin, D. C., & Townes, B. D. (1993). Academic achievement over a decade: A neuropsychological prediction study. *Developmental Neuropsychology, 9*(3–4), 161–176. https://doi.org/10.1080/87565649309540550.

Cockcroft, W. H. (Chairman). (1982). *Mathematics counts* (Report of the Committee of Inquiry into the Teaching of Mathematics in Schools). London, England: Her Majesty's Stationery Office.

Connolly, A. J. (2000). *KeyMath-revised/updated Canadian norms*. Richmond Hill, ON: PsyCan.

De Smedt, B., Noël, M. P., Gilmore, C., & Ansari, D. (2013). How do symbolic and non-symbolic numerical magnitude processing skills relate to individual differences in children's mathematical skills? A review of evidence from brain and behavior. *Trends in Neuroscience and Education, 2*(2), 48–55. https://doi.org/10.1016/j.tine.2013.06.001.

De Smedt, B., Verschaffel, L., & Ghesquière, P. (2009). The reductive value of numerical magnitude comparison for individual differences in mathematics achievement. *Journal of Experimental Psychology, 103*(4), 469–479. https://doi.org/10.1016/j.jecp.2009.01.010.

Desoete, A., Ceulemans, A., De Weerdt, F., & Pieters, S. (2010). Can we predict mathematical learning disabilities from symbolic and non-symbolic comparison tasks in kindergarten? Findings from a longitudinal study. *British Journal of Educational Psychology, 82*(1), 64–81. https://doi.org/10.1348/2044-8279.002002.

Dowker, A. (2016). Factors that influence improvement in numeracy, reading, and comprehension in the context of a numeracy intervention. *Frontiers in Psychology, 7*. https://doi.org/10.3389/fpsyg.2016.01929.

Duncan, G. J., Dowsett, C. J., Claessens, A., Magnuson, K., Huston, A. C., Klebanov, P., et al. (2007). School readiness and later achievement. *Developmental Psychology, 43*(6), 1428–1446. https://doi.org/10.1037/0012-1649.43.6.1428.

Dunn, L. M., & Dunn, L. M. (1997). *Peabody picture vocabulary test–III [Measurement instrument]*. Circle Pines, MN: American Guidance Service.

Dyson, N. I., Jordan, N. C., Beliakoff, A., & Hassinger-Das, B. (2015). A kindergarten number-sense intervention with contrasting practice conditions for low-achieving children. *Journal for Research in Mathematics Education, 46*(3), 331–370. https://doi.org/10.5951/jresematheduc.46.3.0331.

Dyson, N. I., Jordan, N. C., & Glutting, J. (2013). A number sense intervention for low income kindergartners at risk for mathematics difficulties. *Journal of Learning Disabilities, 46*(2), 166–181. https://doi.org/10.1177/0022219411410233.

Fawcett, A., & Reid, G. (2009). Alternative and innovative interventions for dyslexia: A critical commentary. In G. Reid (Ed.), *The Routledge companion to Dyslexia* (pp. 157–174). Abingdon, OX: Routledge.

Frye, D., Barooody, A. J., Burchinal, M., Carver. S. M., Jordan, N. C., & McDowell, J. (2013). *Teaching math to young children: A practice guide* (NNCEE 2014-4005). Washington, DC: US Department of Education and Regional Assistance. Available from https://ies.ed.gov/ncee/wwc/Docs/PracticeGuide/early_math_pg_111313.pdf.

Fuchs, L. S., Fuch, D., Hamlet, C. L., Powell, S. R., Capizzi, A. M., & Seethaler, P. M. (2006). The effects of computer-assisted instruction on number combination skill in at-risk first graders. *Journal of Learning Disabilities, 39*(5), 467–475.

Fuchs, L., Fuchs, D., Compton, D. L., Bryant, J. D., Hamlett, C. L., & Seethaler, P. M. (2007). Mathematics screening and progress monitoring at first grade: Implications for responsiveness to intervention. *Exceptional Children, 73*(3), 311–330. Retrieved from http://go.galegroup.com.proxy1.lib.uwo.ca/ps/i.do?p=AONE&sw=w&u=lond95336&v=2.1&it=r&id=GALE%7CA160331975&sid=summon&asid=37008b001648e22005fe5c7bf8a6da13.

Geary, D. C., Bailey, D. H., Littlefield, A., Wood, P., Hoard, M. K, & Nugent, L. (2009). First-grade predictors of mathematical learning disability: A latent class trajectory analysis. *Cognitive Development, 24*(4), https://doi.org/10.1016/j.cogdev.2009.10.001.

Geary, D. C., Hoard, M. K., Byed-Craven, J., Nugent, L., & Numtee, C. (2007). Cognitive mechanisms underlying achievement deficits in children with mathematical learning disability. *Child Development, 78*(4), 1343–1359. Retrieved from http://www.jstor.org.proxy1.lib.uwo.ca/stable/4620706.

Geary D. C., Hoard, M. K., Nugent, L., & Bailey, D. H. (2013). Adolescents' functional numeracy is predicted by their school entry number system knowledge. *PLoS ONE, 8*(1), https://doi.org/10.1371/journal.pone.0054651.

Gersten, R., & Chard, D. (1999). Number sense: Rethinking arithmetic instruction for students with mathematical disabilities. *Journal of Special Education, 33*(1), 18–28.

Hassinger-Das, B., Jordan, N. C., & Dyson, N. (2015). Reading stories to learn math: Mathematics vocabulary instruction for children with early numeracy difficulties. *The Elementary School Journal, 116*(2), 242–246. https://doi.org/10.1086/683986.

Hicks, M. J., & Devaraj, S. (2017). *The myth and the reality of manufacturing in America*. Muncie, IN: Ball State University.

Holmes, J., Gathercole, S. E., & Dunning, D. L. (2009). Adaptive training leads to sustained enhancement of poor working memory in children. *Developmental Science, 12*(4), F9–F15. https://doi.org/10.1111/j.1467-7687.2009.00848.x.

Hornung, C., Schiltz, C., Brunner, M., & Martin, R. (2014). Predicting first-grade mathematics achievement: The contributions of domain-general cognitive abilities, nonverbal number sense, and early number competence. *Frontiers in Psychology, 5*, https://doi.org/10.3389/fpsyg.2014.00272.

Hulme, C., Bowyer-Crane, C., Carroll, J. M., Duff, F. J., & Snowling, M. J. (2012). The causal role of phoneme awareness and letter-sound knowledge in learning to read: Combining intervention studies with mediation analyses. *Psychological Science, 23*(6), 572–577.

Hume, S., & Hume, T. (2014a). *Pathspan* [Computer software]. Retrieved from https://hume.ca/ix/.

Hume, S., & Hume, T. (2014b). Test Runner [Computer software]. Retrieved from https://hume.ca/ix/.

Institute of Education Sciences. (2014). *What works clearinghouse: Procedures and standards handbook* (Version3.0). Available from http://ies.ed.gov/ncee/wwc/pdf/reference_resources/wwc_procedures_v3_0_standards_handbook.pdf.

Jordan, N. C., & Dyson, N. (2014). *Number sense interventions*. Baltimore, MD: Paul H. Brookes Publishing Company.

Jordan, N. C., & Dyson, N. (2016). Catching math problems early: Findings from the number sense intervention project. In A. Henik (Ed.), *Continuous issues in numerical cognition: How many or how much?* (pp. 59–79). New York, NY: Elsevier.

Jordan, N. C., & Glutting, J. (2012). *Number sense screener (NSS)*. Baltimore, MD: Brookes.

Jordan, N. C., Glutting, J., Dyson, N., Hassinger-Das, B., & Irwin, C. (2012). Building kindergartners' number sense: A randomized controlled study. *Journal of Educational Psychology, 104*(3), 647–660. https://doi.org/10.1037/a0029018.

Jordan, N. C., Glutting, J., & Ramineni, C. (2008). A number sense assessment tool for identifying children at risk for mathematical difficulties. In A. Dowker (Ed.), *Mathematical difficulties: Psychology and intervention* (pp. 45–58). (n.p.): Academic Press.

Jordan, N. C., Glutting, J., & Ramineni, C. (2010a). The importance of number sense to mathematics achievement in first and third grades. *Learning and Individual Differences, 20*(2), 82–88. https://doi.org/10.1016/j.lindif.2009.07.004.

Jordan, N. C., Glutting, J., Ramineni, C., & Watkins, M. W. (2010b). Validating a number sense screening tool for use in kindergarten and first grade: Prediction of mathematics proficiency in third grade. *School Psychology Review, 39*(2), 181–195. Retrieved from http://go.galegroup.com.proxy1.lib.uwo.ca/ps/i.do?p=AONE&sw=w&u=lond95336&v=2.1&it=r&id=GALE%7CA233050811&sid=summon&asid=4b0bb7c47d1bc0864fdfe88656fedac4.

Jordan, N. C., Huttenlocher, J., & Levine, S. C. (1992). Differential calculation abilities in young children from middle- and low-income families. *Developmental Psychology, 28*(4), 644–653. https://doi.org/10.1037/0012-1649.28.4.644.

Jordan, N. C., Kaplan, D., Nabors Oláh, L., & Locuniak, M. N. (2006). Number sense growth in kindergarten: A longitudinal investigation of children at risk for mathematics difficulties. *Child Development, 77*(1), 153–175. https://doi.org/10.1111/j.1467-8624.2006.00862.x.

Jordan, N. C., Kaplan, D., Ramineni, C., & Locuniak, M. N. (2009). Early math matters: Kindergarten number competence and later mathematics outcomes. *Developmental Psychology, 45*(3), 850–867. https://doi.org/10.1037/a0014939.

Krajewski, K., & Schneider, W. (2009). Exploring the impact of phonological awareness, visual-spatial working memory, and preschool quantity-number competencies on mathematics achievement in elementary school: Findings from a 3-year longitudinal study. *Journal of Experimental Child Psychology, 103*(4), 516–531. https://doi.org/10.1016/j.jecp.2009.03.009.

Kroesbergen, E. H., van 't Noordende, J., & Kolkman, M. (2014). Training working memory in kindergarten children: Effects on working memory and early numeracy. *Child Neuropsychology, 20*(1), 23–37. https://doi.org/10.1080/09297049.2012.736483.

Lee, K., & Bull, R. (2015). Developmental changes in working memory, updating, and math achievement. *Journal of Educational Psychology, 108*(6), 869–882. https://doi.org/10.1037/edu0000090.

LeFevre, J.-A., Fast, L., Skwarchuk, S.-L., Smith-Chant, B. L., Bisanz, J., Jamawar, D., et al. (2010). Pathways to mathematics: Longitudinal predictors of performance. *Child Development, 81*(6), 1753–1767. https://doi.org/10.1111/j.1467-8624.2010.01508.x.

Lembke, E., & Forgan, A. (2009). Identifying early numeracy indicators for kindergarten and first-grade students. *Learning Disabilities Research & Practice, 24*(1), 12–20. https://doi.org/10.1111/j.1540-5826.2008.01273.x.

Lyons, I. M., Bugden, S., Zheng, S., De Jesus, S., & Ansari, D. (2017). Symbolic number skills predict growth in nonsymbolic number skills in kindergarteners. *Developmental psychology.* Advanced online publication. https://doi.org/10.1037/dev0000445.

Martin, R. B., Cirino, P. T., Sharp, C., & Barnes, M. (2014). Number and counting skills in kindergarten as predictors of grade 1 mathematical skills. *Learning and Individual Differences, 34,* 12–23. https://doi.org/10.1016/j.lindif.2014.05.006.

Mazzocco, M. M. (2005). Challenges in identifying target skills for math disability screening and intervention. *Journal of Learning Disabilities, 38*(4), 318–323. https://doi.org/10.1177/00222194050380040701.

Mazzocco, M. M., & Myers, G. F. (2003). Complexities in identifying and defining mathematics learning disability in the primary school-age years. *Annals of Dyslexia, 53*(1), 218–253. https://doi.org/10.1007/s11881-003-0011-7.

Mazzocco, M. M., & Thompson, R. E. (2005). Kindergarten predictors of math learning disability. *Learning Disabilities Research & Practice, 20*(3), 142–155. https://doi.org/10.1111/j.1540-5826.2005.00129.x.

McGrew, K. S., Schrank, F. A., & Woodcock, R. W. (2007). *Technical manual: Woodcock-Johnson III normative update.* Rolling Meadows, IL: Riverside Publishing.

McKenzie, B., Bull, R., & Gray, C. (2003). The effects of phonological and visual-spatial interference on children's arithmetical performance. *Educational and Child Psychology, 20*(3), 93–108.

Melby-Lervåg, M., & Hulme, C. (2013). Is working memory training effective? A meta-analytic review. *Developmental Psychology, 49*(2), 270–291. https://doi.org/10.1037/a0028228.

Melby-Lervåg, M., Redick, T. S., & Hulme, C. (2016). Working memory training does not improve performance on measures of intelligence or other measures of "far transfer" evidence from a meta-analytic review. *Perspectives on Psychological Science, 11*(4), 512–534. https://doi.org/10.1177/1745691616635612.

Miyake, A., & Shah, P. (Eds.). (1999). *Models of working memory: Mechanisms of active maintenance and executive control.* Cambridge, England: Cambridge University Press.

Mononen, R., Aunio, P., Koponen, T., & Aro, M. (2015). A review of early numeracy interventions for children at risk in mathematics. *International Journal of Early Childhood Special Education,* 25–54. https://doi.org/10.20489/intjecse.107993.

Morgan, P. L., Farkas, G., & Wu, Q. (2009). Five-year growth trajectories of kindergarten children with learning difficulties in mathematics. *Journal of Learning Disabilities, 42*(4), 306–321. https://doi.org/10.1177/0022219408331037.

Nadler, R. T., & Archibald, L. M. (2014). The assessment of verbal and visuospatial working memory with school age Canadian children. *Canadian Journal of Speech-Language Pathology and Audiology, 38*(3), 262–279. Retrieved from http://cjslpa.ca/files/2014_CJSLPA_Vol_38/CJSLPA_Fall_2014_Vol_38_No_3.pdf#page=8.

Nosworthy, N., Bugden, S., Archibald, L., Evans, B., & Ansari, D. (2013). A two-minute paper-and-pencil test of symbolic and nonsymbolic numerical magnitude processing explains variability in primary school children's arithmetic competence. *PLoS ONE, 8*(7), https://doi.org/10.1371/journal.pone.0067918.

Orsini, A., Grossi, D., Capitani, E., Laiacona, M., Papagno, C., & Vallar, G. (1987). Verbal and spatial immediate memory span: Normative data from 1355 adults and 1112 children. *The Italian Journal of Neurological Sciences, 8*(6), 537–548.

Parsons, S., & Bynner, J. (1997). Numeracy and employment. *Education and Training, 39*(2), 43–51. https://doi.org/10.1108/00400919710164125.

Parsons, S., & Bynner, J. (2005). *Does numeracy matter more?.* London, England: National Research and Development Centre for Adult Literacy and Numeracy.

Passolunghi, M. C., & Lanfranchi, S. (2012). Domain-specific and domain-general precursors of mathematical achievement: A longitudinal study from kindergarten to first grade. *British Journal of Educational Psychology, 82*(1), 42–63. https://doi.org/10.1111/j.2044-8279.2011.02039.x.

Peng, P., Namkung, J., Barnes, M., & Sun, C. (2016). A meta-analysis of mathematics and working memory: Moderating effects of working memory domain, type of mathematics skill, and sample characteristics. *Journal of Educational Psychology, 108*(4), 455–473. https://doi.org/10.1037/edu0000079.

Purpura, D. J., & Ganley, C. M. (2014). Working memory and language: Skill-specific or domain-general relations to mathematics? *Journal of Experimental Child Psychology, 122,* 104–121. https://doi.org/10.1016/j.jecp.2013.12.009.

Raghubar, K. P., & Barnes, M. A. (2017). Early numeracy skills in preschool-aged children: A review of neurocognitive findings and implications for assessment and intervention. *The Clinical Neuropsychologist, 31*(2), 329–351. https://doi.org/10.1080/13854046.2016.1259387.

Ramani, G. B., & Siegler, R. S. (2008). Promoting broad and stable improvements in low-income children's numerical knowledge through playing number board games. *Child Development, 79*(2), 375–394. https://doi.org/10.1111/j.1467-8624.2007.01131.x.

Räsänen, P., Salminen, J., Wilson, A. J., Aunio, P., & Dehaene, S. (2009). Computer-assisted intervention for children with low numeracy skills. *Cognitive Development, 24*(4), 450–472. https://doi.org/10.1016/j.cogdev.2009.09.003.

Rasmussen, C., & Bisanz, J. (2005). Representation and working memory in early arithmetic. *Journal of Experimental Child Psychology, 91*(2), 137–157. https://doi.org/10.1016/j.jecp.2005.01.004.

Reyna, V. F., Nelson, W. L., Han, P. K., & Dieckmann, N. F. (2009). How numeracy influences risk comprehension and medical decision making. *Psychological Bulletin, 135*(6), 943–973. https://doi.org/10.1037/a0017327.

Ritchie, S. J., & Bates, T. C. (2013). Enduring links from childhood mathematics and reading achievement to adult socioeconomic status. *Psychological Science, 24*(7), 1301–1308. https://doi.org/10.1177/0956797612466268.

Romano, E., Babchishin, L., Pagani, L. S., & Kohen, D. (2010). School readiness and later achievement: Replication and extension using a nationwide Canadian survey. *Developmental Psychology, 46*(5), 995–1007. https://doi.org/10.1037/a0018880.

Salminen, J., Koponen, T., Räsänen, P., & Aro, M. (2015). Preventive support for kindergarteners most at-risk for mathematics difficulties: Computer-assisted intervention. *Mathematical Thinking and Learning, 17*(4), 273–295. https://doi.org/10.1080/10986065.2015.1083837.

Savage, R., & Carless, S. (2004). Predicting curriculum and test performance at age 7 years from pupil background, baseline skills and phonological awareness at age 5. *British Journal of Educational Psychology, 74*(2), 155–171. https://doi.org/10.1348/000709904773839815.

Schneider, M., Beeres, K., Coban, L., Merz, S., Schmidt, S. S., Stricker, J., & De Smedt, B. (2017). Associations of non-symbolic and symbolic numerical magnitude processing with mathematical competence: A meta-analysis. *Developmental Science, 20*(3). https://doi.org/10.1111/desc.12372.

Shalev, R. S., Manor, O., & Gross-Tsur, V. (2005). Developmental dyscalculia: A prospective six-year follow-up. *Developmental Medicine and Child Neurology, 47*(2), 121–125.

Simmons, F., Singleton, C., & Horne, J. (2008). Brief report—phonological awareness and visual-spatial sketchpad functioning predict early arithmetic attainment: Evidence from a longitudinal study. *European Journal of Cognitive Psychology, 20*(4), 711–722. https://doi.org/10.1080/09541440701614922.

Slavin, R. E. (2008). Perspectives on evidence-based research in education—what works? Issues in synthesizing educational program evaluations. *Educational Researcher, 37*(1), 5–14. https://doi.org/10.3102/0013189X08314117.

Sowinski, C., LeFevre, J.-A., Skwarchuk, S.-L., Kamawar, D., Bisanz, J., & Smith-Chant, B. (2015). Refining the quantitative pathway of the pathways to mathematics model. *Journal of Experimental Child Psychology, 131,* 73–93. https://doi.org/10.1016/j.jecp.2014.11.004.

Stacy, S. T., Cartwright, M., Arwood, Z., Canfield, J. P., & Kloos, H. (2017). Addressing the math-practice gap in elementary school: Are tablets a feasible tool for informal math practice? *Frontiers in Psychology, 8*(179). https://doi.org/10.3389/fpsyg.2017.00179.

Toll, S. W. M., Kroesbergen, E. H., & Van Luit, J. E. (2016). Visual working memory and number sense: Testing the double deficit hypothesis in mathematics. *British Journal of Educational Psychology, 86*(3), 429–445. https://doi.org/10.1111/bjep.12116.

Toll, S. W. M., & Van Luit, J. E. (2013). Accelerating the early numeracy development of kindergartners with limited working memory skills through remedial education. *Research in Developmental Disabilities, 34*(2), 745–755. https://doi.org/10.1016/j.ridd.2012.09.003.

Toll, S. W. M., Viersen, S. V., Kroesbergen, E. H., & Van Luit, J. E. (2015). The development of (non-)symbolic comparison skills throughout kindergarten and their relations with basic mathematical skills. *Learning and Individual Differences, 38,* 10–17. https://doi.org/10.1016/j.lindif.2014.12.006.

Trick, L. M., Enns, J. T., & Brodeur, D. A. (1996). Life span changes in visual enumeration: The number discrimination task. *Developmental Psychology, 32*(5), 925–932. https://doi.org/10.1037/0012-1649.32.5.925.

Vanbinst, K., Ansari, D., Ghesquière, P., & De Smedt, B. (2016). Symbolic numerical magnitude processing is as important to arithmetic as phonological awareness is to reading. *PLoS ONE, 11*(3). https://doi.org/10.1371/journal.pone.0151045.

Wagner, R. K., Torgesen, J. K., & Rashotte, C. A. (1999). *The comprehensive test of phonological processing [Measurement instrument].* Austin, TX: Pro-Ed.

Watts, T. W., Duncan, G. J., Siegler, R. S., & Davis-Kean, P. E. (2014). What's past is prologue: Relations between early mathematics knowledge and high school achievement. *Educational Researcher, 43*(7), 352–360. https://doi.org/10.3102/0013189X14553660.

Wechsler, D. (2001). *WIAT-II abbreviated: Wechsler individual achievement test [Measurement instrument].* San Antonio, TX: Psychological Corporation.

Woodcock, R. W., & Johnson, M. B. (1989). *Woodcock-Johnson Psycho-Educational battery-revised [Measurement instrument].* Allen, TX: Developmental Learning Materials.

Xenidou-Dervou, I., Molenaar, D., Ansari, D., van der Schoot, M., & van Lieshout, E. C. (2016). Nonsymbolic and symbolic magnitude comparison skills as longitudinal predictors of mathematical achievement. *Learning and Instruction, 50,* 1–13. https://doi.org/10.1016/j.learninstruc.2016.11.001.

Mathematical or Computational Thinking? An Early Years Perspective

Donna Kotsopoulos, Lisa Floyd, Vivian Nelson and Samantha Makosz

Children are born into a digital world, and from a very early age are engaging with technology—from wanting to play on their parents' cell phones to actually working independently on a computer or some form of mobile technology. Countless toys exist that are digital in nature and intended for young children (e.g., robotic toys and basic computing devices). Consequently, many children starting school in kindergarten are already *users* of technology (Blackwell, Lauricella, Wartella, Robb, and Schomburg, 2013).

While a young child's world is inherently digitized, surprisingly, the same may not be true in early learning settings where computers or mobile technology may not be so readily available. Although technology in schools is increasing, there appears to continue to be an underuse of technology in early years' settings (Abu Al Rub, 2015; Blackwell et al., 2013). As Parette, Quesenberry, and Blum (2010) explain, "technology applications are still used less frequently in early childhood education settings than may be the case in real world settings" (p. 336).

The digital realities of most societies and the growing demand for careers that require technology-literate individuals have inspired a need to consider the way in which young children, from the onset of schooling, are also prepared to be future *producers* of technology. Having an understanding of the basics of how computer code works and being able to write programs affords individuals with the ability to create and/or contribute ideas to software applications, rather than simply using

D. Kotsopoulos (✉)
Huron University (at Western University), London, ON, Canada
e-mail: dkotsopo@uwo.ca

L. Floyd
Faculty of Education, Western University, London, ON, Canada

V. Nelson
Faculty of Science, Wilfrid Laurier University, Waterloo, ON, Canada

S. Makosz
Faculty of Education, Wilfrid Laurier University, Waterloo, ON, Canada

K. M. Robinson et al. (eds.), *Mathematical Learning and Cognition in Early Childhood*,
https://doi.org/10.1007/978-3-030-12895-1_6

"apps" that are created by others. While children may start school already using technology, schooling may be lagging behind in terms of supporting an advancement of their skills and understanding to become producers. In the early years, teachers and early childhood educators (ECEs) may require a new and different set of pedagogical skills and content-based competencies to support this type of development.

In this chapter, we describe applied educational psychology research where teachers, ECEs, and researchers explored how to be both more mindful and intentional about *computational thinking* (CT) in a kindergarten play-based setting as a means of potentially supporting children to be both users and future producers of technology. Our objectives are as follows: (1) to advance teachers' and ECEs' thinking about CT in early childhood education and (2) conceptualize CT and MT in relation to existing frameworks.

Simply put, CT, described more fully next, is a collection of cognitive processes and practices that are algorithmic in nature, may or may not use a computer, are drawn from computer science, and are used to help in problem solving or the execution of simple and complex tasks (Gadanidis, Hughes, Minniti, and White, 2016; Yadav, Mayfield, Zhou, Hambrusch, and Korb, 2014). A familiar activity that uses CT is "coding"—or computer programming. Engaging in coding is one vehicle for developing computational thinking, but certainly not the only activity (Grover & Pea, 2018). The integration of computational thinking (CT) into early childhood learning is a field of growing importance; some argue that our digitized world has made CT an essential skill for *all* students (Repenning, Basawapatna, and Escherle, 2016).

Over the course of one school year, we engaged in research that took place in a large urban setting and included kindergarten teachers, early childhood educators, and kindergarten children in six different classrooms from one school. Children in these classes were age three to five, and in this jurisdiction, each kindergarten class has one teacher and one ECE. For the purpose of this chapter, we refer to both of these as "teachers." According to demographic data gathered at the onset of the study, teachers reported having very little comfort with or understanding of CT; indeed, the vast interest in the topic in their school district and beyond was an impetus for agreeing to engage in this research.

What Is CT?

CT can be defined as "the mental activity for abstracting problems and formulating solutions that can be automated" (Yadav et al., 2014, p. 5:2). According to Wing (2006), "computational thinking involves solving problems, designing systems, and understanding human behavior, by drawing on the concepts fundamental to computer science" (p. 33). Essentially, CT is about "thinking like a computer scientist" (Grover & Pea, 2018, 21). Brennan and Resnick (2012) propose that CT consists of both computational concepts and practices. Concepts are tools used in computer programming, including *sequences*, *loops*, *events*, *conditionals*, *operators*, and *data*.

Practices are the context for using the tools and include *experimenting and iterating*, *testing and debugging*, *reusing and remixing*, and *abstracting*.

Though often confused as being limited to computers, CT processes are executed in the daily activities of adults and children alike; CT can be expressed as a subconscious process, but is still present in even the most mundane day-to-day tasks. For example, everyday instances of CT can be as simple as doing laundry or finding a name in an alphabetically sorted list. Each of these examples requires an individual to follow a sequence of steps, where some steps are essential (e.g., add water, go to a specific letter) and other steps may be optional or more efficient (e.g., adding fabric softener, starting the beginning of the alphabet), the steps can be repeated to achieve the same goals, and can be replicated by others. Effective use of CT can enhance such strategies across all subject areas. CT is seen as "a powerful cognitive skill that can have a positive impact on the areas of children's intellectual growth" (Horn, Crouser, & Bers, 2012, p. 380).

CT can be mischaracterized as simply problem solving. Indeed, CT also involves problem solving. Problem solving requires an individual to define the problem, plan, carry out a plan, and check the solution, as well as knowing when to apply these strategies (Polya, 1957). The key difference or enhancement to problem solving is using specifically methods drawn from computer science (e.g., programming, coding, simulation, etc.) that can potentially lead to increased replication or iteration.

We see evidence of widespread acceptance regarding the importance of CT for all children. Increasingly, jurisdictions across the world are implementing mandatory curriculum in CT, recognizing both the cognitive and tangible skills associated with CT and the important growing demand for a highly skilled workforce in all things digital (Berry, 2013; The White House, 2016) England, for example, has now introduced computer programming as part of their nationwide curriculum and it is a mandatory subject from first grade onward (Government of England, 2013). Estonia has demonstrated similar standards since 2013, when coding curriculum starting in first grade was also implemented (SITRA, 2014). North American trends are leaning toward a mandatory curriculum component, but powerhouse countries like the USA have not yet been able to completely integrate CT into all state curriculums (The White House, 2016). In Canada, we have seen mandatory curriculum introduced in several provinces and extensive dialogue about future curriculum across most provinces (British Columbia Government, 2016; Province of New Brunswick, 2016; Province of Nova Scotia, 2015).

CT in a Classroom

As mentioned, although CT is often associated with computer programming, this type of thinking can be developed through other contexts as well, including those without a computer (Grover & Pea, 2018). The Canadian Pediatric Society (2017) recommends limiting screen time use in young children and suggests that sometimes too much screen time can result in less opportunity for teaching and learning

(p. 465). Furthermore, introducing CT along with programming on a computer may add unnecessary complexity and result in children turning away from CT (Yevseyeva & Towhidnejad, 2012). Lu and Fletcher (2009) recommend teaching CT apart from programming languages and separated from computer science to help children to better grasp the key processes and concepts. Presenting CT without technology and in familiar forms can provide a suitable approach for laying the foundation for becoming producers in the digital, computer programming component.

Instances of CT that do not involve a computer are known as "unplugged" experiences or activities (Lamagna, 2015; Taub, Armoni, and Ben-Ari, 2012). Unplugged experiences can be cognitively demanding, but their main pedagogical purpose when used intentionally is to develop a foundational understanding of CT concepts (Kotsopoulos et al., 2017). According to Curzon, McOwan, Plant, and Meagher (2014), unplugged experiences allow for students to witness the processes involved in completing tasks, and this puts CT into a more relatable and familiar context. Furthermore, unplugged experiences or activities make excellent gateways into the world of CT, and could be encouraged by teachers as valuable points of introduction (Kotsopoulos et al., 2017).

Children's free play is already naturally unplugged when not using technology. Free play encompasses artifacts intentionally designed for play, such as toys and puzzles, and other artifacts found in a child's natural environment. Further, and for the most part, intentionally designed learning experiences in the early years may also be predominantly unplugged. It may be that there is also significant CT already embedded in some types of play or teacher-directed activity. The main challenge for educators is in identifying CT instances and then engaging in meaningful provocations or extensions of these instances to support more purposeful play that advances learning, checks for understanding, or reinforces concepts (Kotsopoulos & Lee, 2013, 2014).

The Relationship Between CT and Mathematical Thinking (MT)

One question that was asked frequently during our work with the teachers was whether what they were observing in their students' play was CT, MT, or both. The relationship between the CT and MT became a point of significant discussion in our own work with the teachers. A guiding framework in our work with the teachers was that of Sneider, Stephenson, Schafer, and Flick (2014), who describe differences between CT, MT, and potential overlaps between the two (see Fig. 1). Sneider et al. present an interesting analysis of the "capabilities," that are distinct, those that are shared, and the instances of CT and MT which have the potential to collide.

Sneider et al. (2014) propose that MT occurs when students "approach a new situation with a range of mathematical skills in mind" (p. 54). In contrast, they noted that CT is an "awareness of the many ways that computers can help … visualize systems

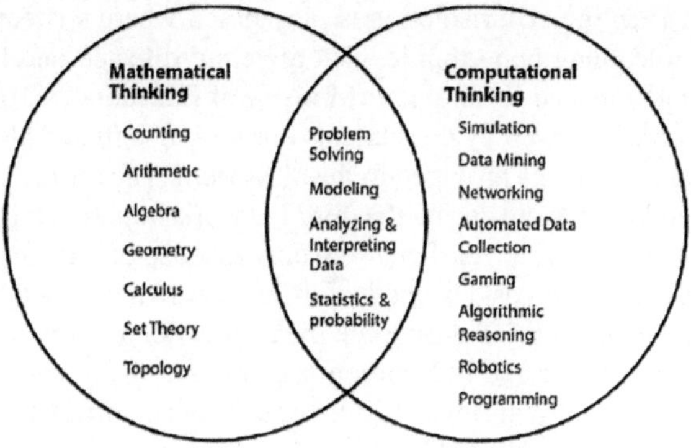

Fig. 1 Venn diagram of mathematical thinking and computational thinking (Sneider et al., 2014)

and solve problems" (p. 54). Understanding whether the CT is simultaneously MT is important. These researchers suggest that multiple capabilities, which we interpret as something that has the potential to be done or completed by an individual, are common to both MT and CT, and this area of overlap is of great importance in facilitating learning of both CT and MT. To fully support learning and to avoid missed opportunities for expanding a child's understanding, a nuanced knowledge of both is required by teachers.

It is easy to understand why differences between CT and MT may be hard to conceptualize. "Computational" in CT implies concepts and processes drawn from computer science, whereas "computational" in a more familiar sense for teachers implies calculations. Indeed, calculations as used for arithmetic, algebra, and other MT identified by Sneider et al. (2014) are also inherently algorithmic, as are those in computer science. While it has been reported that teachers sometimes fail to see the mathematical aspects of children's play (van Oers, 2010), it was nevertheless easier for the teachers in this study to identify the mathematical aspects of the children's play than it was for them to identify the new concept of CT during play. There appeared to be a blurring between concepts that required further contemplation by the researchers and the teachers collectively.

Our Study

The teachers participated in six workshops during the school year, each led by all the authors of this chapter. During the first two professional development (PD) workshops, teachers, along with the research team, explored and learned about CT by reviewing relevant literature (including understanding terms within frameworks explored), engaging in unplugged activities, and engaging in whole-group analyses of the activities and the relationship of these activities to the literature.

Between the second and the third workshop, teachers were asked to capture what they thought were instances of CT in their classrooms using video or images. Kindergarten in our jurisdiction is play-based where students are intended to be involved in learning opportunities that are child-directed and foster problem solving, investigation, and exploration (Ministry of Education, 2016). Consequently, the artifacts were produced by the students during self-initiated play with self-selected materials and resources provided in the classroom. The artifacts they collected were analyzed collaboratively during the third PD workshop. In total, 25 student-generated artifacts were analyzed. Teachers were only asked to capture instances of CT. During our analysis, we then asked whether artifacts were also instances of MT. Focusing teachers' attentions on CT during the data collection allowed us to examine this intersection between CT and MT without a confounding influence of MT from the onset.

Naturally, during the discussions about the artifacts, the question about whether the artifact was CT or MT arose and became persistent. Sneider et al.'s (2014) capabilities (see Fig. 1) were particularly useful in that all the terms were easily understandable. Consequently, collectively, and with the aim of consensus and agreement by the entire group, these artifacts were coded as either CT, MT, or both CT and MT. We used the mathematics curricular documents from our jurisdiction (Ontario Ministry of Education and Training/OMET, 2005) to identify mathematical strands and to assist with understanding the relationship between Sneider and colleagues' framework and the underlying mathematical processes and concepts. So, if something was coded as MT, we also referred to the curriculum documents to further define the mathematics concepts that were apparent. We provide examples of the coding of two artifacts more fully below (see Figs. 2 and 3).

A student-generated artifact that demonstrated both CT and MT is featured in Fig. 2. The student drew a treasure map that included directional arrows and landmarks to indicate the correct path to take to find the treasure. When analyzing this artifact from a CT perspective, there is clear evidence of algorithmic reasoning, in the specific step-by-step instructions provided, but there is also evidence of MT in the form of geometry and topology and in the overlap in the analysis and interpre-

Fig. 2 Treasure map

Fig. 3 Hexagon patterning

tation of the data (Sneider et al., 2014). This area of overlap uses MT capabilities and blends them with CT capabilities to solve a problem (in this example, to find the treasure). This kind of unplugged free play naturally also encourages the use of CT terms. The student, when asked to explain his/her map to the teacher, used words such as "first," "then," "next," and "last." The teacher could have inspired more CT and MT by modeling and then encouraging the student to use Boolean expressions to explain how to continue on the right path after, for example, making one wrong turn (i.e., "IF you accidentally turn right, THEN you will have to turn left three times to make it back to the path").

In Fig. 3, a student's hexagon patterning artifact can also be viewed as evidence of both MT and CT. From the perspective of MT, there is evidence of counting, symmetry, and geometric patterning (Ontario Ministry of Education and Training/OMET, 2005). The CT demonstrated in this artifact included algorithmic reasoning in the form of patterning (Sneider et al., 2014). The student described the structure to the teacher, explaining the need for symmetry in his design and his process of doing one piece of the pattern at a time to be sure.

We did not observe all the capabilities outlined in Fig. 1 in the artifacts collected. Those that were identified in the artifacts included counting (MT), algebra (MT), geometry (MT), typology (MT), problem solving (MT-CT), modeling (MT-CT), analyzing and interpreting data (MT-CT), simulation (CT), and algorithmic reasoning (CT).

Three student artifacts were clearly only CT. For example, one student produced an encrypted message (i.e., an unplugged version of "programming"). Another two students engaged in a "simulation"—one artifact involved bridges and the third artifact, airplanes. An additional artifact was deemed to be neither CT nor MT, despite the teachers thinking otherwise, and this artifact involved students building a random structure with some recycled packaging material available in the classroom referred to by the students as "the doughnut castle."

The teachers who identified "the doughnut castle" were under the impression that the children were engaging in CT. As a group, we arrived at the consensus that it was free play without any underlying CT or MT constructs. The children, for the most part, were simply pushing around some large recycled paper tubing and engaging in some physical play (i.e., rolling it and climbing over it). This was also an important observation that sometimes the child's free play was just that and not all free play could be construed as some type of observable learning. This is not to say that the child was not learning, but the type of learning that was actually internalized was not immediately apparent.

Apart from these four artifacts (one not CT or MT, and three only CT), there was an intersection of MT and CT in the rest of the artifacts ($n = 21$), suggesting an underlying relationship between them. At least four of the MT-only capabilities (i.e., counting (MT), algebra (MT), geometry (MT), and typology (MT)) in Sneider et al.'s (2014) framework may therefore be more appropriately represented as an intersection between MT and CT. Further intersections may also exist but may not have been evident in the samples we explored. This remains a persistent area for further research. It may be, in fact, that there are few areas that do not overlap and those may all be related to technology-based experiences identified in the CT-only part of the framework. This would be an important theoretical contribution from this applied educational psychology research.

There were three mathematical strands from the mathematics curriculum that were particularly evident in artifacts that were deemed to be *only* MT according to the model but also CT based on further analysis. These mathematical strands were number sense and numeration, geometry and spatial sense, and patterning and algebra. With respect to algebra, and given the age range of the children producing the artifacts, early patterning can be considered an early form of algebraic reasoning (National Council of Teachers of Mathematics/NCTM, 2006; Ontario Ministry of Education and Training/OMET, 2005).

The observations by the teachers of the intersection between CT and MT in this study are promising. Teachers were looking at play differently and began to see CT. Simultaneously, they were also paying more attention to the mathematical aspects of the play. It demonstrated to us that the professional development over the school year was not only encouraging a growth in conceptual understanding of CT, but also encouraging greater contemplation about MT and the relationship with CT. The artifacts were all created by the children through unplugged experiences (i.e., without the use of a computer) during free play. Consequently, seeing the intersection between CT and MT may have been easier than had the children been engaged in actual computer coding on computers. Investigating whether the same is true in a "plugged" (i.e., computer) context is an important future direction for research.

Why Is Seeing CT, MT, and the Overlap Between CT and MT Important?

In our classroom-based exploration with teachers in an early years setting, we were inspired by the depth of questioning about CT and MT and the relationship between the two. As mentioned, our task to the teachers at the onset was to identify instances of

CT based on the experiences teachers had in the professional development sessions. The question about the relationship of CT to MT evolved naturally. The importance of beginning to see the connections is important for student learning and so this evolution in the context of the research and in the context of the teachers' learning was transformative for us and for the teachers as well, based on their reports back to us.

Seeing instances of CT, MT, and any potential overlap between CT and MT during children's play is important. First, the perspective that CT can be seen in children's unplugged play and that CT can be developed without a computer is an important perspective for teachers. Some studies suggest that teachers in the early years are particularly anxious and lack professional learning around integrating technology (Jeong & Kim, 2017); moreover, access to technology may not be so prevalent in the early years (Abu Al Rub, 2015; Blackwell et al., 2013; Parette et al. 2010). Thinking about CT through and in free play bypasses these barriers. Second, thinking about play in "unplugged" terms opens opportunities for further CT experiences that may be more complex and/or may also involve technology.

Teachers need to be able to see MT, CT, and instances where they converge in order to support the learning of both versus one or the other. Recognizing MT *in* CT and ensuring simultaneous focus in planned activities by teachers is important because MT is sometimes missed or underemphasized in some CT contexts such as coding (Gadanidis, 2015)—in short, the inability to recognize MT in CT may result in missed opportunities for deeper mathematical learning. The aim, from a mathematics education perspective, is to encourage and support teachers to intentionally design tasks that use CT to also enhance an understanding of MT.

Further Research and Professional Development

While our contributions are partially theoretical, they are also a good example of applied educational psychology research occurring in school settings, with teachers, guiding and informing their thinking and understanding. Teachers need both content- and context-dependent understanding of both MT and CT—with CT being a novel lens for most teachers. This knowledge should develop organically, situated in children's artifacts and naturally occurring events in the classroom.

Over the next decade, there will be a significant need for professional development for teachers that enhances their understanding of CT and CT pedagogy, but also about the ways in which CT intersects with mathematics and even other disciplines. The power and potential of CT is not limited to the computer science realm, but extends to other disciplines as well. Our research demonstrates that oftentimes, instances of CT during young children's free play are in fact *also* instances of MT.

One clear implication of this research is the need to perhaps reconsider the distinction made in the Sneider et al. (2014) model—we found more overlap than what was proposed in their model. Indeed, we propose that more in-depth examination of the intersection between CT and MT is necessary. While the questions regarding the overlap of CT and MT seem to be of mind to researchers and practitioners alike, only Sneider et al.'s framework was found and thus the only one available to us for this analysis of the relationship between the two.

The field of teaching, including in early years education, is sure to see an increased emphasis on computational thinking in coming years, and it is starting to be seen as the core of all STEM fields (Román-González, Pérez-González, and Jiménez-Fernández, 2017). As such, professional development for teachers and ECEs will play a vital role in the introduction of CT in the classroom. Conceptually, unplugged experiences lay an excellent foundation for more complex CT processes that could be fully utilized in a classroom setting prior to digital coding experiences or engagement with computers.

Practically, not all classrooms or even homes may be equipped with technology or sufficient parental support to assist children in engaging in activities such as coding. Unplugged experiences are accessible while building skills that could then be related to more complex processes. These unplugged experiences also lay an important foundation and partner for exploring MT given the level of intersection we observed in our research. Aside from general CT training for teachers and ECEs, an important implication for teachers is that the distinction and connection between CT and MT must be made, so that teachers can appropriately incorporate both realms of learning into their classroom activities but also to be able to recognize and thus support them during play.

CT is also proposed to be widely applicable to other subject areas as well and thus some level of understanding by teachers and ECEs can potentially be translatable to other learning. Finally, more recently Grover and Pea (2018) point out that more recent definitions and elements of CT have also included collaboration and creativity. Consequently, a deeper understanding and application of CT may also have implications for collaborative learning and creativity more broadly.

Acknowledgements Thank you to Wilfrid Laurier University teacher education candidate Nicole Singer for her research support during this study. This research was funded by a Social Sciences and Humanities Research Council of Canada Partnership Grant to the first author (grant co-PI).

References

Abu Al Rub, M. F. (2015). Teachers beliefs and technology use in kindergarten and elementary classrooms. *World Journal on Educational Technology, 3*, 149–156. https://doi.org/10.18844/wjet.v7i3.202.

Berry, M. (2013). *Computing in the national curriculum. A guide for primary teachers*. Bedford, UK: Computing at School.

Blackwell, C. K., Lauricella, A. R., Wartella, E., Robb, M., & Schomburg, R. (2013). Adoption and use of technology in early education. *Computers & Education, 69*(310–319). https://doi.org/10.1016/j.compedu.2013.07.024.

Brennan, K., & Resnick, M. (2012). *New frameworks for studying and assessing the development of computational thinking*. Paper presented at the American Educational Research Association, British Columbia, Canada.

British Columbia Government. (2016). $6 million to help connect students with coding, new curriculum and computers. Retrieved from https://news.gov.bc.ca/releases/2016PREM0065-000994.
Canadian Pediatric Society. (2017). Screen time and young children: Promoting health and development in a digital world. *Paediatrics & Child Health,* 461–468. https://doi.org/10.1093/pch/pxx123.
Curzon, P., McOwan, P., Plant, N., & Meagher, L. (2014). Introducing teachers to computational thinking using unplugged storytelling. In *WiPSCE'14 Proceedings of the 9th Workshop in Primary and Secondary Computing Education* (pp. 89–92).
Gadanidis, G. (2015). Coding as a Trojan horse for mathematics education reform. *Journal of Computers in Mathematics and Science Teaching, 34*(2), 155–173.
Gadanidis, G., Hughes, J. M., Minniti, L., & White, B. J. G. (2016). Computational thinking, grade 1 students and the Binomial Theorem. *Digital Experiences in Mathematics Education.* https://doi.org/10.1007/s40751-016-0019-3.
Government of England. (2013). National curriculum in England: Computing programmes of study. Retrieved from https://www.gov.uk/government/publications/national-curriculum-in-england-computing-programmes-of-study.
Grover, S., & Pea, R. (2018). Computational thinking: A competency whose time has come. In S. Sentance, E. Barendsen, & S. Carsten (Eds.), *Computer science education: Perspectives on teaching and learning in school* (pp. 19–38). London: Bloomsbury Academic.
Horn, M. S., Crouser, R. J., & Bers, M. U. (2012). Tangible interaction and learning: The case for a hybrid approach. *Personal and Ubiquitous Computing, 16*(4), 379–389.
Jeong, H. I., & Kim, Y. (2017). The acceptance of computer technology by teachers in early childhood education. *Interactive Learning Environments, 25*(4), 496–512. https://doi.org/10.1080/10494820.2016.1143376.
Kotsopoulos, D., Floyd, L., Khan, S., Namukasa, I. K., Somanath, S., Weber, J., & Yiu, C. (2017). A pedagogical framework for computational thinking. *Digital Experiences in Mathematics Education.*
Kotsopoulos, D., & Lee, J. (2013). What are the development enhancing features of mathematical play? *An Leanbh Óg: The OMEP Ireland Journal of Early Childhood Studies, 7,* 47–68.
Kotsopoulos, D., & Lee, J. (2014). *Let's talk about math: The LittleCounters® approach to building early math skills.* Baltimore, MD: Brookes Publishing Co., Inc.
Lamagna, E. (2015). Algorithmic thinking unplugged. *Journal of Computing Sciences in Colleges, 30*(6), 45–52.
Lu, J. J., & Fletcher, G. H. L. (2009). Thinking about computational thinking. *ACM SIGCSE Bulletin, 41*(1), 260–264. https://doi.org/10.1145/1539024.1508959.
Ministry of Education. (2016). *The kindergarten program.* Toronto, Ontario: Government of Ontario.
National Council of Teachers of Mathematics/NCTM. (2006). Curriculum focal points: For prekindergarten to grade 8 mathematics. Retrieved from http://www.nctmmedia.org/cfp/full_document.pdf.
Ontario Ministry of Education and Training/OMET. (2005). *The Ontario curriculum Grades 1–8 mathematics, Revised.* Toronto: Queen's Printer for Ontario.
Parette, H., Quesenberry, A., & Blum, C. (2010). Missing the boat with technology usage in early childhood settings: A 21st century view of developmentally appropriate practice. *Early Childhood Education Journal, 37*(5), 335–343. https://doi.org/10.1007/s10643-009-0352-x.
Polya, G. (1957). *How to solve it: A new aspect of mathematical method* (2nd ed.). Princeton, New Jersey: Princeton University Press.
Province of New Brunswick. (2016). *MIddle school technology education.* Author.
Province of Nova Scotia. (2015). *Minister announces coding as a priority during education day.* Retrieved from http://novascotia.ca/news/release/?id=20151021002.
Repenning, A., Basawapatna, A., & Escherle, N. (2016). Computational thinking tools. In *2016 IEEE Symposium on Visual Languages and Human-Centric Computing (VL/HCC)* (pp. 218–222). https://doi.org/10.1109/vlhcc.2016.7739688.

Román-González, M., Pérez-González, J.-C., & Jiménez-Fernández, C. (2017). Which cognitive abilities underlie computational thinking? Criterion validity of the computational thinking test. *Computers in Human Behavior, 72,* 678–691. https://doi.org/10.1016/j.chb.2016.08.047.

SITRA. (2014). *Future will be built by those who know how to code*. Retrieved from http://www.sitra.fi/en/artikkelit/well-being/future-will-be-built-those-who-know-how-code.

Sneider, C., Stephenson, C., Schafer, B., & Flick, L. (2014). Exploring the science framework and NGSS: Computational thinking in the science classroom. *The Science Teacher, 38*(3), 10–15.

Taub, T., Armoni, M., & Ben-Ari, M. (2012). CS unplugged and middle school students' views, attitudes, and intentions regarding CS. *ACM Transactions on Computing Education (TOCE), 12*(2), 8.

The White House. (2016). Computer science for all. Retrieved from https://www.whitehouse.gov/blog/2016/01/30/computer-science-all.

van Oers, B. (2010). Emergent mathematical thinking in the context of play. *Educational Studies in Mathematics, 74*(1), 23–37.

Wing, J. M. (2006). Computational thinking and thinking about computing. *Communications of the ACM, 49,* 33–35.

Yadav, A., Mayfield, C., Zhou, N., Hambrusch, S., & Korb, J. T. (2014). Computational thinking in elementary and secondary teacher education. *ACM Transactions on Computing Education, 14*(1), 1–16.

Yevseyeva, K., & Towhidnejad, M. (2012). Work in progress: Teaching computational thinking in middle and high school. In *Paper presented at the 2012 Frontiers in Education Conference Proceedings*.

Supporting Meaningful Use of Manipulatives in Kindergarten: The Role of Dual Representation in Early Mathematics

Helena P. Osana and Nicole Pitsolantis

Teachers frequently use concrete objects, often called manipulatives, to help young children understand concepts such as number and place value (Carbonneau, Marley, & Selig, 2013; Chao, Stigler, & Woodward, 2000; McNeil & Jarvin, 2007; Moyer, 2001). Typically, the educator's goal for teaching with manipulatives is to make abstract concepts more tangible or concrete (Bruner, 1964; Marley & Carbonneau, 2014a). Base ten blocks, for example, are often used to represent the conceptual structure of the numeration system. Single blocks are intended to represent ones, ten such blocks stuck together in a row are meant to represent one group of 10, and ten such sticks attached represent ten groups of ten (or 100). Teachers often use these blocks to illustrate concepts of place value and to assist students in their computation and problem-solving activities (Dienes, 1964).

Reviews on the effectiveness of manipulatives on student learning in mathematics have revealed mixed results (e.g., Carbonneau et al., 2013; Osana, Przednowek, Cooperman, & Adrien, 2018). Part of the challenge in synthesizing work in this area is that the independent and dependent variables are disparate from one study to the next, rendering comparisons practically impossible. In addition, establishing results that are generalizable requires careful study of the conditions under which manipulatives are effective. As such, research attention has shifted from crude comparisons of manipulatives versus no manipulatives to examining the instructional conditions that are most likely to positively impact students' learning in the classroom (Marley & Carbonneau, 2014b).

One factor that has been investigated at some length is the nature of the instructional guidance provided to students while they engage in activities with manipulatives. Scholars have tested the level of support, the type of support, and the sequencing of specific lessons with manipulatives on a number of different measures, including assessments of conceptual and procedural knowledge in mathematics (Carbonneau & Marley, 2015; Fyfe, McNeil, Son, & Goldstone, 2014;

H. P. Osana (✉) · N. Pitsolantis
Department of Education, Concordia University, Montreal, QC, Canada
e-mail: helena.osana@concordia.ca

K. M. Robinson et al. (eds.), *Mathematical Learning and Cognition in Early Childhood*,
https://doi.org/10.1007/978-3-030-12895-1_7

Osana, Adrien, & Duponsel, 2017). We agree with Sarama and Clements (2009) and others that the degree to which manipulatives are beneficial for learning is contingent on students' interpretations of the objects themselves. If left to their own devices, however, children do not always construct the intended or even appropriate meaning of manipulatives (Osana et al., 2018). Ultimately, students need to see "beyond" the object itself to the abstract referent it is intended to represent.

In this chapter, we describe a study we conducted with kindergarten students that aimed to explore different levels of instructional support on the children's *quantitative interpretations* of base ten blocks. We provided a series of six brief lessons to a group of children in kindergarten. Each lesson focused on a specific prerequisite skill or concept that would support students' interpretations of the blocks as specific quantities. The unit culminated in direct instruction on how to use the objects to read and represent quantities under 100. Our objective was to explore whether students needed direct instruction on the meanings of the blocks or if certain prerequisite skills could be marshaled by the students to construct the meanings for themselves. We also were interested in examining potential differences in the development of *dual representation* as a function of the children's prior numeracy knowledge. Dual representation refers to the notion that manipulatives are objects in their own right as well as symbols that stand for something else (Uttal, Liu, & DeLoache, 2006). Dual representation has been shown to predict children's responsiveness to mathematics instruction (Booth & Siegler, 2008; Fyfe, Rittle-Johnson, & DeCaro, 2012; Ramani & Siegler, 2011).

We begin with a brief overview of the importance of guiding children to make sense of manipulatives, and we include occasional forays, as necessary, into the literature on external knowledge representations more broadly (e.g., pictures, diagrams; Belenky & Schalk, 2014). We then describe the study we conducted and draw conclusions from our data on the instructional factors that are promising for enhancing students' quantitative interpretations of manipulatives.

Supporting Students' Quantitative Interpretations of Manipulatives

Our theoretical framework stems from the work in developmental psychology on children's symbolization (e.g., DeLoache, 2004; Uttal & Yuan, 2014). DeLoache (1987) introduced 2½- and 3-year-old children to a room, complete with items and pieces of furniture, and a scale model of the room that was identical in every way except size. The children watched the experimenter hide a small toy in the model, after which the experimenter asked the children to find the toy in the life-sized room. DeLoache found differences between the two age groups in their ability to use the scale model as a symbol that referred to where a toy was hidden in the actual room, concluding in this and further studies that a developmental shift in children's capacity to use symbols occurs sometime shortly before the age of three (DeLoache, 2000;

Marzolf & DeLoache, 1994; Uttal, Schreiber, & DeLoache, 1995). DeLoache (1995) used the term *dual representation* to describe how by the age of three, the children could represent the model room in two ways—as an object its own right, but also as a symbol for the life-sized room that can be used to solve problems.

In mathematics, symbols are often used to illustrate ideas and procedures. They include standard notation (e.g., numerals, operations, relational symbols, graphs), but also idiosyncratic symbols, such as drawings, maps, and sketches. Uttal, Scudder, and DeLoache (1997) argued that because of their representational role, manipulatives can also be considered symbols (English, 2004; Goldin, 1998). In our work, we have adopted the definition from Uttal and Yuan (2014), namely "A symbol is something that someone intends to stand for something else" (p. 296). Under this definition, base ten blocks can arguably be seen as symbols: A stick of ten single cubes (a "long"), for example, can be considered a symbol for the abstract quantitative referent of one group of ten units. As such, Uttal et al. (2006) proposed that dual representation can be used to explain children's learning mathematics with manipulatives—students do not learn with them if they fail to view them as possessing two "identities," namely objects in their own right and representations of intended conceptual referents.

Dual representation of mathematical symbols, including concrete objects, is not acquired spontaneously, and simply interacting with manipulatives will not guarantee it (Ambrose, 2002; Ball, 1992; Uttal et al., 2006). In a now seminal study, Resnick and Omanson (1987) found that although students could compute with base ten blocks and solve the same problems using the standard algorithm with numerals, they were perfectly content to operate within each of the separate representational contexts without making conceptual connections between them. For example, children were presumably able to see the base ten "long" as a plastic yellow stick, but without dual representation, they did not view it as a representation of a quantity (i.e., one group of ten). Research on external knowledge representations at large has more recently suggested that without conceptual links between representational systems, children's learning tends to be highly proceduralized, tightly connected to the context of learning, and not easily transferred to novel contexts (Belenky & Schalk, 2014; Martin & Schwartz, 2005; Richland, Stigler, & Holyoak, 2012; Vendetti, Matlen, Richland, & Bunge, 2015; Verschaffel, De Corte, de Jong, & Elen, 2010). Indeed, children's difficulties in mathematics have often been ascribed to weak symbol-referent correspondences (e.g., Hiebert, 1992; Janvier, 1987; Martí, Scheuer, & de la Cruz, 2013).

Teachers, therefore, play an important role in helping students understand the quantitative meanings of manipulatives—that is, in supporting the acquisition of dual representation (Brown, McNeil, & Glenberg, 2009; Marley & Carbonneau, 2014b; Wearne & Hiebert, 1988). Much of the research on instructional effects has evaluated the impact of different levels of guidance (e.g., unguided versus explicit instruction) with manipulatives, and external knowledge representations more broadly, on a wide variety of learning and transfer measures in mathematics and science (Alfieri, Nokes-Malach, & Schunn, 2014; Carbonneau & Marley, 2015; Hushman & Marley, 2015; Osana et al., 2017). Overall, the literature suggests that students need some form of guidance when learning mathematics and science with manipulatives. The ideal level of explicitness in instruction remains an open question, but everything else remaining constant, unguided discovery is almost uniformly the least beneficial for students (Alfieri, Brooks, Aldrich, & Tenenbaum, 2011; Kirschner, Sweller, & Clark, 2006).

In contrast to this growing body of research, very little work has examined the effects of different forms of guidance on students' *interpretations of the manipulatives themselves*. In a recent study specifically targeting children's dual representation (Osana et al., 2018), we gave explicit explanations to first graders on the quantitative referents for red and blue chips (blue = 1; red = 10) and showed the students how to represent quantities with the chips in ways that aligned with their intended quantitative referents. Students were exposed to the same chips in two other conditions: In a game-piece condition, the students used the chips as tokens in a board game similar to checkers, and in a free play condition, students were allowed to play with the chips in any way they wished. Students in a fourth condition (control) were not exposed to the chips at all.

After the students were introduced to the chips in all conditions, Osana et al. assessed their dual representation—namely, their ability to read the quantity represented by a display of colored chips and to construct a display given an assigned quantity (Resnick & Omanson, 1987). We found that the students who were told the quantitative referents of the chips were at a considerable advantage on the dual representation tasks compared to all other conditions, a finding that is perhaps not surprising given that the students in the other conditions had encoded the manipulatives in ways that were deliberately non-quantitative. More striking, however, was what happened subsequently. After the administration of the dual representation tasks, all students engaged in small-group activities on addition problems using the chips. After the addition activities, the dual representation tasks were administered a second time. The students in the free play condition did not acquire quantitative meanings for the chips even after using them quantitatively in the addition activity. The students who used the chips as game pieces were also at a disadvantage relative to the quantitative group.

Research also suggests that the extent to which students learn from instruction in mathematics, and the ways in which they benefit from it, is contingent on their prior knowledge in the domain (Booth & Siegler, 2008; Clarke, Ayres, & Sweller, 2005; Cook, 2006; Lee & Chen, 2014; Petersen & McNeil, 2013; Rittle-Johnson, Star, & Durkin, 2009). Only a handful of studies have focused specifically on the impact of prior knowledge on the extent to which children interpret concrete objects as representing quantities. In Osana et al. (2018), we found, for example, that prior encodings of manipulatives greatly influenced the ways children interpreted plastic chips after addition activities. We also found that students' prior perceptions of the objects (as mathematical tools, game pieces, or toys) impacted their interpretations of the chips after the activity.

In another study, Petersen and McNeil (2013) tested the interactive effects of the perceptual salience (i.e., attractiveness) of manipulative counters and the prior knowledge of preschool children on a counting task. The results revealed that children only used the perceptually rich counters as symbols of quantity when their prior knowledge was low. Their established knowledge of what the counter represented (e.g., a giraffe, a zebra) inhibited them from focusing on their known meaning instead of their intended quantitative meaning; without prior knowledge (e.g., of objects that did not cue real-world knowledge), the children were able to use the manipulatives in ways that aligned with their intended quantitative referents. Addressing issues in mathematics learning more generally, Gravemeijer, Doorman, and Drijvers (2010) confirmed that "the meaning of external representations is dependent on the knowledge and understanding of the interpreter" (p. 194).

The Kindergarten Study

In Osana et al. (2018), we concluded that explicit explanation plays an important role in the acquisition of dual representation, but our study left unanswered two important questions. First, we tested the effects of direct instruction on several different skills in the same lesson: the chips' referents, how to count with them, and how to use them to represent quantities. It is unclear, however, how each one of those skills contributes to the development of dual representation over the course of instruction. It is possible, for example, that simply telling students the quantitative referents for the manipulatives is not enough for dual representation to emerge. Practice using the manipulatives in quantitative ways (e.g., grouping them, counting them, representing quantities with them) may be a necessary condition for the development of dual representation. Second, while we were able to conclude that prior encodings and perceptions of the chips themselves were important prerequisites for subsequent interpretations, the role of prior numeracy knowledge in children's interactions with manipulatives remains unanswered.

In the study described in this chapter, we explored the development of dual representation over the course of a six-lesson unit in two kindergarten classrooms. Our objectives were to examine the cumulative effects of a series of lessons, all using direct instruction, that focused on specific prerequisite skills over the course of a five-week instructional unit. We focused on the growth in children's dual representation from lesson to lesson and whether their learning took different trajectories as a function of their incoming number knowledge. We also delivered assessments of how well the children learned each of the prerequisite skills after each lesson; these data served as indices of the specific perquisite skills that might be catalysts for the development of dual representation. We used base ten blocks in this study because at the point during the school year when the data were collected, the children had not yet been exposed to base ten blocks in their class. This minimized possible prior exposure to the blocks interfering with the conclusions we could draw from the data.

Overview

The study design is presented in Fig. 1. Participants were 12 children from two kindergarten classrooms (six students from each class) in a private school in a large cosmopolitan area in Canada. The children were all between the ages of 5 and 6. The second author was the homeroom teacher for both classes. She delivered all six-unit lessons and administered all the learning, dual representation, and transfer tasks with the participants in their classrooms.

The children had not been formally introduced to base ten blocks prior to the study. Prior numeracy knowledge was assessed using the Number Knowledge Test (NKT; Okamoto & Case, 1996), which was administered to all children in both classes. Six children who scored at grade 1/2 or higher were randomly selected for the high knowledge group. Six children who scored at grade level were randomly selected to be in the low knowledge group.

Six lessons were delivered over a five-week period. All lessons included a direct instruction portion that entailed explicit demonstrations of targeted learning objectives, which was then followed by small-group practice activities. Each lesson introduced a distinct skill considered by Resnick and Omanson (1987) as prerequisite to learning how base ten manipulatives map onto their quantitative referents. The first three lessons focused on knowing the referents for each block (Lesson 1); composing and decomposing groups of ten (Lesson 2); and counting groups of ten as units (Lesson 3). The next two lessons focused on explicit demonstration of how to read quantities displayed with blocks ("read a display," Lesson 4) and how to use the blocks to construct displays of given quantities ("construct a display," Lesson 5), also considered by Resnick and Omanson as indices of decimal numeration understanding. Lesson 6 focused on the correspondences between concrete and written representations of quantities. In our previous research, we used measures based on reading and constructing displays as evaluations of dual representation, and as such, these tasks were the target learning objectives for the kindergarteners in our study (Osana, Przednowek, Cooperman, & Adrien, 2013; Osana et al., 2018).

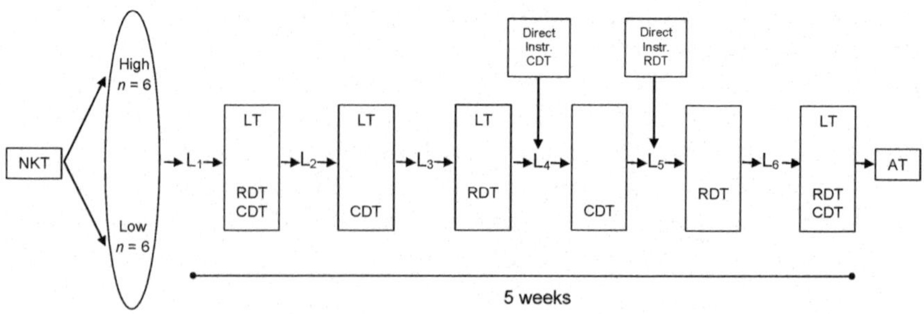

Fig. 1 Design for the kindergarten study. NKT = Number Knowledge Test. LT = learning task. RDT = read a display task. CDT = construct a display task. L = lesson. AT = application task

After each of Lessons 1, 2, 3, and 6, we administered a learning task and at least one dual representation task (i.e., a read a display or construct a display task) individually to all 12 students. Given that direct instruction on the dual representation tasks was provided in Lessons 4 and 5, the dual representation tasks that followed these lessons served as their own learning tasks. A final application task was administered as a test of far transfer after the unit.

Classroom Instruction

Each lesson lasted 25 min and was delivered to all the students in both classes. Each lesson began with direct instruction in a whole-group context, followed by small-group student practice activities that focused on the target idea in each lesson. The teacher circulated through the classroom during the student practice activities to provide them with guidance and feedback.

Lesson 1 focused on the quantitative referents for the unit (block) and the "long" (i.e., stick of ten blocks). The teacher showed the students a unit and a long, named each block, told the students the value of each block, and wrote the corresponding number symbol on the board. Next, each student was given a unit and a long and was asked to hold up each block in response to various teacher instructions (e.g., "if your name starts with the letter A, hold up the ten.") The class was then split into two smaller groups for practice activities. In their small groups, students were given units and longs to work with. Practice activities focused on rehearsing the name of each block and stating its value.

Lesson 2 focused on composing a long from ten units and decomposing a long into ten units. The instruction began with reviewing the ideas from Lesson 1 and then focusing on the concept of "trading" the blocks for equivalent values. The teacher demonstrated various trades (units for longs and longs for units), always focusing on the idea that "ten ones are worth the same as one group of ten; they are both worth ten." Following the demonstration portion of Lesson 2, students were next split into groups of three for practice activities.

Lesson 3 began with a review of the concepts covered in the two prior lessons and then turned to demonstrations on how to count units and how to count groups of ten by skip counting. For example, the teacher counted a group of 19 units and explained that units are counted by ones because they are each worth one. The teacher next repeated the counting demonstration with longs and explained that longs are counted by tens because they are each worth ten. Several more counting exercises were shown, but units and longs were always counted separately. Following the explicit demonstration portion of the lesson, students were given several counting activities to complete in pairs. With a partner, each student counted a total of six sets of blocks: three sets of units and three sets of longs.

Demonstrations were delivered in Lesson 4 on how to construct a display and in Lesson 5 on how to read a display. Both lessons began with a review of the preceding lessons. Next, using the base ten blocks, the teacher showed the students how to construct a display for a numeral presented on an index card (Lesson 4) and how to read a display of pre-arranged blocks (Lesson 5). A new counting method was introduced in Lesson 5: counting across units. In prior lessons that targeted counting (Lessons 3 and 4), units and longs were always counted separately. For example, in Lesson 3, the objective was to learn *how* to count with the blocks—to learn how to count units (by ones) and longs (by tens)—not to tell how many are in a set. In Lesson 4, the objective was to represent a given quantity with the blocks. To do so, one does not need to be able to tell how many there are in the set. To determine the quantity of a given set (i.e., read a display of blocks), however, the student must be able to count across the different units to combine ones and tens. This was the objective of Lesson 5; as such, the teacher explicitly demonstrated counting across units.

After both Lessons 4 and 5, the students were again placed into small groups to complete practice activities. Practice activities involving constructing displays required students to use the base ten blocks to represent one- and two-digit numerals present on index cards. Activities involving reading displays required students to count various pre-arranged sets of blocks, in quantities ranging from 12 to 68, and to record their counts in number symbols on a mini-whiteboard.

Finally, Lesson 6 focused on the correspondence between concrete and written representations for double-digit numerals. Using a place value mat, the teacher demonstrated how to count a set of blocks and place them in the appropriate spot on the mat. For example, the teacher counted a set of 15 blocks then placed the long in the tens column on the mat and explained why it went there (i.e., because it is worth ten). The teacher then placed the units in the ones column on the mat and explained that she did so because units are worth one. Next, the teacher demonstrated how to use the mat as a tool to write the corresponding number symbol by mapping the number symbol to the display on the mat. For example, the teacher wrote 15 on the mat and said, "To write 15, I look and see that I have one group of ten here, so I write a 1 here, and I see that I have five ones here, so I write a 5 here. That's 15 and this is how you write 15." The teacher repeated the demonstration and explanation several more times. The students were then assigned to small groups of three or four to complete similar practice activities.

Learning, Dual Representation, and Application Tasks

Each learning task assessed the skill that was directly addressed in the lesson preceding it. The materials used for the learning tasks included base ten blocks (units and longs), number symbol cards, and paper and pencil for recording written answers. Dual representation was assessed with two tasks based on Resnick and Omanson (1987). The read a display task (RDT) required students to determine the quantity represented by a set of base ten blocks. The construct a display task (CDT) required students to use the blocks to represent double-digit numbers read out loud by the teacher. The dual representation tasks were set in contexts that were familiar to the students. For example, for one CDT task, the teacher told the story of a student who

Fig. 2 Partial setup for the application task

collects hockey cards and brings them to school for show and tell. The child was shown a numeral card corresponding to the number of hockey cards in the collection and was asked to use the base ten blocks to represent that amount.

Finally, the application task (AT) was couched in the context of a pizza shop and was used to assess far transfer. Items on the AT required students to compare quantities that were represented with the blocks, show quantities in different ways, and construct amounts to "pay" for their meals. The students were familiar with the pizza shop because it was part of the dramatic play corner in the classroom. Props in the shop included dishes and cutlery, pizza pans and utensils for making pizza dough, an oven and stove, play food pizza toppings, and menus that were displayed on the walls of the shop. The AT tasks were crafted as a role-play scenario in which the student worked at the pizza shop and the teacher pretended to be a customer visiting the shop. Part of the set up for the AT task is presented in Fig. 2.

The seven tasks on the AT are presented in Table 1. Like the RDT and CDT tasks, the AT tasks required the students to either read a display of blocks or construct a display of blocks in the pizza shop context. The tasks differed from the RDT and CDT tasks, however, in that their objectives required the students to apply what they had learned during the lessons in novel ways. This was achieved by building in goals and constraints that would require the students to transfer their knowledge of dual representation to complete the task. In Task 1, for example, the student was asked to compare the cost of two pizzas presented in base ten blocks. The teacher constructed the two displays prior to the start of the interview and covered them with a sheet of paper. During the interview, the teacher revealed the two displays only briefly (3–4 s) before covering them up again and asking the student to tell which pizza cost more. By giving the student only a few seconds to look at the displays, the student was unable to determine the value of the blocks by counting. This constraint forced the student to estimate which of the two displays had the bigger value by focusing on the type of blocks in each display. That is, the student would have to note that the

Table 1 Items on the application task

Item number	Item type	Item objective	Item description
1	RDT	Comparing only the 10s in a display to tell which is more	The teacher briefly shows two displays, side by side, indicating which is the student's (3 longs and 9 units) and which is the teacher's (7 longs and 0 units), and says, "Your pizza costs this much and my pizza costs this much. Whose pizza costs more?"
2	RDT and CDT	Grouping ones into piles of 10s when counting a large number of units, OR Showing a quantity in two ways	The teacher says, "Your pizza costs this much (shows 47 units in a small bin) but the pizza shop does not want so many ones because the cash register is too full. Can you use the blocks on the table (point to a different bin of blocks on the table containing longs and units) to pay this much another way? Can you find a way to pay this much without so many ones?"
3	RDT	Matching each digit in a numeral to a concrete display	The teacher looks at the menu on the wall and asks for an extra-large cheese pizza. She then says, "My pizza costs this much (shows 6 tens and 6 ones with the blocks). That number looks like this (shows numeral card 66). Which of these blocks mean this much (points to the 6 in the tens place of the numeral)? Which of these blocks mean this much (point to the 6 in the ones place)?"

(continued)

Table 1 (continued)

Item number	Item type	Item objective	Item description
4	CDT	Match number symbol to concrete representation when there is a 0	The teacher says, "Your meal will cost this much (shows numeral card 40). Can you use the blocks to pay this much?" Wait for student to construct the display then asks, "Which of these blocks means this part of the number (point to the 4 in 40)? Which blocks mean this part of the number (points to the 0 in 40)?"
5	CDT	Rebuild a display with more of a denomination	The teacher points to the small pepperoni pizza on the menu and says, "A small pepperoni pizza has 23 pepperonis on it. Can you show how much that is with the blocks?" Allows the student time to construct the display then asks, "Now can you show me a different way to do it?"
6	RDT	Comparing the tens to tell which is more	The teacher briefly shows two displays, side by side, indicating which is the student's (2 longs and 8 units) and which is the teacher's (6 longs and 1 unit), and says, "Your pizza costs this much and my pizza costs this much. Whose pizza costs more?"

(continued)

Table 1 (continued)

Item number	Item type	Item objective	Item description
7	CDT	Regrouping to construct a display	The teacher says, "This meal will cost 17 blocks. These are the blocks you have in your wallet (gives the student 2 tens 3 units). These are the blocks in the bank (points to a bin on the table containing many units and longs). Can you give the shop 17 blocks? You can trade some of yours for some that are in the bank."

display containing more longs was the one with the bigger value because longs are worth more than units.

The teacher administered the learning and dual representation tasks, in that order, during individual interviews conducted in the classroom on the same day the lesson was given. The interviews lasted between seven and 20 min, and all were video recorded. The teacher used base ten blocks (units and longs), numeral cards, and recording sheets for students' written responses. The AT task was administered to all 12 students in a separate interview, at the end of the six-lesson instructional unit over a span of 10 days. The interviews were video recorded and lasted between 12 and 17 min.

Results

In this section, we first describe the development of dual representation across the unit, with a focus on the relationship between children's learning of each lesson's objective and their performance on the dual representation task after each lesson. The next section describes the development of dual representation as a function of the children's learning and their prior numeracy knowledge. Finally, we focus on the relationship between performance on the dual representation tasks across the unit and the children's ability to transfer their knowledge to a task couched in a real-world context. Differences between prior knowledge groups will be addressed in relation to transfer performance. We note that because of the size of the sample, no inferential statistics were performed. As such, any patterns we report are based on descriptive analyses only.

The Development of Dual Representation Across the Sample

The means and standard deviations of the learning and read a display tasks (RDTs) are presented in Table 2. We observed a steady increase in the performance across groups on the RDT from the beginning to the end of the unit, whereas the children's performance on the learning tasks remained relatively constant throughout. These data suggest that, while children were generally able to learn the target idea in each lesson, providing explicit demonstrations on how to read a display of blocks appeared to support students' performance on the RDT. For instance, simply telling students the quantitative referents for the unit and tens blocks (Lesson 1) was not enough for them to use the blocks to read quantitative displays, despite their ability to remember the quantitative referents after the lesson (accuracy was at 92% on the learning task after Lesson 1). After Lesson 3, during which children were shown how to count collections of units by ones and collections of tens by skip counting by 10, the performance on the RDT improved, but it was not until direct demonstrations on reading displays that their performance reached almost 80% accuracy.

A similar pattern was observed for the CDT (see Table 3). The performance on the learning tasks leveled out by Lesson 4, but again, accuracy on the CDT reached 75% (its highest level, up from 33% after the previous lesson) only after direct instruction on the task itself. The students could repeat back to the teacher the quantitative referent for each block (i.e., "this one means one and this one means ten"; performance on the learning task after Lesson 1 reached 92% accuracy), but this was not enough for them to use the blocks in quantitative ways on the CDT (mean performance was 38%). After Lesson 2, which focused on showing children how to

Table 2 Means and (standard deviations) of the learning and read a display tasks after Lessons, 1, 3, 5, and 6

Measure	Lesson			
	1	3	5[a]	6
Learning	0.92 (0.19)	0.77 (0.31)	0.77 (0.33)	0.75 (0.45)
Read a display	0.33 (0.44)	0.53 (0.44)	0.77 (0.33)	0.92 (0.29)

[a]Data on the learning and read a display tasks for Lesson 5 represent performance on the same task

Table 3 Means and (standard deviations) of the learning and construct a display tasks after Lessons, 1, 2, 4, and 6

Measure	Lesson			
	1	2	4[a]	6
Learning	0.92 (0.19)	0.63 (0.38)	0.75 (0.41)	0.75 (0.45)
Construct a display	0.38 (0.43)	0.33 (0.45)	0.75 (0.41)	0.75 (0.38)

[a]Data on the learning and construct a display tasks for Lesson 4 represent performance on the same task

compose and decompose groups of ten with the blocks, students' performance on the CDT hovered at 33%, despite 63% accuracy on the learning task.

Differences in Prior Knowledge

The findings reported above must be interpreted in light of the different patterns we observed in the low and high prior knowledge groups. Children's scores on the learning and dual representation tasks after each lesson are displayed as a function of students' prior knowledge profiles in Table 4 (RDT) and Table 5 (CDT). In each group, dual representation performance increased over the five-week instructional period on both tasks, but the high knowledge group began with DR scores that were considerably higher after the first lesson than the low knowledge group. Specifically, on the RDT, the performance in the high knowledge group ($M = 0.50$, $SD = 0.45$) was three times that of the low knowledge group ($M = 0.17$, $SD = 0.41$) after Lesson 1; on the CDT, the students in the high knowledge group outperformed those in the low knowledge group by a factor of 8. The data also show, however, that while both groups improved on both dual representation tasks, the students in the low knowledge group learned more because of the lower starting point. These findings mirror those of Ramani and Siegler (2011), who found that children who demonstrated lower numerical knowledge at baseline learned more than their higher knowledge counterparts on magnitude estimation, numeral identification, and arithmetic.

Additional group differences were observed in the discrepancies between learning performance and dual representation performance after Lessons 1 and 3. These discrepancies were considerably more pronounced in the low knowledge group than in the high knowledge group on both dual representation tasks. After Lesson 1, for instance, the performance on the learning tasks indicated that students in each group were able to verbally state with 92% accuracy that the unit blocks represented "one" and the tens blocks represented "ten." In contrast, students in the low knowledge

Table 4 Means and (standard deviations) of the learning and read a display tasks by prior knowledge group after Lessons, 1, 3, 5, and 6

Measure	Lesson			
	1	3	5[a]	6
Learning				
Low	0.92 (0.20)	0.71 (0.33)	0.67 (0.44)	0.67 (0.52)
High	0.92 (0.20)	0.83 (0.30)	0.88 (0.14)	0.83 (0.41)
Read a display				
Low	0.17 (0.41)	0.33 (0.42)	0.67 (0.44)	1.00 (0.00)
High	0.50 (0.45)	0.72 (0.39)	0.88 (0.14)	0.83 (0.41)

[a]Data on the learning and read a display tasks for Lesson 5 represent performance on the same task

Table 5 Means and (standard deviations) of the learning and construct a display tasks by prior knowledge group after Lessons, 1, 2, 4, and 6

Measure	Lesson			
	1	2	4[a]	6
Learning				
Low	0.92 (0.20)	0.67 (0.41)	0.54 (0.51)	0.67 (0.52)
High	0.92 (0.20)	0.58 (0.38)	0.95 (0.10)	0.83 (0.41)
Construct a display				
Low	0.08 (0.20)	0.11 (0.27)	0.54 (0.51)	0.61 (0.44)
High	0.67 (0.41)	0.56 (0.50)	0.95 (0.10)	0.89 (0.27)

[a]Data on the learning and construct a display tasks for Lesson 4 represent performance on the same task

group were not able to use the blocks with those quantitative meanings on either the RDT or CDT tasks after the first lesson ($M = 0.17$, $SD = 0.41$; $M = 0.08$, $SD = 0.20$, respectively). Those in the high knowledge group were more successful, however, relative to both the students in the low knowledge group and to their own learning after each lesson. After the second lesson, the gap between performance on the dual representation tasks and the learning task was again considerably less substantial in the high knowledge group than in the low knowledge group. In fact, the mean CDT score in the low knowledge group after Lesson 2 was 83% lower than their mean learning score, but the performance on the learning task and the CDT was comparable in the high knowledge group. Similar patterns were observed after Lesson 3 on learning and RDT performance.

After direct instruction, students in the high knowledge group continued to outperform those in the low knowledge group on the dual representation tasks, although the gap between the two groups narrowed somewhat. Specifically, after Lesson 5, during which direct instruction on reading displays of base ten blocks occurred, the performance in the high knowledge group reached 88% compared to 67% in the low knowledge group. After Lesson 4, which focused on demonstrating how to construct displays of specific quantities, the performance on the CDT reached 96% in the high knowledge group, whereas the performance in the low knowledge group was at 54%.

The apparent effect of direct instruction on dual representation performance was that it was beneficial for both groups, but to different degrees. In the high knowledge group, students' performance on the RDT increased by 22% after direct instruction. In contrast, students in the low knowledge group saw their performance double. On the CDT, students in the high knowledge group increased their scores by 72% after direct instruction; students in the low knowledge group by nearly 400%.

Application Task

Children across both knowledge groups improved their performance on both dual representation tasks over the course of the instructional unit, with mean accuracy levels across the sample of 92% (*SD* = 29%) on the RDT and 75% (*SD* = 38%) on the CDT after Lesson 6. We designed the application task (AT) to evaluate the extent to which the children were able to apply their interpretations of the manipulatives (i.e., as representing quantities) to a setting that simulated a real-world context—in our case, a pizza parlor. The AT means as a function of prior knowledge are presented alongside the RDT means in Table 6 and the CDT means in Table 7.

The mean AT score across groups was 61% (*SD* = 34%), but the performance differed by prior knowledge group (low: $M = 0.40$, $SD = 0.31$; high: $M = 0.83$, $SD = 0.22$). As the data in Tables 5 and 6 illustrate, AT performance of the high knowledge group was consistent with their performance on both the RDT and CDT by the end of the unit. In contrast, the low knowledge group's AT performance dipped substantially compared to their performance on the dual representation tasks after Lesson 6. The children in the low knowledge group reached 100% accuracy on RDT and 61% on the CDT after Lesson 6, but appeared to struggle on the AT, reaching 40% accuracy.

We also observed that the AT score in the low knowledge group, while higher than each of the dual representation scores (RDT and CDT) after both Lesson 1 and Lesson 2, was lower than both dual representation scores in the high knowledge group after the first lesson. This is in spite of the fact that the students in the low

Table 6 Means and (standard deviations) of the read a display and application tasks across the instructional unit by prior knowledge group

PK group	Read a display				AT
	Lesson				
	1	3	5	6	
Low	0.17 (0.41)	0.33 (0.42)	0.67 (0.44)	1.00 (0.00)	0.40 (0.31)
High	0.50 (0.45)	0.72 (0.39)	0.88 (0.14)	0.83 (0.41)	0.83 (0.22)

Note: AT = Application task; PK = prior knowledge

Table 7 Means and (standard deviations) of the construct a display and application tasks across the instructional unit by prior knowledge group

PK group	Construct a display				AT
	Lesson				
	1	3	5	6	
Low	0.08 (0.20)	0.11 (0.27)	0.54 (0.51)	0.61 (0.44)	0.40 (0.31)
High	0.67 (0.41)	0.56 (0.50)	0.95 (0.10)	0.89 (0.27)	0.83 (0.22)

Note: AT = Application task; PK = prior knowledge

knowledge group scored considerably higher on the RDT and at a comparable level on the CDT after Lesson 6 than the students in the high knowledge group did on these same tasks after Lesson 1. Together, these data suggest that, despite having learned how to use the manipulatives in ways that aligned with their quantitative meanings, there was a discrepancy between the two knowledge groups in terms of their ability to transfer their knowledge to a more realistic context.

Discussion

Our objective in this chapter was to take the notion of "manipulatives as symbols" (Uttal et al., 1997) into the kindergarten classroom and to use dual representation, a central aspect of children's symbolization, as a lens to examine its development in the context of mathematics instruction. We worked closely with a small group of children as they learned how to interpret and use base ten blocks in ways that aligned with their intended quantitative meanings. We further investigated the possible role of prior knowledge in the development of children's interactions with the blocks. Although the study was exploratory and replications with larger samples are necessary, our primary aim in this chapter was to provide a fine-grained view of how children learn to interpret manipulatives quantitatively and to address, however tentatively, the conditions under which teachers can foster such interpretations in the classroom.

Overall, the children in our study learned to use base ten blocks in ways that aligned with their quantitative referents, and their performance strengthened over the course of prolonged instruction. Although steady improvement in dual representation was observed from lesson to lesson, substantial increases in performance occurred immediately after explicit demonstrations were provided on the dual representation tasks themselves. Our data also suggested that in an instructional context, the development of children's dual representation may interact in important ways with their prior numeracy knowledge. In our study, for example, the benefits of direct instruction were particularly pronounced for the students with lower numeracy knowledge. With less established knowledge of number, the ability of children in this group to construct meaning for the blocks, and to use them accordingly, may have been compromised as a result. These observations are in line with those of Booth and Siegler (2008), who found that, relative to their peers with stronger knowledge of numerical magnitudes, first graders with weak prior numeracy knowledge were subsequently hindered in learning new mathematical content.

In addition, relative to their high knowledge counterparts, the students with weaker prior knowledge appeared less adept at transferring their learning from each lesson to the dual representation task that followed. This finding is consistent with Goswami's (2004) claim that young children lack the background knowledge to detect the correspondences between two conceptually similar contexts. Specific prerequisite skills, such as declarative knowledge of the blocks' quantitative referents, composing and decomposing groups of ten, and counting collections of blocks, were seemingly not sufficient for the students in the low knowledge group to use the blocks with meaning. It is possible that the deficiencies in prior knowledge of the students in the low knowledge group impeded the development of their dual representation and inhibited transfer to the final application task. The children in the high knowledge group, in contrast, may have been in a better position to marshal their learning on subsequent transfer tasks, which may explain the relatively rapid improvement on the dual representation tasks for this group and superior performance on the application task.

Together, these data may suggest that the students' prior numeracy knowledge served as an important foundation for transferring what they were told about the blocks' quantitative referents to contexts that required them to use the blocks in ways that corresponded to those referents. This line of reasoning is supported by Baroody's (2017) argument that a child's developmental level in a specific mathematical domain is key to rendering manipulatives effective, as development plays a role in the child's ability to connect informal knowledge to concrete experiences and symbolic learning. More specifically, telling the students the quantitative referents for the units and tens blocks appeared to support the development of dual representation in more effective ways for the students in the high knowledge group. The high knowledge students also appeared to capitalize on other prerequisite skills, such as grouping and counting activities (Lessons 2 and 3, respectively). Relative to their peers in the low knowledge group, the naming, counting, and grouping activities may have been more meaningful to the high knowledge students because they had the quantitative structures in place to appropriate the blocks' quantitative meanings during these activities. This, in turn, may have more effectively supported the development of dual representation for those students in the high knowledge group.

In contrast, while students in the low knowledge group learned how to compose and count groups of ten, they could have simply engaged in such activities at a more superficial (i.e., procedural) level relative to their high knowledge peers. This could have contributed to their difficulties in transferring their learning to the dual representation tasks. Our observations are consistent with the findings of McGuire and Kinzie (2013), who found that preschoolers have considerable difficulty grouping and counting by ten, but our study also raises the possibility that students in both groups attended to different aspects of the lessons, despite comparable performance on the learning tasks. Future research is needed to explain how any differences in the children's responses to instruction might explain the development of dual representation and transfer performance (Byrd, McNeil, Chesney, & Matthews, 2015; Lobato, 2012).

The lack of established numerical structures in the low knowledge group may explain only part of their relatively low performance on the final application task, however. Another factor that may account for the variance is the nature of the instruction itself. The lessons we provided consisted exclusively of direct explanations; it is possible that the low knowledge group would have benefited from a combination of different types of activities. Fyfe, DeCaro, and Rittle-Johnson (2014), for example, found that engaging children in exploratory activities before direct instruction was more effective for growth in both conceptual and procedural knowledge of mathematical equivalence than providing explicit explanations first. The authors did not examine these effects as a function of prior knowledge, however, but it is worth considering that different types of instruction and its sequencing may account for at least a portion of the differences observed in the present study. It is also possible that children with lower prior knowledge do indeed benefit from explicit demonstrations, *including* specific demonstrations of how and when to apply new concepts (Kirschner et al., 2006). Additional research is required to tease apart the interrelated factors of student characteristics, such as prior knowledge, and instructional conditions, including the type and sequencing of lessons and the nature of the specific representations used (see Osana et al., 2017).

Furthermore, our line of research promises to contribute to current understandings of the role of prior knowledge and instruction when children engage in typical classroom numeration activities. Indeed, grouping and counting activities with manipulatives emphasize *unitizing*—working with groups of ten as new units—the hallmark of place value understanding (Carpenter, Fennema, Franke, Levi, & Empson, 2014; Carpenter, Franke, Jacobs, Fennema, & Empson, 1998; Fosnot & Dolk, 2001), and grouping activities have been found to be positively associated with children's growth in place value and numeration (e.g., Pagar, 2013; Peled, Meron, & Rota, 2007). Our data contribute to this literature by suggesting a more nuanced interpretation of current understandings of children's numeration development in the early school years; at the very least, our work provides the impetus for more research on the nature of children's manipulatives use in classroom contexts.

Clearly, there are limitations to the current study that must be acknowledged before any prescriptions for practice are made. For one, the sample was small and further reduced by examining the relative effects of the two prior knowledge groups. In addition, the psychometric properties of the dual representation tasks are unknown; a fruitful avenue for further research would be on the construction of valid and reliable assessments of children's interpretations of mathematical representations. Despite the study's limitations, we contend that our results allow us to consider a number of possible implications for practice. For one, our data appear to suggest that children can learn specific facts and procedures, but have considerably more difficulty transferring their knowledge to solve new problems. This finding is not new, of course (e.g., see Barnett & Ceci, 2002; Lobato, 2012), but our research highlights the notion that rich, quantitative interpretations of the symbols' meanings likely take time to develop. Moreover, the differences we observed in the students with stronger and weaker prior knowledge may highlight that appropriate use of manipulatives requires more than declarative knowledge of quantitative referents or procedural fluency in counting objects. While a wide variety of grouping and counting activities with manipulatives is likely necessary, children appear to need sufficient prior knowledge to give meaning to their physical actions with the objects (see also Martin & Schwartz, 2005). Whether or not kindergarten teachers incorporate base ten manipulatives and concepts into their classrooms, focusing on foundational numeracy skills appears critical in the early years (see also Duncan et al., 2007).

References

Alfieri, L., Brooks, P. J., Aldrich, N. J., & Tenenbaum, H. R. (2011). Does discovery-based instruction enhance learning? *Journal of Educational Psychology, 103,* 1–18.

Alfieri, L., Nokes-Malach, T. J., & Schunn, C. D. (2014). Learning through case comparisons: A meta-analytic review. *Educational Psychologist, 48*(2), 87–113.

Ambrose, R. C. (2002). Are we overemphasizing manipulatives in the primary grades to the detriment of girls? *Teaching Children Mathematics, 9,* 16–21.

Ball, D. L. (1992). Magical hopes: Manipulatives and the reform of mathematics education. *American Educator, 16*(2), 14–18, 46–47.

Barnett, S. M., & Ceci, S. J. (2002). When and where do we apply what we learn? A taxonomy for far transfer. *Psychological Bulletin, 128*(4), 612–637.

Baroody, A. J. (2017). The use of concrete experiences in early childhood mathematics instruction. In J. Sarama, D. H. Clements, C. Germeroth, & C. Day-Hess (Eds.), *Advances in child development and behavior: The development of early childhood mathematics education* (pp. 43–87). Cambridge, MA: Academic Press.

Belenky, D. M., & Schalk, L. (2014). The effects of idealized and grounded materials on learning, transfer, and interest: An organizing framework for categorizing external knowledge representations. *Educational Psychology Review, 26,* 27–50.

Booth, J. L., & Siegler, R. S. (2008). Numerical magnitude representations influence arithmetic learning. *Child Development, 79,* 1016–1031.

Brown, M., McNeil, N., & Glenberg, A. (2009). Using concreteness in education: Real problems, potential solutions. *Child Development Perspectives, 3,* 160–164.

Bruner, J. (1964). The course of cognitive growth. *American Psychologist, 19*(1), 1–15.

Byrd, C. E., McNeil, N. M., Chesney, D. L., & Matthews, P. G. (2015). A specific misconception of the equal sign acts as a barrier to children's learning of early algebra. *Learning and Individual Differences, 38,* 61–67.

Carbonneau, K. J., & Marley, S. C. (2015). Instructional guidance and realism of manipulatives influence preschool children's mathematics learning. *Journal of Experimental Education, 83*(4), 495–513.

Carbonneau, K. J., Marley, S. C., & Selig, J. P. (2013). A meta-analysis of the efficacy of teaching mathematics with concrete manipulatives. *Journal of Educational Psychology, 105,* 380–400.

Carpenter, T. P., Fennema, E., Franke, M. L., Levi, L., & Empson, S. B. (2014). *Children's mathematics: Cognitively guided instruction* (2nd ed.). Portsmouth, NH: Heinemann.

Carpenter, T. P., Franke, M. L., Jacobs, V. R., Fennema, E., & Empson, S. B. (1998). A longitudinal study of invention and understanding in children's multidigit addition and subtraction. *Journal for Research in Mathematics Education, 29,* 3–20.

Chao, S.-J., Stigler, J. W., & Woodward, J. A. (2000). The effects of physical materials on Kindergartners' learning of number concepts. *Cognition and Instruction, 18*(3), 285–316.

Clarke, T., Ayres, P., & Sweller, J. (2005). The impact of sequencing and prior knowledge on learning mathematics through spreadsheet applications. *Educational Technology Research and Development, 53*(3), 15–24.

Cook, M. P. (2006). Visual representations in science education: The influence of prior knowledge and cognitive load theory on instructional design principles. *Science Education, 90*(6), 1073–1091.

DeLoache, J. S. (1987). Rapid change in the symbolic functioning of very young children. *Science, 238,* 1556–1557.

DeLoache, J. S. (1995). Early understanding and use of symbols: The model model. *Current Directions in Psychological Science, 4*(4), 109–113.
DeLoache, J. S. (2000). Dual representation and young children's use of scale models. *Child Development, 71,* 329–338.
DeLoache, J. S. (2004). Becoming symbol-minded. *Trends in Cognitive Sciences, 8*(2), 66–70.
Dienes, Z. P. (1964). *Building up mathematics* (2nd ed.). London, England: Hutchinson of London.
Duncan, G. J., Dowsett, C. J., Claessens, A., Magnuson, K., Huston, A. C., Klebanov, P., … Japeli, C. (2007). School readiness and later achievement. *Developmental Psychology, 43*, 1428–1446.
English, L. D. (2004). Mathematical and analogical reasoning. In L. English (Ed.), *Mathematical and analogical reasoning of young learners* (pp. 1–22). Mahwah, NJ: Erlbaum.
Fosnot, C. T., & Dolk, M. (2001). *Young mathematicians at work: Constructing number sense, addition, and subtraction*. Westport, CT: Heinemann.
Fyfe, E. R., DeCaro, M. S., & Rittle-Johnson, B. (2014a). An alternative time for telling: When conceptual instruction prior to problem solving improves mathematical knowledge. *British Journal of Educational Psychology, 84*(3), 502–519.
Fyfe, E. R., McNeil, N. M., Son, J. Y., & Goldstone, R. L. (2014b). Concreteness fading in mathematics and science instruction: A systematic review. *Educational Psychology Review, 26,* 9–25.
Fyfe, E. R., Rittle-Johnson, B., & DeCaro, M. S. (2012). The effects of feedback during exploratory mathematics problem solving: Prior knowledge matters. *Journal of Educational Psychology, 104*(4), 1094–1108.
Goldin, G. A. (1998). Representational systems, learning, and problem solving in mathematics. *The Journal of Mathematical Behavior, 17,* 137–165.
Goswami, U. (2004). Commentary: Analogical reasoning and mathematical development. In L. English (Ed.), *Mathematical and analogical reasoning of young learners* (pp. 169–186). Mahwah, NJ: Erlbaum.
Gravemeijer, K., Doorman, M., & Drijvers, P. (2010). Symbolizing and the development of meaning in computer-supported algebra education. In L. Verschaffel, E. De Corte, T. de Jong, & J. Elen (Eds.) (2010). *Use of representations in reasoning and problem solving* (pp. 191–208). New York, NY: Routledge.
Hiebert, J. (1992). Mathematical, cognitive, and instructional analyses of decimal fractions. In G. Leinhardt, R. Putnam, & R. Hattrup (Eds.), *Analysis of arithmetic for mathematics teaching* (pp. 283–322). Hillsdale, NJ: Erlbaum.
Hushman, C. J., & Marley, S. C. (2015). Guided instruction improves elementary student learning and self-efficacy in science. *The Journal of Educational Research, 108*(5), 371–381.
Janvier, C. E. (Ed.). (1987). *Problems of representation in the teaching and learning of mathematics*. Mahwah, NJ: Erlbaum.
Kirschner, P. A., Sweller, J., & Clark, R. E. (2006). Why minimal guidance during instruction does not work: An analysis of the failure of constructivist, discovery, problem-based, experiential, and inquiry-based teaching. *Educational Psychologist, 41*(2), 75–86.
Lee, C. Y., & Chen, M. J. (2014). The impacts of virtual manipulatives and prior knowledge on geometry learning performance in junior high school. *Journal of Educational Computing Research, 50*(2), 179–201.
Lobato, J. (2012). The actor-oriented transfer perspective and its contributions to educational research and practice. *Educational Psychologist, 47,* 232–247.
Marley, S. C., & Carbonneau, K. J. (2014a). Theoretical perspectives and empirical evidence relevant to classroom instruction with manipulatives. *Educational Psychology Review, 26,* 1–7.
Marley, S. C., & Carbonneau, K. J. (2014b). Future directions for theory and research with instructional manipulatives [Special issue]. *Educational Psychology Review, 26,* 91–100.
Martí, E., Scheuer, N., & de la Cruz, M. (2013). Symbolic use of quantitative representations in young children. In B. M. Brizuela & B. E. Gravel (Eds.), *Show me what you know: Exploring student representations across STEM disciplines* (pp. 7–21). New York, NY: Teachers College Press.

Martin, T., & Schwartz, D. L. (2005). Physically distributed learning: Adapting and reinterpreting physical environments in the development of fraction concepts. *Cognitive Science, 29,* 587–625.

Marzolf, D. P., & DeLoache, J. S. (1994). Transfer in young children's understanding of spatial representations. *Child Development, 65*(1), 1–15.

McGuire, P., & Kinzie, M. B. (2013). Analysis of place value instruction and development in pre-kindergarten mathematics. *Early Childhood Education Journal, 41*(5), 355–364.

McNeil, N. M., & Jarvin, L. (2007). When theories don't add up: Disentangling the manipulatives debate. *Theory Into Practice, 46,* 309–316.

Moyer, P. S. (2001). Are we having fun yet? How teachers use manipulatives to teach mathematics. *Educational Studies in Mathematics, 47,* 175–197.

Okamoto, Y., & Case, R. (1996). Number knowledge test. *Monographs of the Society for Research in Child Development, 61*(1–2), 27–58.

Osana, H. P., Adrien, E., & Duponsel, N. (2017). Effects of instructional guidance and sequencing of manipulatives and written symbols on second graders' numeration knowledge. *Education Sciences, 7*(2), 52–74.

Osana, H. P., Przednowek, K., Cooperman, A., & Adrien, E. (2013). *Making the most of math manipulatives: Play is not the answer*. Presented at the annual meeting of the American Educational Research Association (AERA), San Francisco, CA.

Osana, H. P., Przednowek, K., Cooperman, A., & Adrien, E. (2018). Encoding effects on first-graders' use of manipulatives. *Journal of Experimental Education, 2,* 154–172.

Pagar, D. (2013). *The effects of a grouping by tens manipulative on children's strategy use, base ten understanding and mathematical knowledge*. Doctoral dissertation. Available from ProQuest Dissertations and Theses database. (UMI No. 3560909).

Peled, I., Meron, R., & Rota, S. (2007). Using a multiplicative approach to construct decimal structure. In Woo, J. H., Lew, H. C., Park, K. S. & Seo, D. Y. (Eds.). *Proceedings of the 31st Conference of*
the International Group for the Psychology of Mathematics Education (Vol. 4, pp. 65–72). Seoul, Korea: PME.

Petersen, L. A., & McNeil, N. M. (2013). Effects of perceptually rich manipulatives on preschoolers' counting performance: Established knowledge counts. *Child Development, 84,* 1020–1033.

Ramani, G. B., & Siegler, R. S. (2011). Reducing the gap in numerical knowledge between low- and middle-income preschoolers. *Journal of Applied Developmental Psychology, 32,* 146–159.

Resnick, L. B., & Omanson, S. F. (1987). Learning to understand arithmetic. In R. Glaser (Ed.), *Advances in instructional psychology* (Vol. 3, pp. 41–95). Hillsdale, NJ: Erlbaum.

Richland, L. E., Stigler, J. W., & Holyoak, K. J. (2012). Teaching the conceptual structure of mathematics. *Educational Psychologist, 47*(3), 189–203.

Rittle-Johnson, B., Star, J. R., & Durkin, K. (2009). The importance of prior knowledge when comparing examples: Influences on conceptual and procedural knowledge of equation solving. *Journal of Educational Psychology, 101,* 836–852.

Sarama, J., & Clements, D. H. (2009). 'Concrete' computer manipulatives in mathematics education. *Child Development Perspectives, 3*(3), 145–150.

Uttal, D. H., Liu, L. L., & DeLoache, J. S. (2006). Concreteness and symbolic development. In L. Balter & C. S. Tamis-LeMonda (Eds.), *Child psychology: A handbook of contemporary issues* (2nd ed., pp. 167–184). Philadelphia, PA: Psychology Press.

Uttal, D. H., Schreiber, J. C., & DeLoache, J. S. (1995). Waiting to use a symbol: The effects of delay on children's use of models. *Child Development, 66,* 1875–1889.

Uttal, D. H., Scudder, K. V., & DeLoache, J. S. (1997). Manipulatives as symbols: A new perspective on the use of concrete objects to teach mathematics. *Journal of Applied Developmental Psychology, 18,* 37–54.

Uttal, D. H., & Yuan, L. (2014). Using symbols: Developmental perspectives. *Wiley Interdisciplinary Reviews: Cognitive Science, 5*(3), 295–304.

Vendetti, M. S., Matlen, B. J., Richland, L. E., & Bunge, S. A. (2015). Analogical reasoning in the classroom: Insights from cognitive science. *Mind, Brain, and Education, 9*(2), 100–106.

Verschaffel, L., De Corte, E., de Jong, T., & Elen, J. (Eds.). (2010). *Use of representations in reasoning and problem solving*. New York, NY: Routledge.

Wearne, D., & Hiebert, J. (1988). A cognitive approach to meaningful mathematics instruction: Testing a local theory using decimal numbers. *Journal for Research in Mathematics Education, 19*(5), 371–384.

Kindergarteners' and First-Graders' Development of Numbers Representing Length and Area: Stories of Measurement

Serife Sevinc and Corey Brady

Introduction

Models-and-modeling perspective is a tradition in mathematics education research that has investigated the nature and growth of conceptual systems of problem-solvers across a wide variety of ages, disciplines, and settings (Lesh, 2006, Lesh & Doerr, 2003a). To both stimulate and document problem-solving processes, modeling researchers have developed a genre of activities known as model-eliciting activities (MEAs) based on six design principles that have proven robust over decades of work (Lesh, Hoover, Hole, Kelly, & Post, 2000). While much prior modeling research has involved the design and implementation of MEAs with adult or teenage learners, recent work suggests MEAs may be applied productively with learners in the early primary grades (i.e., K-2) (English, 2010; Lehrer & Kim, 2009; Lesh, English, Riggs, & Sevis, 2013). In this chapter, we share two K-1 level story-based MEAs for which we present a sample instructional cycle that supported the implementation in K-1 classrooms and through which we characterized young learners' models on measurement.

Models-and-Modeling Perspective

The first fundamental question to be re-examined in light of primary grades' applications of the modeling perspective concerns the nature of models and modeling.

S. Sevinc (✉)
Department of Mathematics and Science Education, Middle East Technical University, Ankara, Turkey
e-mail: sserife@metu.edu.tr

C. Brady
Department of Teaching and Learning, Vanderbilt University, Nashville, TN, USA

K. M. Robinson et al. (eds.), *Mathematical Learning and Cognition in Early Childhood*,
https://doi.org/10.1007/978-3-030-12895-1_8

Building upon work by Lesh and colleagues, we adopt the following definition of the term "model":

> Models are conceptual systems (consisting of elements, relations, operations, and rules governing interactions) that are expressed using external notation systems, and that are used to construct, describe, or explain the behaviors of other system(s)—perhaps so that the other system can be manipulated or predicted intelligently. (Lesh & Doerr, 2000, p. 10)

Proceeding from this notion of models, the models-and-modeling perspective asserts that real-world problem-solving processes are iterative, involving multiple cycles in which learners *describe* a problematic situation, *express* "draft" of solution approaches, *test* these approaches, and then *revise* or refine their ways of thinking (i.e., their models) (Lesh & Lehrer, 2003). Idea development in these cycles occurs through conceptual reorganization along multiple continua such as concrete-versus-abstract, simple-versus-complex, intuitive-versus-formal, situated-versus-decontextualized, and specific-versus-general (Lesh & Doerr, 2003a; Lesh & Lehrer, 2003; Lesh & Yoon, 2004). However, it is not the case that an idea that is further toward the "right-hand side" of each of these continua is always more useful than the one that is further to the "left-hand side."

Models developed in this complex manner suggest both a challenge and an opportunity for research. On the one hand, it shows that the essence of idea development is not captured in the final structure of learners' models, albeit understanding the *growth* of these structures is a critical part of understanding their *nature*. Moreover, under normal circumstances, it is not easy to identify the conceptual systems that are active in learners' thoughts and actions, as these models are often tacit (c.f. Borromeo Ferri & Lesh, 2013). On the other hand, because model development involves expressing iterative drafts of increasingly sophisticated ways of thinking, it may be possible to design a learning environment in which these cycles produce "thought-revealing artifacts" (Lesh et al., 2000), which can serve as a record of the idea development process. Indeed, MEAs, as a genre of learning environments, were developed precisely for this purpose (Lesh, 2006; Lesh & Doerr, 2003b; Lesh & Harel, 2003; Lesh, Cramer, Doerr, Post, & Zawojewski, 2003).

Early design research on MEAs has produced the following six design principles that guide much current work in the field: (a) model-construction principle, (b) reality or personal meaningfulness principle, (c) self-assessment principle, (d) construct-documentation principle, (e) construct-shareability and reusability principle, and (f) effective prototype principle (Lesh et al., 2000). These six principles are essential for the design and implementation of MEAs, ensuring that learners will develop mathematically significant, generalizable, transferable, and reusable models (Lesh et al., 2000). Moreover, we used these principles as an analytical frame to design an instructional cycle emphasizing particular instructional practices that support the implementation of MEAs in kindergarten and first-grade classrooms.

Modeling at Early Years

In the last decade, researchers have increasingly argued that modeling should start in preschool and become a core component of mathematics learning at all levels. As argued, this is *possible* because the modeling process does not necessarily require high-level mathematics (Greer & Verschaffel, 2007; Greer, Verschaffel, & Mukhopadhyay, 2007). Moreover, it is *valuable* because modeling activities have potential to help children develop "the skills and concepts necessary to be a competent user of mathematics" (Usiskin, 2007, p. 263) and support them in "using mathematics as a powerful personal tool for the analysis of issues important in their personal lives and in society" (Verschaffel, 2002, p. 76).

Like adults or upper elementary and secondary school students, K-2 students are capable of developing models and revising them iteratively (Biembengut, 2007). However, the traditional mathematics curriculum might mask this capability in primary school students by posing stereotypical problems which (a) do not include more information than needed (in contrast with the overabundance of information in real-life situations), (b) do not require decisions about what information is relevant to solving the problem, and (c) admit only a single answer, as opposed to the range of possible solutions that may be appropriate under different assumptions or conditions (Biembengut, 2007; Verschaffel, de Corte, & Lasure, 1994).

Verschaffel et al. (1994) analyzed textbook word problems, showing that although these exercises aimed to connect mathematical ideas with real-life situations, most did not require real-world knowledge or skills beyond the procedural and computational level. Moreover, only a small percentage of students' responses (17%) to such problems reflected realistic considerations (Verschaffel et al., 1994). In other words, students are often presented with artificial problems in schools and, not surprisingly, often give "school answers" which involve using simple arithmetic techniques and ending up with non-realistic answers. Greer et al. (2007) hypothesized that this tendency of elementary students might be exacerbated by teachers' instructional decisions, specifically their selection of questions and their own perceptions of the problems. Worse still, another research suggests that as children progress in school, they tend increasingly to concentrate on mathematics that they have already learned, opting to solve new problems with familiar mathematical rules and procedures rather than taking on different and alternative perspectives involving interdisciplinary or real-life knowledge (Biembengut, 2007).

On the other hand, there are also indications that changes in instruction can support changes in students' problem-solving behaviors. Verschaffel and de Corte (1997) investigated the influence of instructional practices and classroom cultures that valued real-world and commonsense knowledge resources on students' modeling behavior. They compared a fifth-grade treatment classroom that encouraged authentic problem solving with a pair of sixth-grade comparison classrooms working on stereotypical word problems. Pre- to post-test results showed a substantial increase in students' use of realistic considerations in their solutions to word problems in the treatment classroom. Such studies offer hope that introducing innovative problem-solving activities in early schooling could provide a point of leverage for changing children's later problem-solving readiness.

This chapter contributes to an emerging belief that modeling activities can be effective contexts for both learning of and research with young children. Thus, we present two story-based modeling activities and discuss possible instructional practices that support the implementation of them in early primary grades. With the help of these rich contexts, we aimed to understand the characteristics of young learners' models on length and area measurement.

Story-Based MEAs for Young Learners

Here, we present two of the seven story-based MEAs implemented in a first-grade and two kindergarten classrooms during spring semesters of three consecutive years by the same teacher. The implementation of each MEA took approximately one week, sometimes occupying additional time depending on the group.

All three classrooms were from the same rural midwestern school, where over fifty percent of the students qualify for free or reduced lunch. This school placed special emphasis on developing writing skills and welcomed MEAs in part because they asked students to document their thinking by writing short letters to present their solutions to the fictional characters, which has commonly been used in modeling research as a technique to ensure particularly two of the six design principles—construct-documentation and construct-shareability and reusability (Doerr & Lesh, 2011; Lesh et al., 2003).

Two story-based MEAs, The Proper Hop and Fussy Rug Bugs, were designed to support the development of young learners' models on measurement in two different dimensions; namely, length—1-dimensional measurement and area—2-dimensional measurement. In these MEAs, measurement constructs were built on counting knowledge and skills that students already possess, but in both cases, students needed to use numbers in ways that go beyond mere counting. Moreover, the activities asked students to operationally define concepts involving maximization (e.g., maximal area estimates) and minimization (e.g., shortest paths).

The Proper Hop MEA

This MEA is based on a story called The Proper Hop from the book titled *Mathematics around us: Skills and applications; kindergarten* (Lesh & Nibbelink, 1978). It deals with a community of frogs in Sugar Swamp, including the main character, Beauregard. In the story, when communicating about the locations of lily pads where Beauregard's friends live, he observes a problem that emerged because of differences in the length of hops of different frogs. So, Beauregard attempts to establish a constructed conventional unit of linear measure that he calls a "proper hop." Following

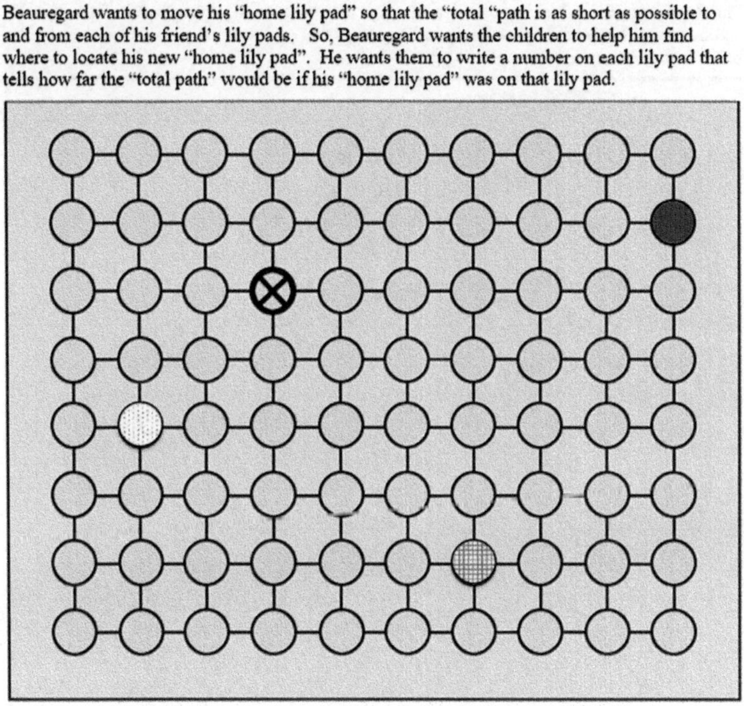

Fig. 1 The Proper Hop MEA

this story, students were given a rectilinear lattice whose dots represent lily pads. The distance between lattice-connected dots was the length of a proper hop; that is, the frogs are not allowed to hop diagonally. As seen in Fig. 1, three points on the lattice (i.e., one dotted-white, one solid-black, and one squared-gray) were marked to indicate where Beauregard's three best friends live that he likes to visit every day.

Beauregard's central problem was to determine where he should build his house so that the sum of the distances to his three friends would be the smallest. In this MEA, students used numbers to represent and optimize distance measurements that involve both direction and magnitude.

Fussy Rug Bugs MEA

This MEA is based on a story called Fussy Rug Bugs from *Mathematics around us: Skills and applications; kindergarten* (Lesh & Nibbelink, 1978). In this MEA, ten-footed rug bugs wish to establish their own town and need to determine the arrangement of their rug-dwellings depending on the shape of the rug bugs. Following the story, students were given post-it notes and a habitat area defined by a closed

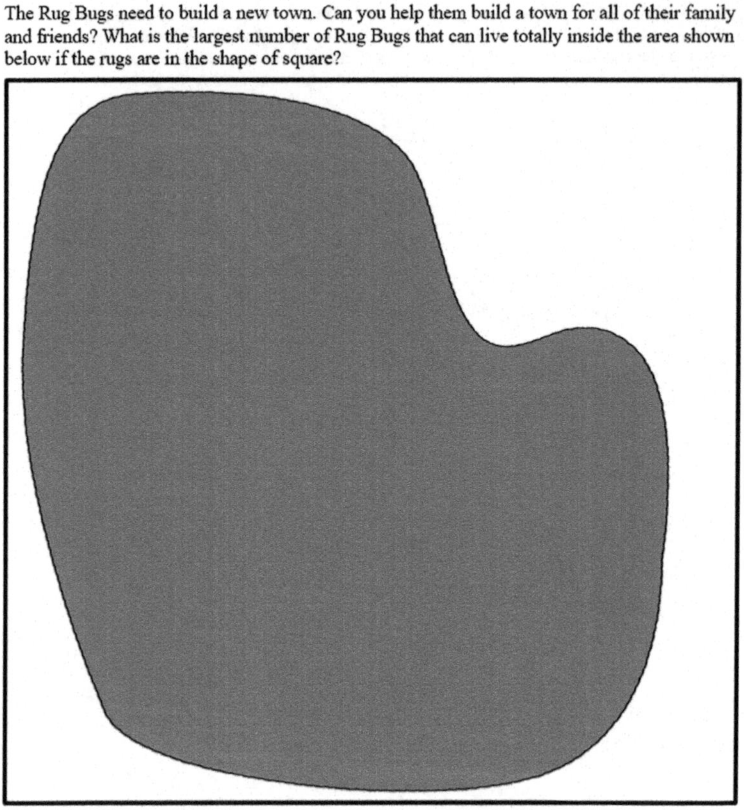

Fig. 2 Fussy Rug Bugs MEA

rectilinear or curvilinear region, as shown in Fig. 2. They were asked to find the largest number of rug bugs that could fit completely inside the habitat. Fussy Rug Bugs MEA provided an occasion to describe a quantity—the amount the floor space in different shapes—using a non-conventional proxy for the area (square, triangle, or circle post-it notes). The Fussy Rug Bugs' requirements—that their dwellings should (a) fit entirely inside the habitat, (b) fit together as closely as possible without gaps, (c) but not overlap—prompted students to develop optimal strategies for "tiling" or "packing" the habitat regions with rug shapes.

Data Collection Methods in K-1 Classrooms

As the teacher implemented these MEAs in our K-1 classrooms, we collected data through regularly photographing students' work, which was based on Rieger's systematic visual measurement, particularly used to understand social change: "Photography is well-suited to this process because of its capacity to record a scene with far greater speed and completeness than could ever be accomplished by a human observer taking notes" (1996, p. 144). We took Rieger's premise relating to a social change and extended it further to a cognitive change because photographic sequences of students' work provided us with a rich data source to analyze their cognitive change. The photograph sequences showing students' work on The Proper Hop and Fussy Rug Bugs MEAs were useful not only for us, researchers, but also for the teacher to understand the model development of young kids who are not fully capable of expressing their thinking clearly. The teacher also used these photographs in her class web blogs to communicate with parents.

Hence, our data collection method, photo-documentation, is followed by coding and analysing the changes in the visual content of the photos. Our analysis allowed us to display an instructional cycle and strategies which influenced young learners' model development and to articulate the characteristics of students' models about the measurement of length and area.

Instructional Cycle for the Implementation of MEAs in Early Grades

The use of MEAs in K-1 classrooms required special attention to encourage model development because young students needed careful scaffolding to develop a model as a solution to the problems of the characters in these mathematically rich stories. Our classroom implementations of MEAs unfolded a three-phased instructional cycle: (a) narrative introduction, (b) model development, and (c) model sharing.

In the narrative introduction phase, the teacher read the story to the whole class, introduced the characters and the problem situation through cartoon-like animated PowerPoint presentations to motivate students toward the problem, and asked several short questions to activate their initial approaches. She engaged the group in a whole-class role-playing activity where students re-enacted aspects of the story to deepen their understating. In the model development phase, students worked in groups of three using hands-on materials to express, test, and revise their thinking. During this period, they also wrote response letters involving their solutions to the problems of the characters in the stories. Finally, in model sharing phase , groups presented their solutions, using a karaoke microphone and standing in front of the class. Figure 3 presents the three-phased instructional cycle for implementing story-based MEAs in K-1 classrooms.

While the narrative introduction phase is important for motivating students to solve the problems involving numbers and measurement, the model development phase encompasses young learners' engagement in mathematical thinking and logico-mathematical actions, in Piagetian terms. The model sharing phase of the instructional cycle intended not only to encourage students to share their final products with others but also to lead them to articulate and refine their models by making some

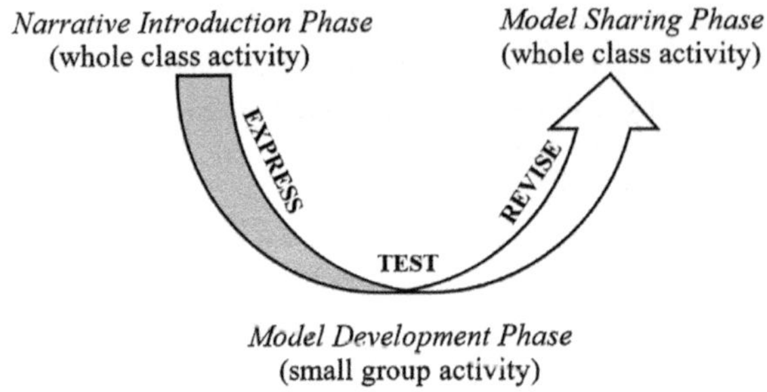

Fig. 3 Three-phased instructional cycles for implementing MEAs to young learners

of that process public and visible to others and contending with the approaches of other groups. In that sense, this final phase of the instructional cycle is as important as the first two phases for helping young learners' developing models on numbers representing length and area measurement.

Embedded within this three-phased instructional cycle, the following five instructional practices were employed by the teacher:

1. using fanciful stories as contexts of MEAs,
2. role-playing and using concrete manipulatives within supportive social interactions,
3. capturing the model development of young learners by photograph sequences,
4. using parallel whole-class and small-group activities, and
5. creating a class weblog about students' work on MEAs.

In the next section, we present how each of these practices facilitated young learners' models-and-modeling processes.

Using Stories as Contexts of MEAs

We have identified the power of fanciful stories in The Proper Hop and Fussy Rug Bugs MEAs in terms of supporting students in seeing the problem context as meaningful. Also, the characters of those stories offered mathematically relevant perspectives to the problem situations. The stories of our two MEAs had fairly "thin" plots; their characters were vivid and idiosyncratic. Their foibles and drives corresponded to mathematically significant aspects of the MEA worlds. Beauregard's need for imposing order, regularity, and equivalency was a useful perspective for defining a system of standard measurement; and the rug bugs' fear of gaps between their dwellings made the notions of tiling, tessellation, packing, and the measurement of an area more concrete for young learners. Furthermore, these characters at the MEAs offered means of anchoring perspectives that could serve as prototypes for future problem-solving.

We also found that MEAs' narratives amplified the self-assessment of students by providing the core information that they needed to develop their models. This core information was composed of the answers to the following three questions: (a) who are the main characters in the story that have a problem?, (b) why (in the worldview of the characters) a solution is needed for this problem? and (c) what would constitute a solution?

Role-Playing and Using Concrete Manipulatives Within Supportive Social Interactions

The careful use of physical activity and concrete manipulatives in supportive social settings was consistent with the assertion that knowledge and abilities are organized not only around abstractions but also around personal experiences (Lesh, English, Riggs et al., 2013; Lesh, Hamilton, & Kaput, 2007). As an example of this, during the narrative introduction phase of The Proper Hop, students role-played Beauregard and practiced moving according to "proper hops" on a large mat. The student playing Beauregard carried a stuffed animal puppet frog and experimented with paths between the red lily pad (at one corner of the mat) and the yellow lily pad (at the opposite corner of the mat). Other students in the class observed and helped: counting hops, reminding Beauregard of the rules of proper hopping and, with the teacher's support, building number sentences reflecting Beauregard's movement (e.g., 3 hops + 4 hops = 7 hops).

Physically acting out proper-hop movement rules supported students in thinking about the constraints of length measurement (e.g., counting hops, not pads, and only vertical and horizontal hops being allowed) and how they affected travel distances. For instance, when the student wanted to hop diagonally, some observer students said "She can't go like that [*diagonally*]. She can only go this way [*horizontal*] or that way [*vertical*]." The teacher further probed students' thinking with the question "why is not this hop [*diagonal hop*] a proper hop?" Most students responded that the story of Beauregard said so. Students were then asked to compare the length of the horizontal, vertical, and diagonal hops, which made students realize that the diagonal hop was longer (i.e., they needed to widen their legs more in the diagonal hop). Thus, students who both physically acted out the hops and were in the observer role gained important insights about the problem situation, particularly about mathematical aspects of the actions (i.e., diagonal hops are not at the size of proper hops).

In addition, role-playing invited students to enter into the worldview of the characters who needed their help. For example, physically acting out Beauregard's hops promoted empathic feelings which became a powerful resource for developing a solution which would meet the need of Beauregard. In contrary to the observations of Greer et al. (2007), young learners' solutions did not reflect the teacher's perspective in relation to the instructional goal; rather, they were sensible within the contexts of the stories and meaningful for the characters. Further, both stories involved students emotionally in *finding a suitable home* for characters. Adopting a caring-for attitude toward the characters supported children in maintaining a focus on the requirements for satisfactory solutions.

Capturing Model Development of Young Learners by Photograph Sequences

In modeling research, students are often asked to present their solutions by letters. For students with more experience with writing letters to an unseen reader, this approach provides a stimulus to document the rationale for their solution and for their final draft of thinking (Chamberlin & Coxbill, 2012; Diefes-Dux et al., 2004; Lesh, Amit, & Schorr, 1997). In implementing MEAs with young learners, we also used letter-writing to encourage students to document their thinking. However, it is not surprising that the younger learners did not have a stable understanding of the genres of the letters in literacy or skills to express their thinking through writing. To support the consolidation of the construct-documentation principle, the teacher periodically captured the states of students' work while they were working during the model development phase of the instructional cycle.

These time-sequenced images captured the variety of different approaches used by either the same or different groups of students over the course of the activity. On the one hand, this was useful for the teacher to decide (a) whether she needed another day for the model development phase and (b) which groups to call on to share their work during model sharing phase. On the other hand, these photographs offered us, as researchers, insights into the students' model development and the existence of modeling cycles. For instance, the image sequences discussed in a later section allowed both the teacher and us to understand that the three local strategies let to different global outcomes (i.e., the overall appearances of the tiling solutions).

Using Parallel Whole-Class and Small-Group Activities

As mentioned above, the narrative introduction phase of the instructional cycle was supported by a whole-class role-playing activity, which then fed into a period of small-group work where students were again supported with similar physical manipulatives and social structures. During the model development phase, groups were given a lily pad board and a toy frog to enact Beauregard's proper hops on a smaller scale better suited for recording path lengths. They calculated proper-hop distances between candidate locations for Beauregard's house and the houses of three of his friends. Using a toy frog similar to the puppet frog that bridged from the whole-class to the small-group experience also helped students to focus on counting actions (i.e., hops) instead of endpoints (i.e., lily pads). In this sense, we found that use of parallel activities and manipulatives contributed to the consolidation of the model-construction principle for K-1 students because when students started their small-group work, they could immediately continue to explore and develop the mathematical insights that they initiated in the whole-class activity.

Creating Class Weblog for MEAs

The last instructional practice was publishing student work, which the teacher captured in photograph sequences, in classes' web blogs. Through these blogs, parents were able to follow their children's work, achieving two goals. First, this indicated to the students that their modeling work was a significant mathematical activity to share with their parents. Second, it promoted conversation at home between students and their parents about the MEAs. Extending the modeling activity outside of the school setting also produced a possibility of a conversation about other similar situations in which mathematical constructs could apply. So, it encouraged students to think of their models as constructs that have applicability outside of the classroom. While we could not follow how often this school-to-home connection occurred, it is a suggestive area for future research design. Regarding these aspects, the teacher's writing on the class weblog consolidated the construct-documentation and construct-shareability and reusability principles.

In the next section, we present our articulation of the characteristics of young learners' models on measurement, which would not be possible to elicit without the aforementioned instructional cycle incorporating the five instructional practices.

Characteristics of Young Learners' Models on Measurement

The two story-based MEAs allowed us to see that young learners could use numbers in non-standard but mathematically interesting situations to describe locations (i.e., the coordinates of the lily pads), measure lengths or distances (i.e., numbers of proper hops), and estimate areas (i.e., numbers of rugs or post-it notes). Their thinking built upon basic counting strategies but went beyond merely counting concrete objects.

In The Proper Hop MEA, students described absolute and relative locations (i.e., hops), and distance measures involving horizontal and vertical components. In Fussy Rug Bugs, they used numbers to evaluate tilings and estimate areas (see details in Lesh, English, Sevis, & Riggs, 2013). As expected, their solutions were informal, did not explicitly reference these mathematical constructs, and did not exhibit a high degree of mathematical formalism. Their models were more situated in the narrative problem setting and grounded in concrete, socially negotiated actions and measurements. However, we observed that their models were personally meaningful and mathematically significant. Moreover, as we explain below, students were able to transfer "big ideas" developed in these MEAs to other situations experienced in their daily lives and remember the story contexts and their solutions, even many weeks later. Thus, this study revealed three characteristics of young learners' models: They tended to be tacit and intuitive; situated and embodied; and social and collaborative.

Tacit and Intuitive Models

A distinctive feature of story-based MEAs was consistently providing the learners the opportunity and means to express their thinking through physical actions. However, there were still insights and ideas which could not be fully captured by the teacher and the researchers. Modeling research has, in fact, argued that *all* models are composed of both implicit and explicit dimensions. Even in settings when models are expressed, tested and revised several times to reach n^{th} draft of it, they still retain implicit features which were called "visible tips of icebergs" (Borromeo Ferri & Lesh, 2013, p. 59). Borromeo Ferri and Lesh (2013) also argued that "implicit models are especially important for younger children, or for older students who are at early stages in the development of specific models, or for students who are functioning at lower cognitive levels" (p. 60). Therefore, special attention was required both in the design and in the implementation of MEAs in K-1 classrooms.

To support students in using their tacit knowledge resources to build toward more explicit articulations of their ideas, they were asked to share their reasoning in front of the whole class. Furthermore, the teacher used verbal interaction techniques including re-voicing and a patient use of wait time to enable groups to express their models. Table 1 presents the transcript of one group's presentation at the end of Fussy Rug Bugs MEA and the photograph sequence of this presentation.

As seen in the transcript, this group primarily attended to the boundary of the habitat space, focusing on the shape of the post-it rugs that are in common with features of that contour, and that constrained their placement. In the course of their presentation, and with the support of the teacher, their approach became more concrete and fully articulated.

The students' packing strategy involved filling the most constrained areas first. Their habitat contour included two squared peninsulas that each accommodated a post-it rug so that it touched the border on three of its sides. They filled these first, reasoning that the rug shape was directly suggested by the space and that the two peninsulas were "exactly the same"—both the same as each other and the same as the post-it rug. The mathematical notion of *congruence* lies behind this initial strategy. The group then proceeded to the next-most constrained setting, the corners of the habitat. Here, their attention was drawn to the right angles of their habitat as compared to the right angles of the square post-it notes. Relaxing the notion of congruence and attending to this common right-angle property, they "easily" placed rugs in the corners. As with congruence, the identification of right angles in the contour and in the rug shape was intuitive and tacit; in other words, the formal notion of a right angle was not explicitly referenced. Together, the first and second strategies for filling the habitat involved *attending to the boundary and moving from the most constrained spaces to less-constrained spaces.*

Table 1 Students' presentation of their work on Fussy Rug Bugs MEA

T (Teacher): How you guys decided what you did for your rug bugs town. What'd you do first? G (Girl): We, um, put this one right there (see Fig. 4a) and that one (see Fig. 4b). We did the corners first (see Fig. 4c) and then we didn't know where to put, em, because we didn't want to overlap T: Good idea. Let me see if I understand. Can you hold up your map? Your rug-bug town? They said they did this part (see Fig. 4b) and this part first (see Fig. 4a). And I'm going to peel that off so you can see what it looks like. See how it sticks out? They thought that one rug bug could stay there, which is pretty smart B: And that one was exactly the same as that one (see Fig. 4b) T: Ok. So they're the exact same …. Okay, good idea. And then you said you did the corners, right? And what'd you do after you did those corners? B: um then we did um. Then we did these ones (see Fig. 5) T: These ones? How come? Why? (B hands microphone to G) G: Because, because these were um easy and those weren't
T: Ok. Do you see any problems with your rug bug town? (G nods) Ok. Tell me what the problems are B: um there's a, there're lines right there (indicating open spaces in Fig. 6a); right there; right there, and right there, and right there, and right there (see Fig. 6a, b) T: and so what would happen to the rug bugs? B: I know. They would fall off T: They would fall off? What else? B: They would hurt their back T: Ok all right (G) wants to say something, so let him have a turn G: Because, because if they … if one of, em, had to go to their friend's house and they um, and there wasn't another one (indicating open space) they would um fall T: They would fall, they would trip, they might land on their backs, and they would not be happy, would they? Boys and girls, give them a round of applause (class claps) (B indicates he wants to talk) T: (to B) ok, you want to say something? B: um … these um squares um. They couldn't fit right there and right there T: mm-hmm … Not big enough, is it? (B shakes head)
B: and, we used ten all T: Oh, ten in all … B: 1 … 1, 2, 3, 4, 5, 6, 7, 8, 9, 10 (see Fig. 7). T: Wow. How many did you guys think you were going to need, for this? B: thirteen G: thirty B: thirty (laughing) T: Thirty. But you really only used how many? B: Ten T: Wow, which is more, ten or thirty? B: Thirty T: That's right. Ok, give 'em a round of applause …

(a) (b) (c)

Fig. 4 **a** Squared peninsula #1 considered in the model. **b** Squared peninsula #2 considered in the model. **c** Corners considered in the model

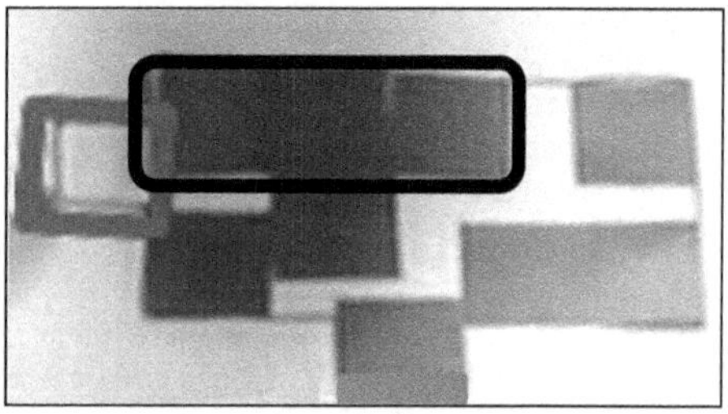

Fig. 5 Lining up with one side of the rugs

(a) (b)

Fig. 6 **a** Spot 1 indicating a limitation in students' model. **b** Spot 2 indicating a limitation in students' model

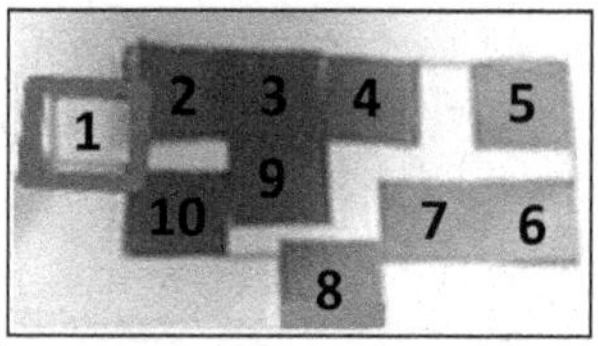

Fig. 7 Number of rugs in the model

They completed their tiling of the space by juxtaposing new rugs to rugs that had already been placed. Here, they might have been guided by the characters' story; the rug bugs' desire for neighbors whose dwellings line up perfectly with one side of their own rug. The result was that their packing strategy built from the boundary and toward the middle of the habitat, leaving gaps in the center. While they recognized that there were shortcomings in their solution, within the model that they developed, it was a reasonable one. As seen in the above exchange between students and the teacher, features of the narrative (e.g., bugs' hurting themselves when there are gaps or overlaps between rugs) contributed substantially to ongoing model development, supported students in externalizing aspects of their implicit models, and encountered the inner logic behind alternative perspectives.

Other groups developed models that attended less exclusively to the boundary of the habitat. As seen in Fig. 8a, one group started with filling the central region without holes and then expanding it by adding layers around it, while another group started by filling one side of the boundary and then adding rows below (see Fig. 8b).

The third group also started with the boundary, but not just one side; rather they first made one row along with the boundary and then added rows below it (see Fig. 8c). These strategies were partly triggered by the teacher, who gave different groups differently shaped habitat contours to work with. To better understand the differences between models of the two groups which started from the boundary (Fig. 8b, c), we analyzed the image sequences. The group in Fig. 8c started by placing post-it notes around the inside perimeter, attending primarily to the alignment between the outer borders of their "rugs" and the boundary line. They then made the second and third rows below the upper boundary row to reach a packing of the whole area. In contrast, the group in Fig. 8b attended more closely to *pairs* of constraining boundaries. Thus, in placing rugs to fill the top row from left to right, they negotiated a tight fit not only with the habitat boundary but also the last-placed rug. This approach supported them in tiling the available area from top to bottom. Hence, the two local strategies attending in a different way to boundary constraints led to different global outcomes (i.e., the overall appearances of the tiling solutions).

Situated and Embodied Models

The students' models were also highly situated with respect to the narratives. Engaging with the problems was supported by the whole-class introductory activities, where students not only listened to the story and discussed but also enacted aspects of the

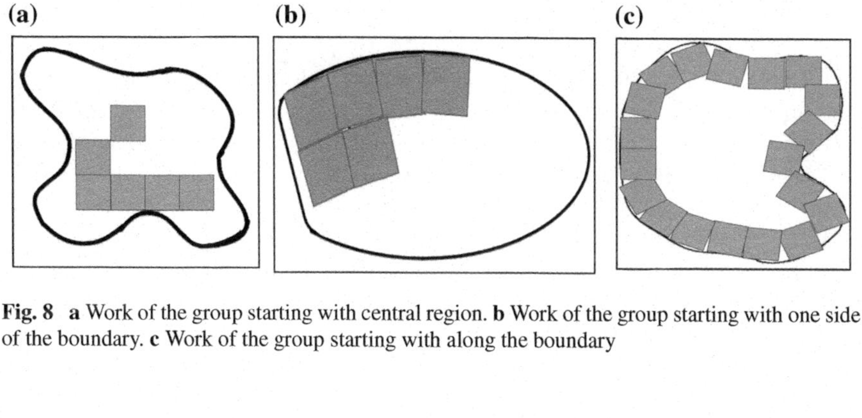

Fig. 8 **a** Work of the group starting with central region. **b** Work of the group starting with one side of the boundary. **c** Work of the group starting with along the boundary

core problem situation. Students' identification and empathy with the characters of the MEA were fostered by the embodied and performative (i.e., theatrical) nature of these role-playing activities.

The modeling literature suggests implementing model-extension activities following the MEAs in order to provide students with the opportunities of testing their models *outside of a specific real-world context*, often using dynamic mathematics software (Doerr & Lesh, 2011; Lesh, 2010; Lesh, Carmona, & Moore, 2009). Such activities contribute to the educational goal of supporting students in constructing generalizable and reusable models. In our case with young learners, we saw instances in which the situated nature of their models actually contributed to the "re-shareability" and reusability of these models. Indeed, some students in the class showed evidence of believing that their ways of thinking could be useful in understanding situations beyond the classroom. For instance, the teacher reported that three weeks after The Proper Hop activity, one student made the comment that her apartment complex was like Sugar Swamp, with a lot of lily pads. Her mother had told her that she could not go on her own to visit a friend who lived in an apartment on the other side of the complex. The student interpreted this in terms of proper hops: Crossing the complex would be like Beauregard having to take twenty hops or more! This student also reflected that she would like to be able to choose the location of her home, as Beauregard had done so that her friends' apartments would be within reach.

Adapting solutions to new situations requires students to *decenter* their own ways of thinking. From a psychology perspective, Piaget (1962) described this process as "shifting one's focus and comparing one action with other possible ones, particularly with the actions of other people, leads to an awareness of 'how' and to true operations" (p. 13). In the above example, the student shifted from the story context to a real-life situation, from the locations of lily pads to the locations of apartments, and from the concept of optimizing sums of proper-hop path lengths to ideas about distances between next-door and across-the-complex neighbors. Hence, we found this example promising for the reusability of the situated and embodied models of young learners.

Social and Collaborative Models

Our analysis revealed features of kindergarteners' and first-graders' social and collaborative ways of knowing. In itself, this was not a surprise: We would claim that *all* knowledge is irreducibly social. What was noteworthy was *when* and *how* young learners in these MEAs drew upon social resources.

In modeling research, small-group work has always been suggested to allow problem-solvers to negotiate and combine their ideas (Lesh & Fennewald, 2010; Lesh & Yoon, 2004, 2007). As Piaget articulated with an educational psychology perspective, "… peer interaction can be a fruitful means of stimulating natural cognitive conflicts that can generate accommodation to the views of others and evaluation of one's own concepts" (Wadsworth, 1971, p. 128). In our case, students' interactions with the teacher and their peers significantly contributed to them placing measurement units within the constraints of the measurement process and counting the hops and rugs to find the measure of the length and area. In this sense, the social and collaborative nature of the models revealed the important role of social interaction on young learners' conceptual development of measurement, which, in fact, is not easy to achieve for K-1 students.

Given the limitation of communication and collaboration skills of the young learners, the teacher was encouraged to take a hands-off approach and focus on observing the patterns of idea development. This was because teacher intervention could collapse the diversity of thinking of the groups and promote a return to "school answers." Therefore, in our case, the teacher took an active role in tapping into the social resources of the classroom as a whole to increase the diversity of ways of thinking to which her students were exposed. When the teacher called for different groups to share their ways of thinking, she made collaborative models developed within small groups accessible by other groups and so made the diversity of approaches visible. Hence, we articulate that the social and collaborative nature of young learners' models were the result of the leverage created by the teacher's practices. And this leverage was referred by Piaget as "the potential effect of the 'right' experience at the 'right' time" (Wadsworth, 1971, p. 118) on the cognitive development of a child. In our case, the teacher's practices created a potential cognitive conflict and triggered accommodation(s) in the children's measurement schemes, which were developed collaboratively.

Conclusions

In this chapter, we considered kindergarten and first-grade implementations of two story-based MEAs dealing with measurement. Our goal was to present a sample instructional cycle that supported the implementation of two MEAs in kindergarten and first-grade classrooms and characterize young learners' models on measurement. The conclusions that we drew from this case focus on (a) educational implications, (b) conceptual understanding of measurement, and (c) implications for educational psychology.

Concerning Educational Implications

The implementations of two story-based MEAs that we presented in this chapter were supported by a three-phased instructional cycle. During the instructional phases, the teacher employed five instructional practices that offered (a) pedagogical support—facilitating the implementation of MEAs in K-1 classrooms, (b) affective support—encouraging young learners' model development, (c) cognitive support—helping students in constructing, testing, and revising their models, and (d) theoretical support—contributing to the consolidation of six principles of MEAs for young learners.

These instructional practices within the instructional cycle were the main resources for the model development and elaboration in early primary grades, which also supports the claim of Verschaffel and de Corte (1997) about the role of instructional strategies on children's problem-solving behaviors. In this regard, this study illustrates that the support of instructional practices should not be underestimated by researchers during the design of MEAs for young learners and by teachers during the implementation of the MEAs in K-1 classrooms. We claim that teachers have a distinct role in scaffolding students' modeling processes either by design decisions (e.g., distributing different habitats in the Fussy Rug Bugs MEA) or by facilitation choices during the whole-class discussion. Indeed, during the model sharing phase, she enabled the class as a whole to engage with this diversity productively. Although sharing sessions gave groups occasions to develop and clarify their models, occasionally, insights appeared in those sessions that could not be fully processed "on the fly." For example, in the transcript in Table 1, students mentioned that there were problems with their solutions owing to the spaces between post-it rugs. Because of these gaps, the rug bugs would "fall off" and could "hurt their backs." While the "boundary-focused" approach that this group followed did not immediately suggest a remedy for this problem, other groups who pursued "interior-focused" approaches might have been used to suggest productive ways of improving the presented solution. However, managing these opportunities for cross-fertilization of ideas is a formidable teacher facilitation challenge.

Concerning Conceptual Understanding of Measurement

In this chapter, we exemplified that early primary learners were capable of developing models on their own, with the teacher's effective scaffolds, which is similar to other research investigating children's modeling behaviors (e.g., English, 2010; Lesh, English, Sevis, et al., 2013). Furthermore, our case investigation revealed three features of kindergarten and first-grade students' models: tacit and intuitive; situated and embodied; and social and collaborative. Even though tacit, intuitive, and situated models might be seen as low-level cognitive constructs, they provided rich information about young learners' mathematical thinking which is not easy to understand in other ways.

In The Proper Hop and Fussy Rug Bugs MEAs, young learners iterated the equal units (i.e., proper hops and rug shapes) and counted the number of iterated units that covered the distance and the space to be measured. Therefore, they could recognize numbers as representations of length and area even though they do not formally name these concepts mathematically. Instead, they contextually labeled them as the number of lily pads on the path of which Beauregard Frog traveled to visit his friends and the largest number of rugs that could be placed to build a town for rug bugs. For those scholars who may not interpret these constructions as the development of knowledge of length and area measurement, we remind readers about Piaget and his colleagues' work investigating a child's understanding of measurement. Piaget, Inhelder, and Szeminska (1960) described measurement as "transferring a succession of changes of position" (p. 121) where a subdivision was not needed and where footsteps or fingers could be used for successive placement. In our K-1 level story-based MEAs, this applied as the successive placement of proper hops along the distance between lily pads and rug shapes without gaps and overlaps to cover the town of rug bugs. Although young learners' measurement of length and area were based on their conceptions of numbers and counting skills, their models convinced us that they understood the need of an invariant measurement unit (i.e., proper hop) and the need for iteration without gaps and overlaps in order to measure.

Concerning Educational Psychology Implications

As mentioned earlier, one of the characteristics of young learners' models is that they are tacit and intuitive, but this should not lead us to underestimate the power of these students' models. The level of consciousness or awareness does not always occur immediately as students engage in activities: "In fact, it is a very general psychological law that the child can do something in action long before he really becomes 'aware' of what is involved—'awareness' occurs long after the action" (Gruber & Vonèche, 1977, p. 731). Although students' models were still rooted in their intuitions at the model sharing phase; expressing their reasoning for others helped students become aware of their actions.

Another characteristic of the young learners' models was situatedness. The story-based MEAs being simple, but not simplistic, allowed students to easily comprehend the problem, start developing a way of thinking about the solution, and easily "package" the situation and the solution method to reuse in similar situations. Fussy Rug Bugs MEA could be packed into "placing rugs as close as possible" while The Proper Hop MEA could be packed into "finding the shortest distance with proper hops." As modeling researchers, we know that weeks or even months after learners engage with an MEA, they often find themselves using that experience as a lens for looking at new problems that exhibit similar mathematical structures (Lesh et al., 2000; Lesh, English, Riggs et al., 2013). This is the essence of what it means for an activity to serve as a prototype.

Both the simplicity of story-based MEAs and the characters in the narratives offered a means to anchor perspectives that could serve as prototypes for future problem-solving. More specifically, a young learner who has made the connection between her apartment complex and Beauregard's sugar swamps could productively approach many new situations by asking, "What would Beauregard do here?" or "What would the rug bugs think about this?" Although there may have been only one student who could see the similarity between locations of lily pads and apartment complexes, and who could extend her model to finding the location for her house to be closest to her best friends that she wanted to visit frequently, it exemplified that even young learners have the potential to develop decontextualized models given enough appropriate experiences.

We claim that this potential is the result of three types of experiences that the young learners had as they worked on two story-based MEAs: physical experience (Piaget, 1970), logico-mathematical experience (Piaget, 1970), and social experience (Vygotsky, 1934/1986). Students acted in situations and coordinated their actions by counting the hops and rugs and by publicizing their actions as they interacted with the whole class and the teacher. With the help of these experiences and the narratives, students had prototypical experiences, which they could become aware of and potentially recall weeks later. We suggest that mathematics education researchers capitalize on cognitive psychology and the psychology of mathematics education because these psychological aspects of concept development are rich sources for a better understanding of cognitive development and the evolution of mathematical ideas in the child's mind (Fischbein, 1999).

Concluding Remarks

Being reminded of Bruner's famous claim that "any subject can be taught effectively in some intellectually honest form to any child at any stage of development" (1960, p. 33), we presented in this chapter that young learners can develop powerful models of length and area measurement. These models were initiated by their intuitions, situated in the story narrative, developed collaboratively, and expressed socially. This chapter also contributes to the modeling literature by articulating the cognitive and psychological characteristics of young learners' models and by presenting the conditions under which these models occur (i.e., three-phased instructional cycle and supporting instructional practices). Well, why is it useful to understand the characteristics of young learners' models and the instructional practices mediating the model development of K-1 students? With the current results in mind, our response to this question is that we may design better interactions for young learners to increase their potential conceptual development. As such, we may prepare children for the world in the twenty-first century that "is increasingly governed by complex systems that are dynamic, self-organizing, and continually adapting" (English, 2007, p. 121) and that requires individuals to make legitimate decisions in organizing their lives and developing different approaches to complex real-life problems.

Acknowledgements We would like to thank Chanda Riggs, the teacher of K-1 students, and Emeritus Professor Richard Lesh for their contributions to this research.

References

Biembengut, M. S. (2007). Modelling and applications in primary education. In W. Blum, P. L. Galbraith, H. -W. Henn, & M. Niss (Eds.), *The 14th ICMI study: Modelling and applications in mathematics education* (pp. 451–456). New York, NY: Springer.

Borromeo Ferri, R., & Lesh, R. (2013). Should interpretation systems be considered to be models if they only function implicitly? In G. A. Stillman, G. Kaiser, W. Blum, & J. P. Brown (Eds.), *Teaching mathematical modelling: Connecting to research and practice* (pp. 57–66). Netherlands: Springer.

Bruner, J. S. (1960). *The process of education*. Cambridge, MA: Harvard University Press.

Chamberlin, S. A., & Coxbill, E. (2012). *Using model-eliciting activities to introduce upper elementary students to statistical reasoning and mathematical modelling*. Retrieved from http://www.uwyo.edu/wisdome/_files/documents/chamberlin_coxbill.pdf.

Diefes-Dux, H., Follman, D., Imbrie, P. K., Zawojewski, J., Capobianco, B., & Hjalmarson, M. (2004, June). Model eliciting activities: An in-class approach to improving interest and persistence of women in engineering. In *Proceedings of the 2004 American Society for Engineering Educating Annual Conference & Exposition*, Salt Lake City, UT.

Doerr, H. M., & Lesh, R. (2011). Trends in teaching and learning of mathematical modelling. In G. Kaiser, W. Blum, R. B. Ferri, & G. Stillman (Eds.), *Trends in teaching and learning of mathematical modelling, international perspectives on the teaching and learning of mathematical modelling* (pp. 247–268). Dordrecht: Springer.

English, L. (2007). Cognitive psychology and mathematics education: Reflections on the past and the future (monograph 2). In *The Montana mathematics enthusiast* (pp. 119–126).

English, L. D. (2010). Modeling with complex data in the primary school. In R. Lesh, P. L. Galbraith, C. R. Haines, & A. Hurford (Eds.), *Modeling students' mathematical modeling competencies* (pp. 287–299). New York, NY: Springer.

Fischbein, E. (1999). Psychology and mathematics education. *Mathematical Thinking and Learning, 1*(1), 47–58. https://doi.org/10.1207/s15327833mtl0101_2.

Greer, B., & Verschaffel, L. (2007). Modelling competencies—Overview. In W. Blum, P. L. Galbraith, H. -W. Henn, & M. Niss (Eds.), *The 14th ICMI study: Modelling and applications in mathematics education* (pp. 219–224). New York, NY: Springer.

Greer, B., Verschaffel, L., & Mukhopadhyay, S. (2007). Modelling for life: Mathematics and children's experience. In W. Blum, P. L. Galbraith, H. -W. Henn, & M. Niss (Eds.), *The 14th ICMI study: Modelling and applications in mathematics education* (pp. 89–98). New York, NY: Springer.

Gruber, H. E., & Vonèche, J. J. (Eds.). (1977). *The essential Piaget*. New York: Basic Books.

Lehrer, R., & Kim, M. J. (2009). Structuring variability by negotiating its measure. *Mathematics Education Research Journal, 21*(2), 116–133.

Lesh, R. (2006). New directions for research on mathematical problem solving. In P. Grootenboer, R. Zevenbergen, & M. Chinnappan (Eds.), *Identities, cultures and learning spaces, Proceedings of the 29th Annual Conference of the Mathematics Education Research Group of Australasia, Canberra* (Vol. 1, pp. 15–34). Adelaide: MERGA.

Lesh, R. (2010). Tools, researchable issues & conjectures for investigating what it means to understand statistics (or other topics) meaningfully. *Journal of Mathematical Modelling and Application, 1*(2), 16–49.

Lesh, R., & Doerr, H. M. (2000). Symbolizing, communicating, and mathematizing: Key components of models and modeling. In P. Cobb, E. Yackel, & K. McClain (Eds.), *Symbolizing and communicating in mathematics classrooms: Perspectives on discourse, tools, and instructional design* (pp. 361–384). Mahwah, NJ: Lawrence Erlbaum Associates.

Lesh, R., & Doerr, H. (2003a). In what ways does a models and modeling perspective move beyond constructivism? In R. Lesh & H. Doerr (Eds.), *Beyond constructivism: A models and modeling perspective* (pp. 519–556). Mahwah, NJ: Lawrence Erlbaum Associates.

Lesh, R., & Doerr, H. (2003b). Foundations of a models and modeling perspective on mathematics teaching, learning, and problem solving. In R. Lesh & H. Doerr (Eds.), *Beyond constructivism: A models and modeling perspective* (pp. 3–33). Mahwah, NJ: Lawrence Erlbaum Associates.

Lesh, R., & Fennewald, T. (2010). Introduction to part I modeling: What is it? Why do it? In R. Lesh, P. L. Galbraith, C. R. Haines, & A. Hurford (Eds.), *Modeling students' mathematical modeling competencies (ICTMA 13)* (pp. 5–10). New York: Springer.

Lesh, R., & Harel, G. (2003). Problem solving, modeling, and local conceptual development. *Mathematical Thinking and Learning, 5*(2–3), 157–189.

Lesh, R., & Lehrer, R. (2003). Models and modeling perspectives on the development of students and teachers. *Mathematical Thinking and Learning, 5*(2–3), 109–129.

Lesh, R. A., & Nibbelink, W. H. (1978). *Mathematics around us: Kindergarten*. Glenview, IL: Scott Foresman & Co.

Lesh, R., & Yoon, C. (2004). Evolving communities of mind—In which development involves several interacting and simultaneously developing strands. *Mathematical Thinking and Learning, 6*(2), 205–226.

Lesh, R., & Yoon, C. (2007). What is distinctive in (our views about) models & modelling perspectives on mathematics problem solving, learning, and teaching? In W. Blum, P. L. Galbraith, H. -W. Henn, & M. Niss (Eds.), *The 14th ICMI Study (New ICMI Study Series: Vol. 10). Modelling and applications in mathematics education* (pp. 161–170). New York: Springer.

Lesh, R., Amit, M., & Schorr, R. (1997). Using "real-life" problems to prompt students to construct conceptual models for statistical reasoning. In I. Gal & J. B. Garfield (Eds.), *The assessment challenge in statistics education* (pp. 65–83). Amsterdam: IOS Press.

Lesh, R., Carmona, G., & Moore, T. (2009). Six sigma learning gains and long term retention of understandings and attitudes related to models & modeling. *Mediterranean Journal for Research in Mathematics Education, 9*(1), 19–54.

Lesh, R., Cramer, K., Doerr, H. M., Post, T., & Zawojewski, J. S. (2003). Model development sequences. In R. Lesh & H. Doerr (Eds.), *Beyond constructivism: A models and modeling perspective* (pp. 35–58). Mahwah, NJ: Lawrence Erlbaum Associates.

Lesh, R., English, L., Riggs, C., & Sevis, S. (2013, January). Problem solving in the primary school (K-2) (Special issue). *The Mathematics Enthusiast, Special Issue: International Perspectives on Problem Solving Research in Mathematics Education, 10* (1 & 2), 35–60.

Lesh, R., English, L., Sevis, S., & Riggs, C. (2013). Modeling as a means for making powerful ideas accessible to children at an early age. In S. Hegedus & J. Roschelle (Eds.), *The SimCalc vision and contributions: Democratizing access to important mathematics* (pp. 419–436). Berlin: Springer.

Lesh, R., Hamilton, E., & Kaput, J. (Eds.). (2007). *Foundations for the future in mathematics education*. Mahwah, NJ: Lawrence Erlbaum Associates.

Lesh, R., Hoover, M., Hole, B., Kelly, A., & Post, T. (2000). Principles for developing thought-revealing activities for students and teachers. In A. Kelly & R. Lesh (Eds.), *Handbook of research design in mathematics and science education* (pp. 113–149). Mahwah, NJ: Lawrence Erlbaum Associates.

Piaget, J. (1962). *Supplement to L. Vygotsky, thought and language*. Cambridge, MA: MIT Press.

Piaget, J. (1970). *Genetic epistemology* (E. Duckworth, Trans.). New York: W. W. Norton.

Piaget, J., Inhelder, B., & Szeminska, A. (1960). *The child's conception of geometry*. New York: Basic Books.

Rieger, J. (1996). Photographing social change. *Visual Sociology, 11*(1), 5–49.

Usiskin, Z. (2007). The arithmetic operations as mathematical models. In W. Blum, P. L. Galbraith, H. -W. Henn, & M. Niss (Eds.), *The 14th ICMI study: Modelling and applications in mathematics education* (pp. 257–264). New York, NY: Springer.

Verschaffel, L. (2002). Taking the modeling perspective seriously at the elementary school level: Promises and pitfalls (Plenary lecture). In A. Cockburn & E. Nardi (Eds.), *Proceedings of the 26th Annual Conference of the International Group for the Psychology of Mathematics Education*

(Vol. 1, pp. 64–82). UK: School of Education and Professional Development, University of East Anglia.

Verschaffel, L., & de Corte, E. (1997). Teaching realistic mathematical modeling in the elementary school: A teaching experiment with fifth graders. *Journal for Research in Mathematics Education, 28*(5), 577–601.

Verschaffel, L., de Corte, E., & Lasure, S. (1994). Realistic considerations in mathematical modeling of school arithmetic word problems. *Learning and Instruction, 4*(4), 273–294.

Vygotsky, L. S. (1934/1986). *Thought and language* (A. Kozulin, Trans. and Ed.). Cambridge, MA: MIT Press.

Wadsworth, B. J. (1971). *Piaget's theory of cognitive development: An introduction for students of psychology and education.* New York: David McKay Company.

Serife Sevinc is Assistant Professor at Mathematics Teacher Education Program at Middle East Technical University, Turkey. She received her doctoral degree in the Mathematics Education Program at Indiana University, USA, with a minor degree in inquiry methodology. Her main research focus is mathematics teacher education with a particular interest in models-and-modeling perspectives. She investigates the problem-solving processes of a wide range of learners through modeling perspectives and is involved in projects investigating the mathematical thinking of elementary and middle school students and teachers. She believes that communication with and about mathematics with different learners is her main resource as a researcher and a teacher educator.

Corey Brady is an Assistant Professor of the Learning Sciences at Vanderbilt University. His research focuses on mathematical and computational modeling, with a particular emphasis on supporting and understanding the collective learning of classroom groups. Prior to entering Academia he has worked in industry, leading the design and development of Texas Instruments' TI-Navigator system (2001–2006). He has also been a classroom teacher at the middle school, high school, and community college levels. He holds degrees in Pure Mathematics (M.S.), English Literature (M.A.), and Mathematics Education (Ph.D.).

Young Children's Patterning Competencies and Mathematical Development: A Review

Nore Wijns, Joke Torbeyns, Bert De Smedt and Lieven Verschaffel

Introduction

The past twenty years have witnessed a strong growth in research on young children's early mathematical competencies. Although several reasons can be given for the growing interest of researchers in this topic (Clements & Sarama, 2007; New & Cochran, 2007), one major argument is the impact of these competencies on children's future academic achievement (Claessens & Engel, 2013; Nguyen et al., 2016). In an attempt to assess the mathematical difficulties children encounter as early as possible, various indicators of early mathematics-related competencies have been proposed. These indicators include subitizing (Schleifer & Landerl, 2011), counting (Geary, Bow-Thomas, & Yao, 1992), transcoding a number from one representation to another (Göbel, Watson, Lervåg, & Hulme, 2014), comparing numerical magnitudes (De Smedt, Noël, Gilmore, & Ansari, 2013), and positioning numerical magnitudes on an empty number line (Siegler & Booth, 2004). More recently, these *ability-oriented* measures have been complemented with measures that address the *dispositional* side of children's early numerical development, particularly children's spontaneous focusing on numerosities (SFON; Hannula-Sormunen, Lehtinen, & Räsänen, 2015), their spontaneous focusing on number symbols (SFONS; Rathé, Torbeyns, De Smedt, & Verschaffel, 2017), or, more generally, their spontaneous attention to number (Baroody, Li, & Lai, 2008). Although both researchers and educational practitioners acknowledge the importance of the aforementioned indicators of children's later mathematical development (Clements & Sarama, 2007; Frye et al., 2014), they have been criticized because of their limited—and more partic-

N. Wijns (✉) · J. Torbeyns · L. Verschaffel
Centre for Instructional Psychology and Technology, KU Leuven,
Dekenstraat 2, Postbox 3773, 3000 Leuven, Belgium
e-mail: nore.wijns@kuleuven.be

B. De Smedt
Parenting and Special Education Research Unit, KU Leuven, Leuven,
Belgium

K. M. Robinson et al. (eds.), *Mathematical Learning and Cognition in Early Childhood*,
https://doi.org/10.1007/978-3-030-12895-1_9

ularly: too exclusively numerical—scope (Mulligan et al., 2018; Starkey, Klein, & Wakeley, 2004; Verschaffel, Torbeyns, & De Smedt, 2017). As stated by Clements and Sarama (2007), "(t)he mathematical world of young children is much richer" (p. 537).

The aim of this review is to give an overview of the available literature concerning one particular topic that has recently been proposed as an interesting complementary avenue for further research in early mathematics, namely early patterning (Björklund & Pramling, 2014; English & Mulligan, 2013; New & Cochran, 2007; Nguyen et al., 2016; Rittle-Johnson, Fyfe, McLean, & McEldoon, 2013; Sarama & Clements, 2004; VanDerHeyden et al., 2011; Verschaffel et al., 2017). Early patterning comprises various activities (e.g., copying, creating, extending) that children in preschool and the first years of elementary school can do with any type of discernible regular arrangement of objects in their environment (e.g., repeating patterns like ABABAB, growing patterns like 1–3–5, spatial structure patterns like :::). The motivation for promoting patterning activities from an early age onwards appears to be evident. First, patterning activities seem feasible and appropriate for young children, given that many of them spontaneously create patterns during free play (Fox, 2005; Piccolo & Test, 2010; Seo & Ginsburg, 2004) and that activities with patterns (mainly extending or copying a repeating pattern) are already common in many preschool settings (Cross, Woods, & Schweingruber, 2009; Economopoulos, 1998). Second, the critical value of patterning for (school) mathematics has been acknowledged for decades. Already in the late 1980s, Steen (1988) stated that "(m)athematics is the science of patterns" (p. 616, see also Wittmann & Müller, 2007), and, as will be discussed below, older studies have already empirically documented that early patterning competencies contribute to children's later mathematical development (Herman, 1973; Threlfall, 2005). Early childhood teachers are also increasingly convinced that patterning is an important mathematical topic (Sarama, Clements, Starkey, Klein, & Wakeley, 2008; Waters, 2004). This conviction is reinforced by the creation of a patterns and (pre)algebra strand as part of the early mathematics curriculum in a growing number of countries (e.g., acara, 2015; Ontario Ministry of Education and Training (OMET), 2005). The name and content of this strand reveal a fourth rationale underlying the importance attached to these patterning activities, namely their potential as a route to algebra (English & Warren, 1998; Hargreaves, Shorrocks-Taylor, & Threlfall, 1998; National Research Council, 2001; NCTM, 1989, 2000; Smutny, 1998; Taylor-Cox, 2003). Nevertheless, the question remains whether the patterning activities that are nowadays common in preschool optimally exploit this potential.

In sum, current mathematics education research, policy, and practice suggest the implementation of patterning in the early school years, but questions arise about how these activities are currently implemented and enacted. In an attempt to enhance our understanding of early mathematical development and to support early educational practices, several researchers have, since the start of the twenty-first century, been focusing on early patterning. This review summarizes and discusses the recent literature on patterning in children in the transition from preschool to formal schooling by considering four questions: (1) How is patterning defined and operationalized in the context of early mathematics education? (2) How does patterning develop in the early school years? (3) What is the relation between early mathematical patterning and other competencies? (4) What is the impact of interventions aimed at the early stimulation of patterning?

While the above introduction might suggest that the attention for patterning in early mathematics education is a recent trend, it should be acknowledged that research on early patterning did not start this century. Several older studies have already looked into the nature of patterning, its link with other cognitive variables, and its development and stimulation in young children (see Orton, 1999, for a partial review of this older literature). Because this older research receives little or no attention in the current work on the topic, we will start by giving a short summary of the contribution of this older research to each of our four questions. It is, however, important to keep in mind that this summary is not intended to be exhaustive, but only serves to delineate the historical background for the more recent literature.

A Short History of Research on Early Mathematical Patterning Competencies

In older research, a pattern is often not explicitly defined, but if so, it is described as the abstract representation (Sternberg & Larson, 1976) or the rule (Close & Glennon, 1977) of an indefinite sequence. Most studies focus on repeating patterns (e.g., ABABAB; Close & Glennon, 1977; Herman, 1973; Sternberg & Larson, 1976; Threlfall, 2005), and some studies also consider growing patterns (e.g., ABAABAAAB; Mckillip, 1970a). Several tasks are used to operationalize patterning, with the most common ones being copying (i.e., "make the same pattern"), interpolating (i.e., "fill in missing elements"), extending (i.e., "continue the given pattern"), generalizing (i.e., "make the same pattern with these different materials"), and identifying the pattern (i.e., "find the same pattern") (see Table 1 for our categorization of patterning tasks).

With respect to the development of patterning, it is difficult to draw conclusions from this older work, as longitudinal studies with large groups of children are missing. Some indications for development can be found in cross-sectional studies with repeating patterns, which showed that copying or describing a repeating pattern is easier than extending or interpolating it (McKillip, 1970a, 1970b). Other authors concluded that gaining insight into the unit of repeat (i.e., the smallest sequence of elements that constitutes a pattern, also referred to as the repetend) is a crucial step in children's mathematical development (Economopoulos, 1998; Threlfall, 2005).

The relationship between patterning abilities and general mathematical development was never explicitly assessed in the older literature, but an intervention study by Herman (1973) suggested an association between early patterning and early numerical abilities. During that intervention, kindergarteners had to copy, extend, or interpolate repeating patterns in a series of 24 lessons. At the end of the school year, the children from the intervention group performed better on a standardized test that

Table 1 A classification of pattern tasks

Name also referred to as (by)	Instruction	Example
Copy Duplicate (Clements & Sarama, 2014) Level 1 (Rittle-Johnson et al., 2013)	Make the same pattern	★ ■ ★ ■ ★ ■
Create	Make a pattern using these blocks	★ ■
Extend Level 2 (Rittle-Johnson et al., 2013)	What element comes next?	★ ■ ★ ■ ★ ...
Generalize Extra-variable transfer (Close & Glennon, 1977) Translate (Lüken, 2016; Threlfall, 2005) Abstract, Level 3 (Rittle-Johnson et al., 2013)	Make the same pattern using different materials	Make ★ ■ ★ ■ ★ ■ Using (several) ● ▲
Identify the pattern unit Level 4 (Rittle-Johnson et al., 2013)	Which is the smallest part of the pattern that repeats?	★ ■ ★ ■ ★ ■
Recognize (Sternberg & Larson, 1976)	Which pattern is the same?	Model pattern ★ ■ ★ ■ ★ ■ ★ ■ ★ ■ ★ ■ or ★ ★ ■ ★ ★ ■
Interpolate Extrapolate (Lüken, 2016) Fix (Clements & Sarama, 2014) Understand (Kidd et al., 2013)	Which element is missing?	★ ■ ★ ... ★ ■
Recognize	Where do you see a pattern? Is this a pattern?	★ ■ ★ ★ ★ ■

assessed number knowledge compared to those from a comparison group. Unfortunately, children were not randomly allocated to the intervention or comparison group, and there was no pretest to indicate whether both groups were comparable in their performance before the intervention.

The impact of a patterning intervention was researched in the study by McKillip (1970a), in which three- to five-year-olds performed a sequence of activities with repeating and growing patterns that were assumed to increase in difficulty, starting from free play, through copying and describing, to interpolating and extending. A growth in patterning abilities was observed for three and four-year-olds during the implementation of that intervention, but unfortunately the effect could not be properly assessed as there was no control group.

Besides the methodological queries mentioned above, the design of the intervention studies by Herman (1973) and McKillip (1970a) did not allow them to determine whether the interventions actually stimulated any insight into the unit of repeat, which was increasingly being considered to be *the* quintessence of patterning skill. According to Threlfall (2005), perceiving the unit of repeat may occur naturally, but most often it must be explicitly taught. He proposed several patterning activities to teach this awareness, such as explicitly defining the unit of repeat or generalizing a given pattern toward other materials (which he calls translating, see Table 1). Remarkably, Close and Glennon (1977) found their generalization (which they termed extra-variable transfer) and pattern identification tasks to be easier than interpolation and extension. According to these authors, establishing a superficial one-to-one correspondence strategy between the model pattern and the pattern to be generalized or identified might be sufficient to complete a generalization or identification task, while a child must recognize the relation among the elements of the pattern to some level for interpolation and extension.

Recent Research on Early Mathematical Patterning Competencies

Definition and Operationalization

Although the word "pattern" is often used in daily interactions, defining a pattern is not as easy as it might seem. There are two key features, namely regularity (or order) and predictability, that are explicitly present in most current definitions (Lüken, 2010; Mulligan & Mitchelmore, 2009; Rittle-Johnson et al., 2013), and that can also be found in older research (Close & Glennon, 1977; Sternberg & Larson, 1976), albeit in more implicit terms. Furthermore, several contemporary researchers make a distinction between a pattern, which refers to the sequence itself, and its structure, which refers to the organization of or rule behind the pattern (Lüken, 2012; Mulligan & Mitchelmore, 2009). They also differentiate between different types of patterns. The most common are repeating patterns, growing patterns, and spatial structure patterns (Papic & Mulligan, 2007). A repeating pattern contains a constant "unit of repeat" that reoccurs indefinitely (e.g., ABABAB, ABCABCABC), whereas a growing pattern increases or decreases systematically (e.g., ABAABAAAB, 1–3–5). Spatial structure patterns are invariant and describe the organization of individual elements in a two- or three-dimensional space (e.g., :::, , Papic & Mulligan, 2007).

In order to conceptualize patterning, Mulligan and associates introduced the construct "Awareness of Mathematical Pattern and Structure" (AMPS; Mulligan & Mitchelmore, 2009), which is assumed to comprise two components: a cognitive one (i.e., knowledge of structure) and a metacognitive one (i.e., tendency to search for patterns and analyze them). So, their conceptualization of AMPS seems to involve both an ability and a dispositional component. Lüken proposed the concept "Early Structure Sense" to refer to a collection of abilities "to easily and flexibly operate with mathematical pattern and structure" (2012, p. 263). Kidd and colleagues prefer the term "sequencing" over patterning and defined it as the "ability to understand abstract relationships" (2013, p. 255). Not surprisingly, these different definitions and interpretations of patterning have led to different ways of measuring patterning competencies.

To measure AMPS, Mulligan and Mitchelmore (2009) constructed the Pattern And Structure Assessment (PASA), an instrument consisting of several tasks in which children have to identify, visualize, represent, or replicate elements of pattern and structure (i.e., repeating, growing, and spatial structure patterns, but also multiplicative structures, measurement units, and data representation). For each task, a child's answer is categorized into five levels: pre-structural stage, emergent stage, partial structural stage, stage of structural development, and advanced stage of structural development. Different versions of the PASA exist for children in the foundational years (five-year-olds), first grade, and second grade. Later, a similar instrument was developed to measure patterning in four-year-olds: the Early Mathematical Patterning Assessment (EMPA; Papic, Mulligan, & Mitchelmore, 2011) consisting of several diagnostic tasks in which children have to copy, extend, create, interpolate, and represent repeating, spatial, and growing patterns.

To measure early structure sense, Lüken (2010, 2012) created an instrument consisting of six task categories in which children have to conceive, reproduce, copy from memory, use, extend, and create repeating and spatial structure patterns presented in a visual, audio, or tactile mode. While the repeating patterns (e.g., a ten chain with five red and five blue beads) in this study were rather unusual, Lüken (2016) focused in a more recent study only on the more common repeating patterns (i.e., AB, ABC, AAB patterns) presented visually. In one-on-one interviews, children had to copy, extend, interpolate (in her terminology: extrapolate), and generalize (in her terminology: translate) repeating patterns. Similarly, Rittle-Johnson and associates (2013) created a construct map with four levels, namely (1) copy (or duplicate), (2) extend, (3) generalize (or abstract), and (4) unit recognition, which formed the foundation for their repeating pattern assessment tasks.

For Kidd and colleagues (2013, 2014), the current focus of early patterning research on simple repeating alternations of the type AB or AAB does not cover the full potential of these young children. To properly assess first graders' patterning skills, they used an interpolation task presented in a multiple choice format with four options in which children were confronted not only with repeating patterns but also with other patterns, which the authors consider to be more complex. The patterns consisted of letters, numbers, clock faces, shapes, or objects and had increasing or decreasing value, rotated, repeated, or were symmetrical.

From the overview above, it is clear that patterning comprises a broad range of competencies that can be measured with several *patterning tasks* and *types of patterns.* Throughout the years, a relatively consistent classification of different types of patterning tasks has been distinguished, although the terms for these types may vary by author (see Table 1). The most common tasks are copying (i.e., making the same pattern), extending (i.e., finding the next element), interpolating (i.e., finding the missing element), and generalizing (i.e., making the same pattern with different materials) (e.g., Lüken, 2016; Rittle-Johnson et al., 2013). In addition, several new theoretical concepts covering a broader interpretation of patterns and structures (e.g., AMPS, early structure sense) have been introduced, together with new instruments (e.g., PASA, EMPA). Quite often these instruments contain similar tasks (e.g., the interpolation task in Kidd et al., 2013; extending spatial structure patterns in Mulligan, Mitchelmore, & Stephanou, 2015), but new tasks have also been introduced (e.g., structuring counters in Lüken, 2012).

Moreover, the studies reviewed above have used diverse types of patterns, with repeating patterns, growing patterns, and spatial structure patterns being the most common ones. Most researchers have focused on repeating patterns. The need for more integrated research, in which multiple tasks with different types of patterns are compared across ages, has already been urged by Clements and Sarama (2014, 2009). Making a valid comparison between types of patterns is difficult, however, because implementing different types of patterns in the same task can yield pattern-specific differences that are difficult to control.

A final issue is the almost exclusive focus on the ability side of patterning competence, thereby largely neglecting its dispositional aspects. With the term disposition, we refer to children's spontaneous attention to or feeling for mathematical patterns and structures (Mulligan et al., 2018; Verschaffel et al., 2017). As stated above, during the past decade, researchers have started to explore children's spontaneous tendencies to focus on mathematical elements such as numerosities, number symbols, or quantitative relations (e.g., Hannula & Lehtinen, 2005). Importantly, these spontaneous tendencies are not about what children think and do when they are *guided* to the mathematical elements or relations in the situation by a researcher, teacher, or parent, but rather they refer to what they *spontaneously* focus on in informal everyday situations. Several researchers have recently made a plea for systematically investigating young children's spontaneous focusing on mathematical patterns and structures in particular (e.g., Sarama & Clements, 2009; Verschaffel et al., 2017). A few observational studies have documented children's spontaneous engagement in patterning activities (Fox, 2005; Garrick, Threlfall, & Orton, 2005), including children from disadvantaged backgrounds (Seo & Ginsburg, 2004). Several decades ago, Mckillip (1970b) already suggested that children exhibiting spontaneous engagement in patterning activities might have more advanced patterning abilities than children with no such spontaneous interest in patterns, simply because such a spontaneous engagement will create more opportunities for "deliberate practice" (Lehtinen, Hannula-Sormunen, McMullen, & Gruber, 2017). One could argue that the metacognitive component of Mulligan and Mitchelmore's AMPS construct, which they describe as "a tendency to seek and analyze patterns" (Mulligan & Mitchelmore, 2009, p. 38), somehow coincides with the notion of spontaneous focus on patterns. However, in their actual research, the cognitive and metacognitive components of patterning are conceived as one single construct that can be measured by the same assessment tool wherein the instructions and tasks clearly guide the children toward patterning. Therefore, it is very hard to disentangle the nature and development of the dispositional side of patterning on the basis of their research.

Development

In line with the older literature summarized in the introduction, more recent research on the development of early patterning has mainly considered repeating patterns. In an exploratory longitudinal study, Lüken (2016) found repeating patterning abilities to grow between the ages of three and four in German kindergarteners. Three-year-old children were not able to copy, interpolate, or extend a simple AB pattern. However, one year later, most four-year-olds were able to do so not only with AB, but also with ABC and AAB patterns. The generalizing task (termed by Lüken as translating) was only completed by four-year-olds and appeared to be the most difficult task for them.

Rittle-Johnson and colleagues (2015, Study 1, 2013) observed a growth in repeating patterning abilities of four-year-old preschoolers. In the fall, most four-year-olds were capable of duplicating and extending patterns, whereas almost one-third of them could also generalize a pattern. Only a few children were able to identify the unit of repeat (e.g., "What is the smallest tower you could make and still keep the same pattern as this?", Rittle-Johnson et al., 2013, p. 382). A growth in duplicating, extending, and generalizing was observed between fall and spring scores, whereas there was no improvement yet in pattern unit recognition. A decrease in the types of errors that did not involve patterns was seen between fall and spring.

The results of these studies (Lüken, 2016; Rittle-Johnson et al., 2015, Study 1, 2013) are in line with the developmental progression that Clements and Sarama (2014) propose in their *learning trajectory for Pattern and Structure.* According to this trajectory, which is mainly based on their studies for the Building Blocks program (Sarama & Clements, 2004) and the TRIAD project (Sarama et al., 2008), three-year-olds are able to recognize a simple pattern, whereas four-year-olds can interpolate (referred to as fix), copy (referred to as duplicate), and extend repeating patterns (first AB patterns, later other repeating patterns like ABB patterns). Six-year-olds are assumed to recognize a pattern unit (i.e., identify the unit of repeat and generalize), and seven-year-olds should be able to work with growing patterns and describe these patterns numerically. Although this learning trajectory appears to be a useful tool for early childhood educators, the authors acknowledged their main focus on repeating patterns and suggested more research is needed to extend this limited interpretation of patterns and structures (Sarama & Clements, 2009). Moreover, the empirical background of this (hypothetical) learning trajectory remains somewhat obscure.

Within the context of development, one important element is the difficulty level of a pattern item, which is assumed to mainly—but not exclusively—depend on two aspects: the nature of the pattern (e.g., repeating, growing) and the type of patterning task being used (e.g., copying, extending, interpolating). Concerning the nature of the pattern, the five-year olds in the study of Skoumpourdi (2013) were better in extending or copying a repeating pattern, than extending a growing pattern. To the best of our knowledge, no other, more systematic, comparison between the difficulty levels of different types of patterns has been made for kindergartners. Studies in primary school children have shown that activities with growing patterns were considerably more difficult than similar activities with repeating patterns (Banerji & Ferron, 1998; Gadzichowski, 2012b; Kyriakides & Gagatsis, 2003). It is, however, unclear whether these researchers controlled for possible pattern-specific differences within a task. Concerning the repeating patterns, a rational analysis suggests that the more elements and the more different elements constitute a pattern unit, the more difficult it is to fulfill a task (Simon, 1972). So, an AB pattern appears to be the easiest repeating pattern and an AAB pattern is easier than an AABB pattern or an ABC pattern (Close & Glennon, 1977). No research so far has looked into a similar progression in difficulty level for growing or other types of patterns.

There is also a progression in difficulty level between patterning tasks. There seems to be agreement among researchers on the easiest task, namely copying, which is manageable by the age of three or four, and the most difficult task, namely identifying the unit of repeat, which may remain even difficult for nine-year-olds (Warren & Cooper, 2007). Extending and interpolating are also considered as rather easy tasks that are already feasible for some four-year-olds (Brownell, Chen, Ginet, & The Early Math Collaborative Erikson, 2014; Lüken, 2016; Rittle-Johnson et al., 2013; Sarama & Clements, 2009). As most research so far has focused on repeating patterns, it remains unclear whether children are able to fulfill the same tasks with other types of patterns (e.g., growing patterns) at the same age.

In most theoretical accounts for the observed differences in difficulty of the distinct types of patterning tasks, a distinction is made between two levels of thinking, namely *recursive* versus *functional thinking*. This move from recursive, local thinking toward functional, global thinking has been suggested to be an important milestone in the development of early patterning abilities in older (Economopoulos, 1998; Threlfall, 2005) and more recent studies (Lüken, 2012; Papic & Mulligan, 2007; Rittle-Johnson et al., 2013). Children who think recursively only see the relationship between consecutive elements in a pattern and therefore can only predict the next one (i.e., the +1th), whereas those who are able to think functionally can see the underlying structure of a pattern and therefore predict any (i.e., the *n*th) element of a sequence. Consider, for example, an extending task involving a simple AB repeating pattern. In order to think functionally, children should acquire insight into the unit of repeat (Economopoulos, 1998; Threlfall, 2005), instead of simply following a recursive rule (e.g., "B follows A and A follows B"). For growing patterns, the transition should be made from creating and applying a recursive rule (e.g., $X_n = X_{n-1} + 3$) to a functional rule (e.g., $X_n = 2 + n * 3$) (English & Warren, 1998). Finding a functional rule for a growing pattern appears to be difficult, even at the end of primary school (National Research Council, 2001).

While most researchers make a distinction between recursive and functional thinking, the question remains which patterning tasks require the latter type of thinking. Moreover, children can use their understanding of the unit of repeat to solve tasks that do not necessarily require functional thinking, such as copying, extending, or interpolating. Making use of a variety of data they collected, Collins and Laski (2015) suggested that the four-year-olds in their study solved a generalization task using functional thinking (which they termed: a relational similarity strategy), whereas copying and extending were approached by a more basic recursive strategy (which they termed: a one-to-one appearance matching strategy). Performance on the unit identification task was very low; therefore, it remains unclear which type of thinking is required for good performance on the unit identification task. More research is necessary to unravel the reasoning processes underlying young children's good and weak performance on the various patterning tasks.

As was already hinted at, the nature of the pattern and the type of patterning task together do not fully account for the level of difficulty of a specific patterning task. This difficulty is also affected by several additional factors related to the way in which a given task is precisely implemented. For example, the difficulty level of the interpolation task is also known to depend on the number of elements missing and the number of options in the multiple choice response option (Gadzichowski, 2012a). Likewise, in the generalization task, the difficulty level—and the extent to which it requires functional thinking—also depends on the relationship between the features of the model pattern and those of the to-be-generalized pattern. Consider the following examples. In the study of Close and Glennon (1977), children were given a pattern with colored blocks they had to copy using colored houses. In other studies (Collins & Laski, 2015; Rittle-Johnson et al., 2013, 2015), children were given a pattern with colored shapes they had to copy using either colored blocks (colors different from the model) or shapes (different from the model) in one color. The former generalization task seems to be much easier than the latter, which might explain why generalizing appeared to be easier than extending in the study of Close and Glennon (1977), whereas the opposite was the case in the study of Rittle-Johnson and associates (2013). Such inconsistent results may of course also be due to subject characteristics, such as the amount of children's experiences with various types of patterns in their home or (pre)school environment.

Relationship with Other Cognitive Abilities

Several studies have explored the relationship of early patterning with other domain-general and domain-specific cognitive abilities. We will consider the relationship with mathematical abilities, reading abilities, and general cognitive abilities.

Indications of a relationship between early patterning abilities and *mathematical abilities* have been found in numerous longitudinal studies, again most of which focused on repeating patterns. In a first study with five- and six-year-olds, repeating patterning abilities at the start of the school year (i.e., identify the unit of repeat, extend, interpolate, and create a repeating pattern) predicted general mathematical abilities at the end of the school year (Warren & Miller, 2013). Lüken, Peter-Koop, and Kollhoff (2014) found a correlation between repeating patterning abilities (i.e., copy, extend, and explain an ABCC pattern) in preschool (i.e., four-year-olds) and mathematical abilities in preschool, kindergarten (i.e., five-year-olds) and first grade. Rittle-Johnson and colleagues (2017) showed that patterning abilities of low-income children in preschool (i.e., four-year-olds) predicted mathematics achievement in fifth grade and that this effect was mediated by the patterning abilities in first grade.

Indications of a relationship between patterning abilities and mathematical achievement have also been found with other types of patterns. A correlation between patterning abilities in fall and mathematical achievement in spring suggested that being able to interpolate growing number sequences at the start of first grade might support mathematical achievement at the end of first grade (Pasnak et al., 2016). Interestingly, the authors argued that interpolating these sequences might "reflect fluid reasoning and that fluid reasoning is related to mathematics concepts" (p. 644). Similar patterning abilities (i.e., interpolating a range of patterns) uniquely predicted mathematical abilities, above executive functioning, in first graders (Schmerold et al., 2017). Lüken (2012) found a correlation between early structure sense in first grade and general mathematical ability in kindergarten (i.e., five-year-olds) and in second grade. Variance in mathematical ability in Grade 2 was predicted by early mathematical abilities, early structure sense, and the interaction between both. The majority of variance, however, remained unexplained. Additional qualitative analyses showed that children with low patterning abilities (or, as Lüken states, low early structure sense) had no difficulties perceiving the external aspects of a pattern (i.e., the different colors or visual gaps that group sub-structures), but had difficulties connecting these external aspects with their mathematical aspects, more particularly, their numerical structure. Children with high patterning abilities, on the other hand, showed an understanding of the unit of repeat, which Lüken considered as a milestone in the development of their patterning abilities.

Much less is known about the relationship between early patterning abilities and *reading abilities*, although one research group did systematically look at this relationship in several cross-sectional studies in first grade (Bock et al., 2015; Pasnak et al., 2016; Schmerold et al., 2017). In all these studies, first graders completed a pattern interpolation task and a reading task. Bock and colleagues (2015) observed a correlation between this patterning ability and early reading abilities in the fall of first grade. Pasnak and colleagues (2016) found a correlation between patterning abilities in the fall and word reading abilities in the spring of first grade, as well as between reading abilities in the fall and patterning abilities in the spring. According to the latter authors, two mechanisms may explain their findings: (1) reading and patterning may have a reciprocal relationship, or (2) a third (untested) variable, such as executive functioning, may underlie the relationship between patterning and reading. Schmerold and colleagues (2017) reported, however, that patterning ability remained a unique predictor of word reading abilities after controlling for executive functioning.

Research on the relationship between early patterning abilities and *general cognitive abilities* can be divided into two sets. A first set of studies consists of two cross-sectional studies with four-year-olds in which a relation was found between cognitive variables and a repeating patterning instrument based on the four-level construct map of Rittle-Johnson and colleagues (2013). Collins and Laski (2015) explored the relationship between these tasks and several cognitive measures (i.e., visuospatial short-term memory, verbal short-term memory, working memory, and inhibition). General accuracy on the four patterning tasks was predicted by visuospatial short-term memory and working memory, but patterning abilities did not correlate with inhibition or verbal short-term memory. Accuracy on extending and identifying the pattern unit was predicted by visuospatial short-term memory, whereas accuracy on generalizing was predicted by working memory.

Miller, Rittle-Johnson, Loehr, and Fyfe (2016) explored the relationship between the patterning instrument as a whole and several cognitive measures on two different days. On the first day, children completed a pretest patterning instrument, as well as an inhibition task and a set-shifting task. They also received brief instruction on six generalization items that same day and on four more items the following day. On the second day, they completed a posttest patterning assessment, a relational knowledge task, and a working memory measure. Relational knowledge, working memory, and set shifting were unique predictors of pretest repeating patterning abilities, whereas inhibition was not. Moreover, working memory was a predictor for performance on the posttest patterning task, suggesting that working memory might support learning to generalize patterns.

A second set of studies consists of two cross-sectional studies in which first graders completed an interpolation activity with a range of patterns. Bock and colleagues (2015) found a correlation between this patterning activity and cognitive flexibility, while Schmerold and colleagues (2017) found a correlation between this patterning activity and both cognitive flexibility and working memory, but not inhibition. Only cognitive flexibility—not working memory—appeared to be a predictor of patterning abilities and the effect of cognitive flexibility on both reading and mathematics was completely mediated by patterning (Schmerold et al., 2017).

To sum up, there is evidence for a relationship between patterning abilities, on the one hand, and mathematical, reading, or cognitive abilities, on the other hand. Not surprisingly, most studies on early patterning have investigated the relationship with mathematical abilities. In these studies, patterning performance in preschool or first grade predicted general mathematical achievement up to fifth grade (Lüken, 2012; Lüken et al., 2014; Pasnak et al., 2016; Rittle-Johnson et al., 2017; Schmerold et al., 2017; Warren & Miller, 2013). Additional qualitative analyses suggested that it was the children who had insight into the unit of repeat who were most likely to score better on a general mathematics achievement test (Lüken, 2012; Warren & Miller, 2013). The scarce evidence that is available on the relationship with reading abilities used only one type of task, namely interpolating (Bock et al., 2015; Pasnak et al., 2016; Schmerold et al., 2017). So far, it is unclear whether this relationship is consistent over other patterning tasks. A few studies have also looked into general cognitive abilities that might support patterning abilities, namely visuospatial and verbal short-term memory, working memory, cognitive flexibility, relational knowledge, and set shifting, but different tasks seem to rely on different cognitive abilities (Collins & Laski, 2015; Miller et al., 2016). Certainly, more theoretical and empirical research is needed to unravel the complex relationship between patterning abilities and these general cognitive abilities (Verschaffel et al., 2017).

Intervention

The research reviewed in the previous sections has also led to a variety of interventions on mathematical patterns and structures, some being implemented and tested in broader research programs consisting of various studies. We will present four sets of intervention studies, each time considering the nature of the intervention, how it was implemented and tested, and its effects.

A first set of interventions focused on first graders and their performance on an interpolation task with a range of patterns. During the intervention, children had to find the missing element in a pattern with scaffolding of the teacher or the computer, and the same interpolation task—without scaffolding—was used both as a pre- and posttest. In a first study (Hendricks et al., 1999), four first graders received individual instructional sessions of 15–20 min, 4 days per week, until the child had mastered all items (>400). The intervention yielded an improvement in children's patterning abilities, IQ, and academic achievement. In a more systematic study in first graders (Hendricks, Trueblood, & Pasnak, 2006), children received a 15 min intervention for four days a week over a period of five months in groups of three. They were randomly assigned to a patterning or control condition (i.e., extra lessons in general subject matters). Children in the patterning intervention scored higher on the posttest patterning assessment and a broad posttest measure of academic achievement. A similar patterning intervention was used in three more recent studies (Kidd et al., 2013, 2014; Pasnak et al., 2015). In all three studies, children were randomly assigned to four conditions (i.e., patterning, reading, mathematics, and social studies) in which they individually received lessons of 15 min three times a week over a period of six to seven months. In each study, children in the patterning condition performed afterward equally well or better than children in one of the other conditions on patterning tasks similar to the intervention, as well as on transfer tasks using patterns that were different from those used in the intervention. Remarkably, children following the patterning intervention scored as least as well or even better on mathematical tasks than children following a mathematical intervention. In two of these studies, these children scored as least as well on reading tasks compared to children following the reading instruction (Kidd et al., 2014; Pasnak et al., 2015). In one study, it was shown that these transfer effects of patterning instruction on reading and mathematics were fully mediated by patterning abilities (Kidd et al., 2014).

A second set of intervention studies comprised of different implementations of the Pattern and Structure Mathematics Awareness Program (PASMAP) developed by Mulligan and colleagues. The program consists of several activities (similar to activities of the assessment tool PASA) through which a child is encouraged to represent pattern and structure (Mulligan, English, Mitchelmore, & Robertson, 2010). Teachers using PASMAP are provided with materials and activities they can implement in their classroom. In one study with a large group of low-achieving kindergarten (i.e., four- to six-year-olds) and primary school students, the PASMAP was implemented over the course of nine months (Mulligan, Prescott, Papic, & Mitchelmore, 2006). An improvement in PASA scores and measures of general mathematical achievement suggested possible learning and transfer effects. Because there was no control group in this design, it is hard to derive conclusions concerning the program's impact. In a second study—again without a control group—a small group of low-achieving four- to six-year-olds (kindergartners) following the PASMAP for one hour a week during 15 weeks improved their scores on the PASA (Mulligan, Mitchelmore, Marston, Highfield, & Kemp, 2008). In a later intervention study with a control group, kindergarten teachers were asked to implement either the standard mathematics curriculum or PASMAP during three school terms (Mulligan, English, & Mitchelmore, 2013; Mulligan et al., 2010). All teachers followed a one-day training session and received professional support by weekly visits of the research team. Teachers in the intervention condition could implement the program at their own pace, leading to different implementation times from class to class (i.e., 50 min–5 h a week). At the start of the intervention, there were no differences in PASA or mathematical competencies between children in both conditions. Posttest results confirmed the impact of PASMAP on PASA scores, but no differences were found in general mathematical abilities.

The same research group tested a different, but similar, intervention with repeating and spatial structure patterns (Papic et al., 2011). Over a period of six months, five-year-olds were confronted—individually and in small groups—with several pattern-eliciting tasks (similar to the EMPA) and their teachers were encouraged to "patternize" their regular school program. A similar kindergarten class was tested pre- and post-intervention, serving as a contrast group. Posttest data suggested a learning effect on patterning as well as transfer effects toward another type of patterns, namely growing patterns, and general numerical abilities. According to the authors, this transfer toward growing patterns might be caused by an acquired tendency to look for patterns in the children who received the experimental program. Additional qualitative analyses during the intervention and the posttest indicated that a solid understanding of the unit of repeat helped children in tasks with repeating patterns.

A third set of interventions assessed the impact of a short (20–30 min) feedback session on four-year-olds' repeating patterning abilities (Fyfe, McNeil, & Rittle-Johnson, 2015; Rittle-Johnson et al., 2015, study 3, 2013). In such a feedback session, children were given a few examples of generalization items (similar to those in their assessment instrument) and they received feedback from a researcher on these items or on a few generalization items they had to solve themselves. This feedback session had a positive influence on children's accuracy on generalization items, suggesting the potential of this instructional technique for further stimulation in generalizing patterns (Rittle-Johnson et al., 2015, study 3, 2013). When comparing the origin of feedback, there were no differences in accuracy whether the instructor provided the feedback, the child explained the pattern himself, or when a combination of both was used (Rittle-Johnson et al., 2015, study 3). However, when comparing the kind of feedback the instructor provided, results indicated that the use of abstract labels (e.g., A–B–A–B) yields better results on generalizing than the use of concrete labels (e.g., red–blue–red–blue) (Fyfe et al., 2015). According to the authors, abstract labels are better since they draw children's attention to the structure of the pattern and can be shared across patterns.

Finally, Kampmann and Lüken (2016) designed an intervention study for first graders that involved repeating, spatial structure, and growing patterns. The intervention group followed 13 lessons on patterning, which were implemented within the regular mathematics curriculum and were given by one of the researchers, whereas the control group received the regular mathematics program. During the patterning intervention, children could create, recognize, use, describe, and explain patterns and structures. At the start of the intervention, there were no differences between the two groups in intelligence or mathematical abilities, whereas differences in mathematical abilities were found post-intervention, suggesting a transfer effect toward mathematical abilities in first-grade students. Moreover, it was the low-achieving children in particular who benefited from the intervention.

In addition to these four groups of interventions specifically focusing on patterning, various scholars have also designed, implemented, and tested general programs for early mathematics that included patterning activities (e.g., Building Blocks, Clements, Sarama, Spitler, Lange, & Wolfe, 2011; Big Math for Little Kids, Greenes, Ginsburg, & Balfanz, 2004; TRIAD, Sarama et al., 2008; Pre-K Mathematics intervention, Starkey et al., 2004). While there is evidence for the effectiveness of all these programs in general, the design of these studies does not allow us to disentangle the contribution of the patterning activities to the positive impact of these intervention programs on their mathematical development in general and the growth of their patterning competencies in particular (Frye et al., 2014).

To sum up, there is a wide range of interventions focusing on early patterning abilities that have been developed and tested for both preschoolers and primary school students, some in the context of short and strictly controlled psychological experiments and others in the context of ecologically valid educational settings. Interventions aimed at four-year-olds (Fyfe et al., 2015; Rittle-Johnson et al., 2015, study 3, 2013) or five-year-olds (Mulligan et al., 2013, 2010) all yielded learning effects toward patterning. A transfer effect toward numerical abilities was only observed in one study (Papic et al., 2011), but multiple interventions for first graders yielded transfer effects toward mathematics (Hendricks et al., 1999, 2006; Kampmann & Lüken, 2016) and reading (Hendricks et al., 1999, 2006). Extensive and detailed descriptions of the intervention programs are often missing. The above-mentioned programs are also relatively broad and include multiple components, which is valuable from an educational perspective but makes it hard to determine which of these components caused the observed learning and transfer effects. Therefore, it is difficult to compare the different programs and their effects on a theoretical level.

Educational Implications

From this review, it is clear that there are multiple indications for a relationship between a child's patterning competencies in kindergarten and his or her mathematical competencies later on (Lüken, 2012; Lüken et al., 2014; Pasnak et al., 2016; Rittle-Johnson et al., 2017; Schmerold et al., 2017; Warren & Miller, 2013). Moreover, patterning competencies can be successfully stimulated through well-developed early interventions and this stimulation might also improve young children's concurrent or later mathematical or reading competencies (Hendricks et al., 1999, 2006; Kampmann & Lüken, 2016; Papic et al., 2011). These findings suggest the value of implementing patterning activities in early educational settings. Currently, patterning is already part of the national early childhood curriculum in countries such as Australia and Canada (acara, 2015; Ontario Ministry of Education and Training (OMET) 2005). The educational value of the current patterning approach to algebra has, however, been criticized by a number of researchers, by claiming that the patterning activities that are most often implemented lead children only toward recursive thinking (Carraher & Schliemann, 2007; Hargreaves, Threlfall, Frobischer, & Shorrocks-Taylor, 1999; Macgregor & Stacey, 1992; Radford, 2010; Tall, 1992). Indeed, the main focus of these national curricula is on copying and extending (Clements & Sarama, 2014; Economopoulos, 1998; Mckillip, 1970b) and it is questionable whether these tasks require—and, thus, stimulate—the development of any functional insight. There is, however, only limited empirical evidence that looks into the level of thinking (i.e., recursive or functional) that is required while performing a patterning task. Although several researchers acknowledge that patterning tasks eliciting functional thinking are educationally more valuable (Lüken, 2012; Papic & Mulligan, 2007; Rittle-Johnson et al., 2013), it is unclear whether, and if so to what extent, the intervention programs designed by these same researchers implement tasks to stimulate functional thinking. The diversity in the existing intervention studies makes it hard to provide teachers with good, evidence-based materials, tasks, and techniques for working with patterns in an early childhood setting. Nevertheless, researchers agree that kindergarten teachers should consider implementing more difficult tasks (i.e., tasks that are assumed to involve functional thinking and require insight into the unit of repeat) and more difficult types of patterns (i.e., not only AB repeating patterns).

In most intervention programs so far (Fyfe et al., 2015; Hendricks et al., 2006; Kidd et al., 2013, 2014; Pasnak et al., 2015; Rittle-Johnson et al., 2015, Study 3, 2013), children received extra stimulation either individually or in small groups, which is not always possible in regular educational settings. However, positive effects on patterning (Mulligan et al., 2010, 2013) and mathematical achievement (Kampmann & Lüken, 2016) were also found in studies involving whole classroom interventions.

Finally, patterning interventions were mostly carried out by the researchers themselves (except for PASMAP; Mulligan et al., 2010, 2013). Making available research-based course material for early patterning is an important step toward the solution, but attention should also be given to the quality of teacher education, both in-service and pre-service. Unfortunately, teachers are generally not well prepared to teach early mathematics in general (Ginsburg, Lee, & Boyd, 2008) and early patterning in particular (Verschaffel et al., 2017). As for all other aspects of early mathematics education, the development of teachers' professional knowledge base should involve various elements. Teachers should have sufficient content knowledge (CK) regarding patterning. Research has shown that preschool teachers themselves have difficulty identifying multiple appropriate continuations of repeating patterns (Tsamir, Tirosh, Levenson, Barkai, & Tabach, 2016) and defining repeating patterns (Tirosh, Tsamir, Levenson, Barkai, & Tabach, 2017). Teachers frequently use the word "pattern" (Houssart, 2000), but, as one teacher noticed, a pattern—in a mathematical sense—"is not just a pretty design" (Brownell et al., 2014, p. 88). Teachers should also have sufficient pedagogical content knowledge (PCK) to teach this patterning content to young children. Research has shown that early childhood teachers often do not know how to react when a child spontaneously creates a pattern (Björklund & Barendregt, 2016; Fox, 2005; Hendershot, Berghout Austin, Blevins-Knabe, & Ota, 2016; Waters, 2004). Understanding the learning trajectory (e.g., Clements & Sarama, 2014) for patterning might help them to gain insight into the typical development of patterning competencies, the difficulties young children might encounter, and the typical errors they make (Cross et al., 2009). Besides knowledge of children's learning trajectories, knowing which tasks are appropriate for these young children and knowing how to schematically or symbolically represent and verbally discuss patterns to preschoolers is another important component of teachers' pedagogical content knowledge. Teachers should also believe in the importance of patterning for children's further academic achievement and be convinced that their work in preschool can provide a valuable contribution to children's patterning competences. One program has already been set up to stimulate teachers' pedagogical content knowledge and self-confidence to teach patterning in preschool (Tirosh, Tsamir, Barkai, & Levenson, 2017), but more work is needed.

General Conclusion

The patterning concept covers a broad range of activities, from extension to unit identification, which can be implemented with several types of patterns, with repeating, growing, and spatial structure patterns being the most common ones. Despite the diversity in the definition and operationalization of early patterning, at least three main conclusions can be drawn from the available research. First, children's early patterning abilities grow remarkably over the preschool years. Whereas three- to four-year-olds are generally able to copy or extend an easy repeating pattern, five- to six-year-olds seem to gain some insight into the unit of repeat and are able to generalize a given pattern (Clements & Sarama, 2014; Rittle-Johnson et al., 2015). Second, there is a relationship between children's early patterning abilities and their later mathematical and reading abilities (e.g., Lüken, 2012; Rittle-Johnson et al., 2017; Schmerold et al., 2017). Third, early patterning abilities can be successfully stimulated by means of well-developed interventions (e.g., Fyfe et al., 2015; Kampmann & Lüken, 2016; Kidd et al., 2014; Mulligan et al., 2010; Papic, Mulligan, Highfield, McKay-Tempest, & Garrett, 2015).

Although early patterning is defined and operationalized broadly, the above-reviewed research on early patterning seems to focus only on a limited set of activities and types of patterns. Whereas the main focus so far has been on repeating patterns, future research should also look into spatial structure and growing patterns, thus making a global framework of early patterning possible. One promising starting point for such an endeavor has already been provided by Clements and Sarama's (2014) learning trajectory for pattern and structure. Moreover, the distinction between recursive and functional thinking that has been made by multiple researchers seems like a valuable avenue for further exploration.

Acknowledgements This research was partially supported by Grant KU Leuven project C16/16/001 "Development and stimulation of core mathematical competencies" and Grant DBOF/12/009 "Early mediators of number sense" from the Research Fund KU Leuven, Belgium.

References

acara. (2015). *Mathematics: Sequence of content F-6*. Retrieved November 22, 2017, from https://www.australiancurriculum.edu.au/media/3480/mathematics-sequence-of-achievement.pdf.

Banerji, M., & Ferron, J. (1998). Construct validity of scores on a developmental assessment with mathematical patterns tasks. *Educational and Psychological Measurement, 58*(4), 634–660. https://doi.org/10.1177/0013164498058004007.

Baroody, A. J., Li, X., & Lai, M. (2008). Toddlers' spontaneous attention to number. *Mathematical Thinking and Learning, 10*(3), 240–270. https://doi.org/10.1080/10986060802216151.

Björklund, C., & Barendregt, W. (2016). Teachers' pedagogical mathematical awareness in Swedish early childhood education. *Scandinavian Journal of Educational Research, 60*(3), 359–377. https://doi.org/10.1080/00313831.2015.1066426.

Björklund, C., & Pramling, N. (2014). Pattern discernment and pseudo-conceptual development in early childhood mathematics education. *International Journal of Early Years Education, 22*(1), 89–104. https://doi.org/10.1080/09669760.2013.809657.

Bock, A., Cartwright, K., Gonzalez, C., O'Brien, S., Robinson, M., Schmerold, K., … Pasnak, R. (2015). The role of cognitive flexibility in pattern understanding. *Journal of Education and Human Development, 4*(1), 19–25. https://doi.org/10.15640/jehd.v4n1a3.

Brownell, J. O., Chen, J. -Q., Ginet, L., & The Early Math Collaborative Erikson. (2014). *Big ideas of early mathematics: What teachers of young children need to know* (J. W. Johnston, J. Peters, & A. Hall, Eds.). New Jersey: Pearson Education.

Carraher, D., & Schliemann, A. (2007). Early algebra and algebraic reasoning. In F. Lester (Ed.), *Second handbook on mathematics teaching and learning* (pp. 669–699). Charlotte, NC: Information Age Publishing.

Claessens, A., & Engel, M. (2013). How important is where you start? Early mathematics knowledge and later school success. *Teachers College Record, 115*(6), 1–29.

Clements, D., & Sarama, J. (2007). Early childhood mathematics learning. In F. Lester (Ed.), *Second handbook of research on mathematics teaching and learning* (pp. 461–555). Charlotte, NC: Information Age Publishing.

Clements, D., & Sarama, J. (2014). Other content domains. In D. Clements & J. Sarama (Eds.), *Learning and teaching early math: The learning trajectories approach* (2nd ed., pp. 214–229). New York: Routledge.

Clements, D., Sarama, J., Spitler, M. E., Lange, A., & Wolfe, C. (2011). Mathematics learned by young children in an intervention based on learning trajectories: A large-scale cluster randomized trial. *Journal for Research in Mathematics Education, 42*(2), 127–166.

Close, J. S., & Glennon, V. J. (1977). The development of linear patterning in disadvantaged children of three ethnic groups. *The Irish Journal of Education, 11*(1), 21–32.

Collins, M. A., & Laski, E. V. (2015). Preschoolers' strategies for solving visual pattern tasks. *Early Childhood Research Quarterly, 32*(1), 204–214. https://doi.org/10.1016/j.ecresq.2015.04.004.
Cross, C., Woods, T., & Schweingruber, H. (Eds.). (2009). *Mathematics learning in early childhood.* Washington, D.C.: National Academies Press. https://doi.org/10.17226/12519.
De Smedt, B., Noël, M. -P., Gilmore, C., & Ansari, D. (2013). How do symbolic and non-symbolic numerical magnitude processing skills relate to individual differences in children's mathematical skills? A review of evidence from brain and behavior. *Trends in Neuroscience and Education, 2*(2), 48–55. https://doi.org/10.1016/j.tine.2013.06.001.
Economopoulos, K. (1998). What comes next? The mathematics of pattern in kindergarten. *Teaching Children Mathematics, 5*(4), 230–233. Retrieved from http://www.jstor.org/stable/41197171.
English, L., & Mulligan, J. (Eds.). (2013). *Reconceptualizing early mathematics learning*. Dordrecht: Springer Netherlands. https://doi.org/10.1007/978-94-007-6440-8.
English, L., & Warren, E. (1998). Introducing the variable through pattern exploration. *The Mathematics Teacher, 91*(2), 166–170.
Fox, J. (2005). Child-initiated mathematical patterning in the pre-compulsory years. In H. L. Chick & J. L. Vincent (Eds.), *Proceedings of the 29th Conference of the International Group for the Psychology of Mathematics Education* (Vol. 2, pp. 313–320). Melbourne: PME.
Frye, D., Baroody, A. J., Burchinal, M., Carver, S. M., Jordan, N. C., & McDowell, J. (2014). *Teaching math to young children*. Retrieved from http://whatworks.ed.gov.
Fyfe, E. R., McNeil, N. M., & Rittle-Johnson, B. (2015). Easy as ABCABC: Abstract language facilitates performance on a concrete patterning task. *Child Development, 86*(3), 927–935. https://doi.org/10.1111/cdev.12331.
Gadzichowski, M. (2012a). Examining patterning abilities in first grade children: A comparison of dimension, orientation, number of items skipped and position of the missing item. *Psychology, 3*(12), 1177–1182. https://doi.org/10.4236/psych.2012.312A174.
Gadzichowski, M. (2012b). Patterning abilities of first grade children: Effects of dimension and type. *Creative Education, 3*(5), 632–635. https://doi.org/10.4236/ce.2012.35092.
Garrick, R., Threlfall, J., & Orton, A. (2005). Pattern in the nursery. In A. Orton (Ed.), *Pattern in the teaching and learning of mathematics* (pp. 1–17). London: Continuum.
Geary, D. C., Bow-Thomas, C. C., & Yao, Y. (1992). Counting knowledge and skill in cognitive addition: A comparison of normal and mathematically disabled children. *Journal of Experimental Child Psychology, 54*(3), 372–391. https://doi.org/10.1016/0022-0965(92)90026-3.
Ginsburg, H., Lee, J. S., & Boyd, J. S. (2008). *Mathematics education for young children: What it is and how to promote it* (pp. 3–22). XXII(I): Social Policy Report.
Göbel, S. M., Watson, S. E., Lervåg, A., & Hulme, C. (2014). Children's arithmetic development: It is number knowledge, not the approximate number sense, that counts. *Psychological Science, 25*(3), 789–798. https://doi.org/10.1177/0956797613516471.
Greenes, C., Ginsburg, H., & Balfanz, R. (2004). Big math for little kids. *Early Childhood Research Quarterly, 19*(1), 159–166. https://doi.org/10.1016/j.ecresq.2004.01.010.
Hannula, M. M., & Lehtinen, E. (2005). Spontaneous focusing on numerosity and mathematical skills of young children. *Learning and Instruction, 15*(3), 237–256. https://doi.org/10.1016/j.learninstruc.2005.04.005.
Hannula-Sormunen, M. M., Lehtinen, E., & Räsänen, P. (2015). Preschool children's spontaneous focusing on numerosity, subitizing, and counting skills as predictors of their mathematical performance seven years later at school. *Mathematical Thinking and Learning, 17*(2–3), 155–177. https://doi.org/10.1080/10986065.2015.1016814.
Hargreaves, M., Shorrocks-Taylor, D., & Threlfall, J. (1998). Children's strategies with number patterns. *Educational Studies, 24*(3), 315–331. https://doi.org/10.1080/0305569980240305.
Hargreaves, M., Threlfall, J., Frobischer, L., & Shorrocks-Taylor, D. (1999). Children's strategies with linear and quadratic sequences. In A. Orton (Ed.), *Pattern in the teaching and learning of mathematics* (pp. 67–83). London: Continuum.

Hendershot, S. M., Berghout Austin, A. M., Blevins-Knabe, B., & Ota, C. (2016). Young children's mathematics references during free play in family childcare settings. *Early Child Development and Care, 186*(7), 1126–1141. https://doi.org/10.1080/03004430.2015.1077819.

Hendricks, C., Pasnak, R., Willson-Quayle, A., Trueblood, L., Malabonga, V., & Ciancio, D. (1999). Effects of instruction in sequencing and class inclusion for first graders. *Genetic, Social, and General Psychology Monographs, 125*(3), 297–312.

Hendricks, C., Trueblood, L., & Pasnak, R. (2006). Effects of teaching patterning to 1st-graders. *Journal of Research in Childhood Education, 21*(1), 79–89. https://doi.org/10.1080/02568540609594580.

Herman, M. L. (1973). Patterning before mathematics in kindergarten: The effect of a patterning treatment on the children's ability to learn number concepts later in the year (Doctoral dissertation, Columbia University, 1972). In *Dissertation abstract international: Section A humanities and social sciences* (p. 4060).

Houssart, J. (2000). Perceptions of mathematical pattern amongst primary teachers. *Educational Studies, 26*(4), 489–502. https://doi.org/10.1080/03055690020003665.

Kampmann, R., & Lüken, M. M. (2016). The influence of fostering children's pattern and structure abilities on their arithmetic skills in grade 1. In *13th International Congress on Mathematical Education* (pp. 1–4). Hamburg.

Kidd, J., Carlson, A., Gadzichowski, M., Boyer, C. E., Gallington, D., & Pasnak, R. (2013). Effects of patterning instruction on the academic achievement of 1st-grade children. *Journal of Research in Childhood Education, 27*(2), 224–238. https://doi.org/10.1080/02568543.2013.766664.

Kidd, J., Pasnak, R., Gadzichowski, M., Gallington, D., McKnight, P., Boyer, C. E., et al. (2014). Instructing first-grade children on patterning improves reading and mathematics. *Early Education and Development, 25*(1), 134–151. https://doi.org/10.1080/10409289.2013.794448.

Kyriakides, L., & Gagatsis, A. (2003). Assessing student problem-solving skills. *Structural Equation Modeling: A Multidisciplinary Journal, 10*(4), 609–621. https://doi.org/10.1207/S15328007SEM1004_7.

Lehtinen, E., Hannula-Sormunen, M. M., McMullen, J., & Gruber, H. (2017). Cultivating mathematical skills: From drill-and-practice to deliberate practice. *ZDM Mathematics Education, 49*(4), 625–636. https://doi.org/10.1007/s11858-017-0856-6.

Lüken, M. M. (2010). The relation between early structure sense and mathematical development in primary school. In M. Pinto & T. Kawasaki (Eds.), *Proceedings of the 34th Conference of the International Group for the Psychology of Mathematics Education* (pp. 241–248). Belo Horizonte, Brazil: PME.

Lüken, M. M. (2012). Young children's structure sense. *Journal Für Mathematik-Didaktik, 33*(2), 263–285. https://doi.org/10.1007/s13138-012-0036-8.

Lüken, M. M. (2016). Repeating patterning competencies in 3- and 4-year old kindergartners. In *13th International Congress on Mathematical Education* (pp. 1–4). Hamburg.

Lüken, M. M., Peter-Koop, A., & Kollhoff, S. (2014). Influence of early repeating patterning ability on school mathematics learning. In D. Allan, P. Liljedahl, C. Nicol, & S. Oesterle (Eds.), *Proceedings of the 38th Conference of the International Group for the Psychology of Mathematics Education and the 36th Conference of the North American Chapter of the Psychology of Mathematics education* (pp. 137–144). Vancouver, Canada: PME.

Macgregor, M., & Stacey, K. (1992). A comparison of pattern-based and equation-solving approaches to algebra. In B. Southwell, B. Perry, & K. Owens (Eds.), *Proceedings of the 15th Annual Conference, Mathematics Education Research Group of Australasia* (pp. 362–371). Sydney: MERGA.

McKillip, W. D. (1970a). *Developing and evaluating "Patterns."* Athens.

McKillip, W. D. (1970b). "Patterns"—A mathematics unit for three- and four-year-olds. *The Arithmetic Teacher, 17*(1), 15–18.

Miller, M. R., Rittle-Johnson, B., Loehr, A. M., & Fyfe, E. R. (2016). The influence of relational knowledge and executive function on preschoolers' repeating pattern knowledge. *Journal of Cognition and Development, 17*(1), 85–104. https://doi.org/10.1080/15248372.2015.1023307.

Mulligan, J., & Mitchelmore, M. (2009). Awareness of pattern and structure in early mathematical development. *Mathematics Education Research Journal, 21*(2), 33–49.

Mulligan, J., English, L., & Mitchelmore, M. (2013). An evaluation of the Australian "Reconceptualising early mathematics learning" project: Key findings and implications. In A. M. Lindmeier & A. Heinze (Eds.), *Proceedings of the 37th Conference of the International Group for the Psychology of Mathematics Education* (Vol. 3, pp. 337–344). Kiel, Germany: PME.

Mulligan, J., Mitchelmore, M., & Stephanou, A. (2015). *PASA Teacher guide*. ACER Press.

Mulligan, J., English, L., Mitchelmore, M., & Robertson, G. (2010). Implementing a pattern and structure mathematics awareness program (PASMAP) in kindergarten. In *Proceedings of the 33rd Annual Conference of the Mathematics Education Research Group of Australasia* (pp. 706–803). Fremantle, Western Australia.

Mulligan, J., Mitchelmore, M., Marston, J., Highfield, K., & Kemp, C. (2008). Promoting mathematical pattern and structure in the first year of schooling: An intervention study. In O. Figueras, J. L. Cortina, S. Alatorre, T. Rojano, & A. Sepúlveda (Eds.), *Proceedings of the Joint Meeting of PME 32 and PME-NA XXX* (Vol. 4, pp. 1–8). Morelia, Mexico: PME. Retrieved from http://www2.earlyalgebra.terc.edu/our_papers/2006/Martinez_BrizuelaPME30.pdf.

Mulligan, J., Prescott, A., Papic, M., & Mitchelmore, M. (2006). Improving early numeracy through a pattern and structure mathematics awareness program (PASMAP). In P. Grootenboer, R. Zevenbergen, & M. Chinnappan (Eds.), *Identities, cultures, and learning spaces (Proceedings of the 29th Annual Conference of the Mathematics Education Research Group of Australasia, Canberra)* (pp. 376–383). Adelaide: MERGA. Retrieved from http://www.merga.net.au/documents/RP422006.pdf.

Mulligan, J., Verschaffel, L., Baccaglini-Frank, A., Coles, A., Gould, P., He, S., … Yang, D. - C. (2018). Whole number thinking, learning and development: Neuro-cognitive, cognitive and developmental approaches. In M. Bussi & X. Sun (Eds.), *Primary mathematics study on whole numbers (ICMI Study 23)* (pp. 137–167). New York: Springer. https://doi.org/10.1007/978-3-319-63555-2_7.

National Research Council. (2001). *Adding it up* (J. Kilpatrick, J. Swafford, & B. Findell, Eds.). Washington, D.C.: National Academies Press. https://doi.org/10.17226/9822.

NCTM. (1989). *Curriculum and evaluation standards for school mathematics* (F. J. Crosswhite, J. A. Dossey, & S. M. Frye, Eds.). Curriculum and Evaluation Standards Report. Reston: National Council of Teachers of Mathematics.

NCTM. (2000). *Principles and standards for school mathematics* (J. Carpenter & S. Gorg, Eds.). Portland: Graphics Arts Center.

New, R. S., & Cochran, M. (2007). *Early childhood education: An international encyclopedia* (Vols. 1–4).

Nguyen, T., Watts, T. W., Duncan, G. J., Clements, D. H., Sarama, J. S., Wolfe, C., et al. (2016). Which preschool mathematics competencies are most predictive of fifth grade achievement? *Early Childhood Research Quarterly, 36,* 550–560. https://doi.org/10.1016/j.ecresq.2016.02.003.

Ontario Ministry of Education and Training (OMET). (2005). *The Ontario curriculum, grades 1–8: Mathematics, revised*. Retrieved November 24, 2017, from http://www.edu.gov.on.ca.

Orton, A. (Ed.). (1999). *Pattern in the teaching and learning of mathematics*. London: Continuum.

Papic, M., & Mulligan, J. (2007). The growth of early mathematical patterning: An intervention study. In J. Watson & K. Beswick (Eds.), *Mathematics: Essential research, essential practice (Proceedings of the 30th Annual Conference of the Mathematics Education Research Group of Australasia, Hobart)* (Vol. 2, pp. 591–600). Adelaide: MERGA.

Papic, M., Mulligan, J., Highfield, K., McKay-Tempest, J., & Garrett, D. (2015). The impact of a patterns and early algebra program on children in transition to school in Australian Indigenous communities. In B. Perry, A. MacDonald, & A. Gervasoni (Eds.), *Mathematics and transition to school* (pp. 217–236). Singapore: Springer Singapore. https://doi.org/10.1007/978-981-287-215-9_14.

Papic, M., Mulligan, J., & Mitchelmore, M. (2011). Assessing the development of preschoolers' mathematical patterning. *Journal for Research in Mathematics Education, 42*(3), 237. https://doi.org/10.5951/jresematheduc.42.3.0237.

Pasnak, R., Kidd, J., Gadzichowski, M., Gallington, D., Schmerold, K., & West, H. (2015). Abstracting sequences: Reasoning that is a key to academic achievement. *The Journal of Genetic Psychology, 176*(3), 171–193. https://doi.org/10.1080/00221325.2015.1024198.

Pasnak, R., Schmerold, K., Robinson, M., Gadzichowski, M., Bock, A., O'Brien, S. E., … Gallington, D. (2016). Understanding number sequences leads to understanding mathematics concepts. *The Journal of Educational Research, 109*(6), 640–646. https://doi.org/10.1080/00220671.2015.1020911.

Piccolo, D. L., & Test, J. (2010). Preschoolers' thinking during block play. *Teaching Children Mathematics, 17*(5), 310–316. Retrieved from http://www.jstor.org/stable/41199662.

Radford, L. (2010). Layers of generality and types of generalization in pattern activities. *PNA, 4*(2), 37–62. Retrieved from http://digibug.ugr.es/handle/10481/3505.

Rathé, S., Torbeyns, J., De Smedt, B., & Verschaffel, L. (2017). Young children's spontaneous focusing on Arabic number symbols. In J. Torbeyns (Chair), *Spontaneous mathematical focusing tendencies. Symposium at the Biennial Meeting of the European Association for Research on Learning and Instruction (EARLI)*. Tampere.

Rittle-Johnson, B., Fyfe, E. R., Hofer, K. G., & Farran, D. C. (2017). Early math trajectories: Low-income children's mathematics knowledge from age 4 to 11. *Child Development, 88*(5), 1727–1742. https://doi.org/10.1111/cdev.12662.

Rittle-Johnson, B., Fyfe, E. R., Loehr, A. M., & Miller, M. R. (2015). Beyond numeracy in preschool: Adding patterns to the equation. *Early Childhood Research Quarterly, 31,* 101–112. https://doi.org/10.1016/j.ecresq.2015.01.005.

Rittle-Johnson, B., Fyfe, E. R., McLean, L. E., & McEldoon, K. L. (2013). Emerging understanding of patterning in 4-year-olds. *Journal of Cognition and Development, 14*(3), 376–396. https://doi.org/10.1080/15248372.2012.689897.

Sarama, J., & Clements, D. (2004). Building blocks for early childhood mathematics. *Early Childhood Research Quarterly, 19*(1), 181–189. https://doi.org/10.1016/j.ecresq.2004.01.014.

Sarama, J., & Clements, D. (2009). Other content domains. In *Early childhood mathematics education research* (pp. 319–334). New York: Routledge. https://doi.org/10.4324/9780203883785.

Sarama, J., Clements, D., Starkey, P., Klein, A., & Wakeley, A. (2008). Scaling up the implementation of a pre-kindergarten mathematics curriculum: Teaching for understanding with trajectories and technologies. *Journal of Research on Educational Effectiveness, 1*(2), 89–119. https://doi.org/10.1080/19345740801941332.

Schleifer, P., & Landerl, K. (2011). Subitizing and counting in typical and atypical development. *Developmental Science, 14*(2), 280–291. https://doi.org/10.1111/j.1467-7687.2010.00976.x.

Schmerold, K., Bock, A., Peterson, M., Leaf, B., Vennergrund, K., & Pasnak, R. (2017). The relations between patterning, executive function, and mathematics. *The Journal of Psychology, 151*(2), 207–228. https://doi.org/10.1080/00223980.2016.1252708.

Seo, K. -H., & Ginsburg, H. (2004). What is developmentally appropriate in early childhood mathematics education? Lessons from new research. In D. Clements, J. Sarama, & A. -M. DiBiase (Eds.), *Engaging young children in mathematics* (pp. 91–104). Mahwah, NJ: Lawrence Erlbaum.

Siegler, R. S., & Booth, J. L. (2004). Development of numerical estimation in young children. *Child Development, 75*(2), 428–444. https://doi.org/10.1111/j.1467-8624.2004.00684.x.

Simon, H. A. (1972). Complexity and the representation of patterned sequences of symbols. *Psychological Review, 79*(5), 369–382. https://doi.org/10.1037/h0033118.

Skoumpourdi, C. (2013). Kindergartners' performance levels on patterning. *Hellenic Mathematical Society International Journal for Mathematics in Education, 5,* 108–131.

Smutny, J. F. (Ed). (1998). The nature and role of algebra in the K-14 curriculum. In *Proceedings of a National Symposium*. Washington, D.C.: National Academies Press. https://doi.org/10.17226/6286.

Starkey, P., Klein, A., & Wakeley, A. (2004). Enhancing young children's mathematical knowledge through a pre-kindergarten mathematics intervention. *Early Childhood Research Quarterly, 19*(1), 99–120. https://doi.org/10.1016/j.ecresq.2004.01.002.

Steen, L. A. (1988). The science of patterns. *Science, 240*(4852), 611–616.

Sternberg, L., & Larson, P. (1976). The development of pattern recognition ability in children. *Contemporary Educational Psychology, 1*(2), 146–156. https://doi.org/10.1016/0361-476X(76)90019-9.

Tall, D. (1992). The transition from arithmetic to algebra: Number patterns, or proceptual programming? In *New Directions in Algebra Education, Proceedings of the Research Workshop on Mathematics Teaching and Learning "From Numeracy to Algebra"* (pp. 213–231). Brisbane, Australia: Queensland University of Technology. Retrieved from https://homepages.warwick.ac.uk/staff/David.Tall/pdfs/dot1993f-arith-alg-brisbane.pdf.

Taylor-Cox, J. (2003). Algebra in the early years? Yes! *Young Children, 58*(1), 14–21.

Threlfall, J. (2005). Repeating patterns in the early primary years. In A. Orton (Ed.), *Pattern in the teaching and learning of mathematics* (pp. 18–30). London: Continuum.

Tirosh, D., Tsamir, P., Barkai, R., & Levenson, E. (2017). *Preschool teachers' variations when implementing a patterning task.* Paper presented at the 10th Congress of European Research in Mathematics Education (CERME), Dublin, Ireland. Retrieved from https://keynote.conference-services.net/resources/444/5118/pdf/CERME10_0010.pdf.

Tirosh, D., Tsamir, P., Levenson, E., Barkai, R., & Tabach, M. (2017). Defining, drawing, and continuing repeating patterns: Preschool teachers' self-efficacy and knowledge. In C. Andrà, D. Brunetto, E. Levenson, & P. Liljedahl (Eds.), *Teaching and learning in maths classrooms. Research in mathematics education* (pp. 17–26). Cham: Springer. https://doi.org/10.1007/978-3-319-49232-2_3.

Tsamir, P., Tirosh, D., Levenson, E., Barkai, R., & Tabach, M. (2016). Preschool teachers' responses to repeating patterns tasks. In *13th International Congress on Mathematical Education.* Hamburg.

VanDerHeyden, A. M., Broussard, C., Snyder, P., George, J., Lafleur, S. M., & Williams, C. (2011). Measurement of kindergartners understanding of early mathematical concepts. *School Psychology Review, 40*(2), 296–306. Retrieved from https://eric.ed.gov/?id=EJ936458.

Verschaffel, L., Torbeyns, J., & De Smedt, B. (2017). *Young children's early mathematical competencies: Analysis and stimulation.* Plenure lecture presented at the Congress of European Research in Mathematics Education (CERME), Dublin, Ireland.

Warren, E., & Cooper, T. (2007). Repeating patterns and multiplicative thinking: Analysis of classroom interactions with 9-year-old students that support the transition from the known to the novel. *The Journal of Classroom Interaction Research Library, 4142*(7), 7–17. Retrieved from https://files.eric.ed.gov/fulltext/EJ780288.pdf.

Warren, E., & Miller, J. (2013). Young Australian indigenous students' effective engagement in mathematics: The role of language, patterns, and structure. *Mathematics Education Research Journal, 25*(1), 151–171. https://doi.org/10.1007/s13394-013-0068-5.

Waters, J. (2004). A study of mathematical patterning in early childhood settings. In I. Putt, R. Faragher, & M. MacLean (Eds.), *Proceedings Mathematics Education for the Third Millennium: Towards 2010. The 27th Annual Conference of the Mathematics Education Research Group of Australasia* (Vol. 2, pp. 321–328). Townsville, Queensland, Australia: MERGA. Retrieved from http://www.merga.net.au/documents/RP682004.pdf.

Wittmann, E. C., & Müller, G. N. (2007). Muster und strukturen als fachliches grundkonzept. In G. Walther, M. van den Heuvel-Panhuizen, D. Granzer, & O. Köller (Eds.), *Bildungsstandards für die Grundschule: Mathematik konkret* (pp. 42–65). Berlin: Cornelsen. https://doi.org/10.18452/3121.

Part III
The Elementary School Years

Arithmetic Concepts in the Early School Years

Katherine M. Robinson

Arithmetic Concepts in the Early School Years

A critical task of the early elementary school years is for children to start learning formal mathematics. They begin Grade 1 with varying amounts of knowledge about informal mathematics, and as previous chapters in this volume have examined, children actually have a surprising amount of knowledge about varying aspects of numbers and mathematics that they learn through informal and everyday experiences. Making the transition from informal to formal mathematics can be challenging for children. Children are taught a formal system with symbols attached to specific numbers (e.g., **** objects are represented with the symbol 4) and specific symbols indicating whether to add, subtract, to make equal, and so forth (e.g., +, − =). Further complications arise as problems can be presented in horizontal or vertical formats. In essence, children are being asked to learn a new language that often seems to include more than one dialect. They are also being taught basic arithmetic facts such as "2 and 2 make 4," and eventually, they will be expected to automatically retrieve these facts from memory, so they can free up cognitive resources to deal with the complexity of algebra or calculus problems.

Before problems such as 2 + 2 trigger the answer 4 from long-term memory, children need considerable practice with these problems and also often need to be taught problem-solving strategies or procedures for more difficult problems (Siegler, 1996). For example, counting on your fingers might work well for a problem such as 3 + 4, but what happens when the problem requires more than 10 fingers such as 7 + 8? Beyond the new language of math, learning arithmetic facts, and learning strategies for how to solve more difficult problems, lies a series of complex rules and relationships that need to be understood and applied when solving the new or more challenging problems continuously being introduced to children as

K. M. Robinson (✉)
Department of Psychology, Campion College at the University of Regina,
3737 Wascana Parkway, Regina, SK S4S 0A2, Canada
e-mail: katherine.robinson@uregina.ca

K. M. Robinson et al. (eds.), *Mathematical Learning and Cognition in Early Childhood*,
https://doi.org/10.1007/978-3-030-12895-1_10

they move through their formal schooling. These rules and relationships and what children know about them often get discussed under the umbrella term of "conceptual knowledge of arithmetic." The focus of this chapter is on children's conceptual knowledge of arithmetic—on how children make sense of mathematics in the early school years.

Procedural, Factual, and Conceptual Knowledge of Arithmetic

There are three general types of arithmetical knowledge: factual, procedural, and conceptual (Bisanz & LeFevre, 1990). The first type of knowledge, factual knowledge, is knowing math facts such as $5 + 2$ and 6×8. When children have considerable exposure to these problems and the correct answers, they eventually encode each problem and its associated answer into long-term memory. Then, if they encounter the problem again at some later time, the answer will automatically come to mind (Siegler & Shipley, 1995; Siegler & Shrager, 1984). These math facts become so well learned that it is virtually impossible for individuals to not reflexively think of the answer when they see the problem. For example, when you read the following math problem, try to prevent yourself from automatically retrieving the answer. Here is the problem: 3×3. Most children and adults, once they have learned their basic multiplication facts, cannot prevent themselves from thinking 9 when they see that problem. But how did that problem go from being novel and new to being effortlessly and automatically solved? That is where procedural and conceptual knowledge come into play.

The second type of knowledge, procedural knowledge, is about the problem-solving procedures or problem-solving strategies that children use when they cannot automatically retrieve the answer from memory (Bisanz & LeFevre, 1990; Siegler & Araya, 2005). That is, factual knowledge is not available or is insufficient to help them out (Siegler & Shrager, 1984). In North America, most children and adults have only learned the multiplication tables until the 12 times table. Ask a person to solve a problem such as 6×13 and all of a sudden a simple arithmetic problem is no longer quite as simple as if they had been asked to solve 6×11. So, how to solve the problem of 6×13? This is where procedural knowledge is needed. There are multiple methods, procedures, or strategies that can be used. One person might immediately reach for a calculator, a second might count up by 13 six times (13, 26, 39, 54, 65, 78), another might solve the problem by multiplying 6×3 to get 18 and then 6×10 to get 60 and then add 18 and 60 together, and yet another person might rely on his or her factual knowledge that 6×12 is 72 and then add another 6, and so on. All of these problem-solving strategies should yield the correct answer.

In some instances, some problem-solving strategies may be more attractive than others. If you get told that you will win $1000 if you get the answer correct, you might decide to put your faith in a calculator so that you are absolutely certain that you get the right answer. On the other hand, if you are competing against 10 other people for the $1000 and the fastest person to say the answer wins and you do not have a calculator handy, you might want to rely on a faster solution procedure such as decomposition ($6 \times 12 = 72 + 6$) as automatic retrieval of the first fact plus 6 more is quite fast compared to multiplying 6×3 than 6×10 and then adding the two numbers. Or, you might just completely panic and make a wild guess!

But how come most people know more than one way to solve any given arithmetic problem (LeFevre, Sadesky, & Bisanz, 1996; Robinson, 2001)? Children start learning new problem-solving strategies or procedures before formal schooling and, once they begin formal schooling, are taught several problem-solving procedures. In many cases, children figure out new, more efficient problem-solving procedures on their own without any help from their teachers or parents. In the case of the *min* procedure, most children discover this more efficient problem-solving strategy themselves (Groen & Resnick, 1977; Siegler & Jenkins, 1989). On a problem such as $2 + 6$, if a child does not yet know the answer by heart, he or she will initially use a *count-all* strategy. That is, he or she will painstakingly count the first number (1, 2) and then the second (1, 2, 3, 4, 5, 6) and then put these two numbers together (often by using their fingers). Eventually, they will realize that there is a better way and will start using the *count-on* strategy where they will start with the first number (2) and count up from there (3, 4, 5, 6, 7, 8). This is a faster problem-solving strategy and less prone to errors than the *count-all* strategy. Finally, they will switch to the *min* strategy which is the most efficient counting strategy as it always involves the minimum amount of counting by starting with the largest number (6) and counting up from there (7, 8). This progression is typically something that most children figure out on their own without any assistance or teaching from others. This figuring it out requires children to notice the size of the numbers that they are dealing with, to notice that on some problems such as $8 + 2$, it takes less time to get the answer with the count-on strategy than when the problem is switched to $2 + 8$, and so forth. Children are starting to learn and understand number, counting, and the operation of addition—they are starting to develop conceptual knowledge of arithmetic.

The third type of arithmetical knowledge is conceptual knowledge which will be the focus of the next section and the rest of this chapter. In the meantime, factual, procedural, and conceptual knowledge are deeply intertwined forms of knowledge and they are considered to develop in an "iterative" fashion (Rittle-Johnson & Alibali, 1999; Rittle-Johnson, Schneider, & Star, 2015). Each type of knowledge informs the other types of knowledge, which in turn inform the other types of knowledge, and so on and so forth. This makes it difficult to discuss each of the types of knowledge on their own as they are not completely independent on one another. However, factual and procedural knowledge are relatively straightforward to define and assess. Does a child know the answer to $28 \div 4$ or 7×8 without needing to use a problem-solving procedure? Does a child know three different ways to solve a problem such as $26 - 9$? Conceptual knowledge, on the other hand, has both been challenging for researchers not only to assess but also to define.

Defining Conceptual Knowledge

Conceptual knowledge has been defined in various ways, including as the understanding of mathematical concepts, operations, and the relations between them (Kilpatrick, Swafford, & Findell, 2001). Hiebert and Lefevre (1986) discussed conceptual knowledge as interconnected knowledge within the domain of mathematics. Bisanz and LeFevre (1990) defined conceptual knowledge as concepts or principles needed to understand mathematics. These definitions are somewhat vague, but are perhaps best exemplified by the following from Kilpatrick et al. (2001) in their report for the U.S. National Research Council:

> Students with conceptual understanding know more than isolated facts and methods. They understand why a mathematical idea is important and the kinds of contexts in which it is useful. They have organized their knowledge into a coherent whole, which enables them to learn new ideas to what they already know. Conceptual understanding also supports retention. Because facts and methods learned with understanding are connected, they are easier to remember and use, and they can be reconstructed when forgotten (p. 118).

As Kilpatrick et al. (2001) illustrate, factual, procedural, and conceptual knowledge are deeply intertwined forms of knowledge, so measuring or assessing a child's conceptual knowledge or understanding of arithmetic can be challenging. Although asking children "what do you know about addition?" or "what does the equal sign mean?" can sometimes yield interesting information, children are not always able to communicate or articulate what they know. Sometimes this is because of developmental issues related to language and cognitive capacity, other times it is because they might not understand the question, or sometimes they will have knowledge that they omit to communicate. Therefore, many researchers rely on more indirect routes for assessing conceptual knowledge.

One of the earliest instances of this approach to assessing conceptual knowledge comes from the work by Starkey and Gelman (1982). They presented young children between 3 and 5 years of age with problems such as $2 + 3 - 3$ using concrete objects. They were investigating whether children understood that addition and subtraction are inversely related to one another—a concept first proposed by Piaget (1952) as being critical for children's understanding of number. If children understand that because addition and subtraction are inverse or opposite operations, adding and then subtracting the same number would be redundant. Instead, the answer can easily be solved by ignoring or canceling out the 3s and stating the first number. Thus, children's understanding of the concept of inversion can be accessed via their problem-solving strategies or procedures. Children who do not know that addition and subtraction are inversely related will have to first add the 2 and the 3 together and then subtract the second 3 from the sum of 5. This latter problem-solving procedure takes longer and is more error-prone. Since the work by Starkey and Gelman (1982), researchers have often used information about how children solve problems to infer conceptual knowledge. Although this approach is commonly used, it is not without some controversy (e.g., Prather & Alibali, 2009); nevertheless, it is the one approach to assessing conceptual knowledge that appears to be common to researchers' investigations of many different arithmetic concepts (Crooks & Alibali, 2014).

The Importance of Conceptual Knowledge

Given my brief discussion of the struggles that researchers have faced with both defining and assessing conceptual knowledge, why has research on conceptual knowledge of arithmetic been steadily increasing since Starkey and Gelman (1982)? Kilpatrick et al. (2001), in their report for the U.S. National Research Council on what was known about children's mathematics learning, have already made a compelling case for why conceptual knowledge is important, but Kilpatrick et al. are not alone by any means. The National Council of Teachers of Mathematics (2000) emphasized the teaching of conceptual knowledge as an effective method for increasing students' problem solving performance. The U.S. National Mathematics Advisory Panel (2008) proposed that conceptual knowledge is essential for children to recognize and fix errors when performing mathematical tasks and for generalizing problem-solving strategies to novel mathematical problems and situations. The more expertise and knowledge that a student has in a domain such as mathematics, the more a child's knowledge and skills will become integrated into a coherent knowledge structure (Schneider, Rittle-Johnson, & Star, 2011). Kilpatrick et al. (2001) also noted that conceptual knowledge can result in students actually having *less* to learn because they can extrapolate their current understanding to new types of mathematical problems and concepts.

Following from this, the understanding of arithmetic concepts is a critical precursor to successfully learning how to solve and understand more complex mathematical skills such as algebra (NMAP, 2008). The view that conceptual knowledge of arithmetic is a foundational skill for later mathematical skills is shared widely (Falkner, Levi, & Carpenter, 1999; Kilpatrick et al., 2001; NCTM, 2000; Nunes et al., 2008). Those later mathematical skills are important predictors for attending college or university or, for those who do pursue advanced education, enrolling in degrees in the science, technology, engineering, and mathematics (STEM) disciplines, acquiring basic mathematical skills and knowledge that are so heavily based on understanding the domain of arithmetic is of great concern. In a recent UK report, Hudson, Price, and Gross (2009) concluded that the failure to master basic mathematical skills (such as mental arithmetic) in the early school years had a public cost in the UK of 2.4 billion pounds a year, which they considered to be an underestimate as it did not include factors such as higher social services costs, poorer health, and an increased likelihood to be involved in the criminal justice system, which are all factors correlated with poorer mathematical skills. They proposed that if effective mathematics interventions were put into place when students were 7 years of age, the annual savings would be up to 1.6 billion lb per year in the UK alone. Thus, the development of mathematical knowledge in the early years of formal schooling has important long-term consequences not only for individuals, but also for societies as a whole.

What Concepts Do Children Know When?

So, what conceptual knowledge do children in the early school years possess? Some concepts will be specifically taught in schools, but a surprising number of arithmetic concepts are ones that children discover for themselves (Baroody & Gannon, 1984) in the same way that they typically discover the *min* counting strategy for themselves. Indeed, understanding of concepts may arise from informal, everyday experiences with number rather than via formal instruction (Ginsburg, Cannon, Eisenband, & Pappas, 2006). For the remainder of this chapter, the focus will be on six essential arithmetic concepts (Robinson, Dubé, & Beatch, 2017; Robinson, Price, & Demyen, 2018). There are other concepts that could be considered (e.g., relation to operands, complement, additive composition), but these particular six have all been the focus of a substantive body of research. Each concept will be briefly examined to determine when these concepts may be acquired and whether the instruction is required or not. Finally, data from a recent study on these six concepts with children in Grades 1, 2, and 3 will be presented.

Identity. Using the definition proposed by Gelman and Gallistel (1978), identity here refers to the understanding that when zero is subtracted from a number, the identity or quantity of the original number remains the same, and remains an untransformed quantity. Whether or not children understand identity can be assessed by presenting them with problems of the form $a - 0$. Baroody, Lai, Li, and Baroody (2009) investigated the concept of identity in children from 3 to 7 years of age to determine whether understanding develops before or after the beginning of the early school years. Baroody et al. used concrete objects to show a set of objects and then demonstrated none of them being taken away and then asking children how many objects remained. If children answered correctly on at least 75% of the trials, they were deemed to understand the identity concept. Although only about half of 3-year-olds had good understanding of the identity concept, about three-quarters of 4-year-olds had good understanding, and nearly all 5- to 7-year-olds had good understanding. These findings suggested that before formal schooling even begins, children understand that when zero is subtracted from a number that number remains unchanged.

The children in Baroody and colleagues' (2009) study were presented with concrete objects and, as mentioned earlier in this chapter, children who may have good knowledge of how many objects there are when they are asked to add three cookies and two cookies may struggle when presented with the symbolic version of this problem (i.e., $3 + 2 = ?$). Robinson et al. (2017) investigated how well children understood identity when presented with symbolic problems. They had children in Grades 3, 4, and 5 solve problems such as $28 - 0$, then state their answers, and also explain how they got the answer. If children understand the concept of identity, i.e., that subtracting zero leaves the identity of the first number unchanged, they should be able to explain this. Robinson et al. found that understanding of identity was at approximately 90% in all grades. After correctly answering the identity problems, the typical description from children when asked how they had solved the problem was to state that "when you take zero away from a number, it's like you take away nothing so the answer is just the first number." These findings demonstrate that even on symbolic problems, by Grade 3, understanding of identity is very strong and that understanding of this concept probably develops without the need for formal instruction.

Negation. Negation refers to the understanding that when a number is subtracted from itself it "negates" that initial number, and therefore, there will be nothing, or zero, left (Gelman & Gallistel, 1978). Whether or not children understand negation can be assessed by presenting them with problems of the form $a - a$. Baroody et al. (2009) investigated the concept of negation in children from 3 to 7 years of age to again determine whether understanding develops before or after the beginning of the early school years. Using concrete objects and correct answers as evidence for conceptual knowledge, they found that about a quarter of 3-year-olds, about two-thirds of 4-year-olds, and nearly all 5- to 7-year-olds understand negation. Robinson et al. (2017) followed up Baroody et al.'s work with children in Grades 3–5 using symbolic problems and asked children to report how they had solved the negation problems. Negation had over 90% understanding in all grades. The typical description of children's thinking on these problems was that "taking away the same number just cancels the first number out so you are left with nothing, zero." Overall, it appears that across a wide range of ages and years of formal schooling, children have a solid understanding of both identity and negation before they start formal schooling and that even on symbolic problems, their understanding is strong by Grade 3.

Commutativity. The concepts of identity and negation appear to develop early, require little formal instruction, and may transfer easily from concrete to symbolic problems. Perhaps because these concepts do not seem to pose great difficulties for children, there has been comparatively little research on them and curricula typically do not mention these concepts as ones that need to be explicitly taught to children. On the other hand, other additive concepts have garnered significantly more attention both from researchers and educators. One such concept is the commutativity concept. The commutativity concept involves understanding that the order in which numbers are added together does not affect the answer (Baroody & Gannon, 1984). Whether children understand commutativity can be assessed by presenting them with problems such as $a + b = b + ?$. Children who understand commutativity will often respond that they solved the problem by "flipping" the numbers so that they know that the answer must be a. This knowledge is critical for developing the min strategy discussed previously (Rittle-Johnson & Siegler, 1998). In order for children to add two numbers starting with the larger one, they must realize that even if the larger number is the second one, it will not affect the answer (e.g., $3 + 28 = 28 + 3$).

The consensus is that by the end of kindergarten, many children will understand commutativity (Baroody & Gannon, 1984; Nunes & Bryant, 1998), but understanding might be higher when solving concrete rather than symbolic problems (Canobi,

Reeve, & Pattison, 1998, 2002, 2003). The Common Core Standards (National Governors' Association Center for Best Practices, & Council of Chief State School Officers, 2010) state that, by the end of Grade 1, children are expected to understand commutativity. Canobi found that almost all 7- to 9-year-old children in her study (roughly equivalent to Grade 2–4) understood commutativity and were able to apply it to symbolic problems. In a more recent study, Robinson et al. (2017) found that understanding of commutativity was much weaker. Just over one-tenth of Grade 3 children, one-third of Grade 4 children, and just under two-thirds of Grade 5 children reported using the commutativity concept to solve problems such as $4 + 9 = 9 + ?$. A further study by Robinson et al. (2018) with Grade 5–7 children demonstrated that understanding of commutativity was still not fully understood as only about two-thirds of Grade 5 children and three-quarters of Grade 6 and 7 children reported using the commutativity concept to solve problems. These findings suggest that, despite the assertion that many young children discover commutativity on their own through their informal arithmetic experiences (National Council of Teachers of Mathematics & National Association for the Education of Young Children, 1999), commutativity may require formal instruction, including instruction on what the equal sign means.

Inversion. Another even more researched arithmetic concept is the inversion concept. The inversion concept is the understanding that addition and subtraction are inversely related operations (Piaget, 1952). Starkey and Gelman (1982) used problems of the form $a + b - b$ to evaluate whether children could apply their knowledge of inversion and realize that as both the same number was added and subtracted, the answer to the problem was the first number and that no calculations needed to be performed. One of the reasons that this concept has received so much research attention (e.g., Gilmore & Bryant, 2006; Klein & Bisanz, 2000; Robinson, Ninowski, & Gray, 2006; Siegler & Stern, 1998; Vilette, 2002) is that inversion problems are novel problems so children must apply their conceptual knowledge to formulate a new and efficient problem-solving strategy.

Children who use conceptually based problem-solving procedures—often termed the "inversion shortcut"—on inversion problems (e.g., $4 + 3 - 3$) typically state that the second and third numbers (e.g., $3 - 3$) cancel each other out and the answer is the first number (e.g., 4). Research has shown that even very young children have some understanding of the inversion concept when solving concrete problems (e.g., Rasmussen, Ho, & Bisanz, 2003; Sherman & Bisanz, 2007), but Baroody et al. (2009) found that it took until almost the beginning of formal schooling (i.e., 5 years of age) for a slim majority of children to understand the concept of inversion. On symbolic problems, Gilmore and Bryant (2008) found that almost all 8- and 9-year-olds accurately answered inversion problems of the form $a + b - b$, but it is possible to answer these questions accurately without understanding and/or applying conceptual knowledge. In a study where participants' understanding of inversion was assessed not only by accurate responses, but also by children reporting their problem-solving procedures, Watchorn et al. (2014) classified almost 50% of Grade 2 and 3 children as understanding the inversion concept and 60% of Grade 4 children. This is similar to Robinson and Dubé (2009), who found that children in Grades 2 through 4 applied the inversion shortcut on almost 50% of the problems. These results are also comparable to Robinson et al.'s (2017) slightly older sample of children with about one-third use of conceptually-based problem-solving procedures by Grade 3 students and about 40% use by Grade 4 and 5 students.

Baroody et al. (2009), by comparing performance on identity, negation, and inversion problems, proposed that the concepts of identity and negation precede the concept of inversion. This makes sense, as inversion problems incorporate both negation ($b - b = 0$) and identity ($a - 0 = a$), but few other studies on arithmetic concepts have investigated the relationship among different concepts. Overall, inversion may be similar to commutativity in that even though there are signs that children have an informal understanding of the concept before formal schooling begins, once symbolic forms of the inversion problems are presented, instruction may be required for many children.

Associativity. One concept that has recently started to gain research attention is the associativity concept as applied to problems of the form $a + b - c$. Associativity involves the understanding that the operations of addition and subtraction can be solved in any order (e.g., on a problem such as 8 + 23 − 3, the problem could be solved from left to right, or it could be solved by first subtracting 23 – 3, then adding 8 or even by subtracting 8 – 3, and then adding 23). Originally, associativity was assessed in problems of the form of $a + b + c$ (e.g., Canobi et al., 1998) to investigate whether children knew they could solve the problem either by adding any pair of numbers and then adding the remaining number (e.g., 3 + 4 + 5 could be solved by adding 3 + 5 first and then the 4 or by adding 4 + 5 first and then the 3). Realizing that the numbers could be added in any order could be particularly helpful on problems such as 57 + 38 + 2 where adding the 38 + 2 first is much easier than adding 57 + 38.

However, Klein and Bisanz (2000) noticed that the associativity concept could also be used not only on three-term problems involving only addition but could also be applied on three-term problems involving both addition and subtraction These problems were originally included in studies on the inversion concept as control problems (Bisanz & LeFevre, 1990). If children were using the inversion shortcut on inversion problems such as 8 + 17 − 17, they should make few mistakes as no counting is required, but they should make more errors on problems such as 8 + 17 − 16 which are comprised of numbers of similar magnitudes as in the inversion problems, but calculations are required to get the answer. Any time calculation is involved, errors will be more likely. Klein and Bisanz (2000), in their study of the inversion concept with preschoolers, noted that on the control problems, children sometimes made use of their understanding of the associative relationship between addition and subtraction to simplify their problem solving. Associativity here refers to the fact that on these problems, the numbers can be added or subtracted in any order. Children can make use of this knowledge to make control problems, or what will from now on be called associativity problems, much easier to solve by solving the subtraction component first and then adding the first number (e.g., on the problem of 8 + 17 − 16, children could first perform 17 − 16 and get an answer of 1 and then add 1 on to the first number,

8, for a final answer of 9 which is faster and less error-prone than first adding 8 + 17 and then subtracting 16 from the sum of 25 (Robinson & Dubé, 2009, 2013).

Using concrete problems, Klein and Bisanz (2000) found that the associativity concept was applied by preschoolers on only 5% of problems. Even after several years of formal schooling, associativity use does not increase much. Robinson and Dubé (2009) found that children in Grades 2–4 applied their knowledge of associativity on about 20% of the problems. Robinson et al. (2017) found lower understanding of the associativity concept in Grade 3 through 5 children, who only made use of the associativity concept on approximately 10% of problems.

The results, although they vary somewhat from study to study, consistently demonstrate that the associativity concept appears to be quite a difficult concept for children to grasp. Even studies with older children have shown that understanding of the associativity concept does not increase dramatically even by the end of the middle school years (Robinson et al., 2006). These findings suggest that children need formal instruction about the associative relationship between addition and subtraction.

Equivalence. Like the inversion concept, the equivalence concept has been the focus of intensive research investigation. The equivalence concept involves the understanding that two sides of an equation are equal and interchangeable (McNeil, 2014). It turns out that, although completely obvious to adults, this concept as assessed on problems s of the form $a + b + c = a + ?$ is often very challenging for children (Crooks & Alibali, 2014). Children often interpret the equal sign in these problems as an operation and therefore believe they need to "do something." Usually, they decide that that something should be to add all four of the numbers together (Perry, Church, & Goldin-Meadow, 1988). On a problem such as $3 + 4 + 5 = 3 + ?$, children commonly will add up all of the numbers (3, 4, 5, and 3) and stick the answer into the empty blank (Rittle-Johnson & Alibali, 1999). Other children will add up all of the numbers on the left side of the equation and determine that the sum of those three numbers is what is missing from the right side of the equation.

To adults, these errors seem quite odd and unexpected (Sherman & Bisanz, 2009). Indeed, the meaning of the equal sign is so obvious to us as adults that we overestimate how well children understand the equal sign. Even teachers mistakenly believe that their students have a far greater understanding of the equal sign than the children actually do (Asquith, Stephens, Knuth, & Alibali, 2007; Sherman, 2007). This is potentially concerning as it may mean that, although specifically included in many mathematics curricula (e.g. U.S. Common Core Standards, 2010), teachers may not believe that they need to spend as much time and effort teaching the meaning of the equal sign as is actually needed.

That children need help with the concept of equivalence has been found across a multitude of studies that have assessed children's misunderstanding of the concept as well as attempted to improve children's understanding of the concept (e.g., McNeil & Alibali, 2005; Sherman & Bisanz, 2009). According to McNeil (2014), only about 20% of children between the ages of 7 and 11 are able to correctly solve equivalence problems. Even in older children who were tracked until the end of the middle school years, many Grade 8 children still did not have a solid understanding of the equal sign symbol (Alibali, Knuth, Hattikudur, McNeil, & Stephens, 2007). Robinson et al.

(2017), in a study with Grade 3, 4, and 5 children, found that children in Grade 3 never reported using their understanding of equivalence when solving problems, Grade 4 children used knowledge of equivalence on approximately 10% of problems, while Grade 5 children used their knowledge of equivalence on approximately 33% of problems. Children who did understand the equivalence concept usually reported that they needed to make both sides of the equation the same. Taken together, the results are not encouraging.

Few studies have examined multiple concepts within the same study with a focus on children in Grades 1 to 3—the developmental age range of interest for this section of the present volume. To get a sense of how these concepts develop within the first years of formal schooling requires the cobbling together of results from many different studies using different methods, measures, and ways of assessing conceptual knowledge. What is needed is to investigate these concepts within one single study for a more direct comparison of how concepts relate and develop in conjunction with one another.

Investigating Multiple Concepts in Grades 1–3

As mentioned previously, few studies have investigated more than one or two concepts at a time. As conceptual knowledge is an integrated set of knowledge about operations and the relations among them, more research is needed that directly compares children's understanding of multiple concepts (Robinson et al., 2017, 2018). As discussed previously, Baroody et al. (2009) investigated identity, negation, and inversion within the same study as they proposed that identity and negation are important precursors of inversion. Robinson and colleagues have investigated inversion and associativity together as both of these concepts require paying attention to the entire problem in order to notice that there is an easier way to solve the problem by applying conceptual knowledge versus applying a rote left-to-right problem-solving procedure (Robinson & LeFevre, 2012) and have consistently found that the inversion concept is easier and acquired earlier than the associativity concept (Robinson & Dubé, 2009, 2013; Robinson et al., 2006). Commutativity and equivalence are similar concepts in that they involve the understanding that both sides of the equation should be equal. Despite often being investigated separately, most arithmetic concepts relate to other concepts in some way or another. Children with strong conceptual knowledge are children who understand the ways in which operations and numbers relate to one another and thereby are able to build a coherent, integrated domain of knowledge in mathematics (Schneider & Stern, 2009). In this section of the chapter, the results of a recent study assessing the six concepts described in the previous section are described. No previous research has investigated this age range of children on all of these concepts simultaneously.

The participants in the current study were children in Grades 1, 2, and 3. The study ran in the second and third month of the school year, so the children in Grade 1 had very little formal instruction in arithmetic before the study began. There were

38 children in Grade 1, 30 children in Grade 2, and 30 children in Grade 3 from one school in a mid-sized Canadian city. Children were presented with math problems on sheets of paper and asked to solve the problems without using paper or pencil. Children's answers were recorded, and after each problem, they were asked how they had solved the problem or, if they had difficulties while solving the problem, were asked how they were trying to solve the problems. Based on their verbal reports of their problem-solving procedures, children were classified as having applied conceptual knowledge or not to solve the problem.

On an identity problem such as $8 - 0$, children who were classified as using conceptual knowledge would usually report that the answer was the first number as when nothing is taken away, the first number remains unchanged. On a negation problem such as $8 - 8$, children using conceptual knowledge would usually report that when you take something away from itself, then there is nothing left. On a commutativity problem such as $8 + 2 = 2 + ?$, children credited with having conceptual knowledge would mention that they had "flipped" the problem or that if you had an 8 and a 2 on one side, then you also had to have an 8 and 2 on the other side. On an inversion problem such as $2 + 8 - 8$, children with conceptual knowledge applied the shortcut and would reference that the 8s canceled each other out or that when you both add and subtract the same number, then the first number stayed the same. On an associativity problem such as $2 + 8 - 6$, children who understood the concept would report that they had solved the subtraction part first and then added the result to the first number. On an equivalence problem such as $8 + 2 + 6 = 8 + ?$, children were credited with using their conceptual knowledge if they said they had added all of the numbers on the left side and then subtracted the 8 on the second side or if they said that the answer was the sum of the second and third numbers on the left side of the equation.

As can be seen in the figure below, there were clear grade-related differences as well as differences in conceptual knowledge across the six concepts that were assessed in this study. Across all six concepts, Grade 3 children applied their conceptual knowledge on almost 50% of the problems, closely followed by Grade 2 children who applied their knowledge on just over 40% of their problems, compared to only 20% of problems by Grade 1 children. Consistent with the research previously reported, identity and negation were quite well understood with understanding of 79 and 69%, respectively, but these results are weaker than those obtained by Baroody et al. (2009) using concrete problems. The third most understood concept was the inversion concept with overall understanding just under 33%. The remaining three concepts of commutativity, associativity, and equivalence had weak understanding with all having well under 20% understanding. It appears that in the early school years, understanding of arithmetic concepts leaves considerable room for improvement (Fig. 1).

When each grade is examined individually, the results are somewhat more heartening. Clearly, as seen in the figure, Grade 1 children are struggling on all concepts except for identity, and even then, understanding is just over 50%. There have not been as many investigations of children's understanding of concepts at the beginning of formal schooling as many studies have either focussed on the time before children

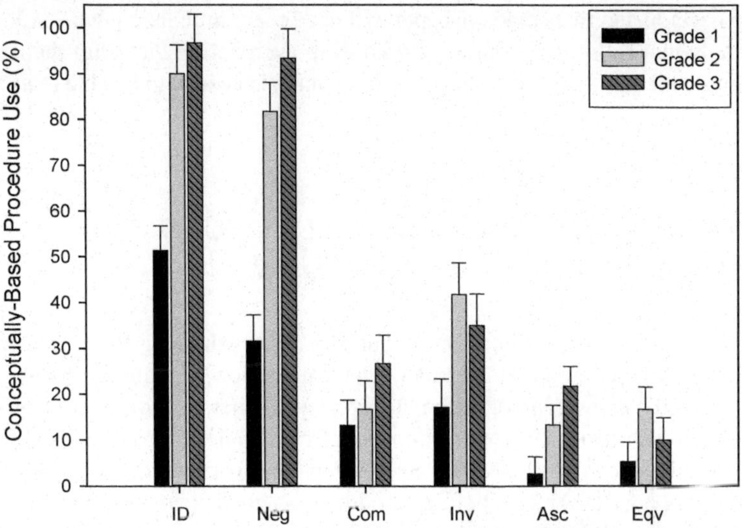

Fig. 1 Percentage use of conceptually based procedures by Grades 1, 2, and 3 children on six concept types

start school or after they have been in school for a few years. As these children were asked to solve these problems within the first few months of formal schooling, it is perhaps not surprising that they struggled as much as they did given that these Grade 1 children had little experience with symbolic arithmetic problem solving and had little time to learn basic arithmetic facts. Not only is their conceptual knowledge weak, but so is their factual and procedural knowledge. As all of these forms of knowledge are thought to develop iteratively, gains in one or more should lead to gains in the other (Rittle-Johnson, Siegler, & Alibali, 2001).

As can also be seen in the figure, by Grade 2, the concepts of identity and negation are well understood, which is more consistent with previous findings. The concept of inversion remains the third most understood concept in Grades 2 and 3, but commutativity, associativity, and equivalence remain very low. Only in Grade 3 does understanding of commutativity and associativity increase to over 20%. In all grades, the understanding of negation and identity is higher than that of inversion, which is compatible with Baroody et al.'s (2009) assertion that understanding of the first two concepts precedes the latter. The inversion concept was always stronger than the associativity concept which is the same pattern as found by Robinson and colleagues (2006), Robinson and Dubé (2009, 2013) and suggests that inversion may precede associativity. Even though understanding of both the commutativity and equivalence concepts was quite poor, understanding of commutativity was more than double than that of equivalence in Grades 1 and 3, but was virtually the same for both concepts in Grade 2. Therefore, there is some evidence that the concept of commutativity may precede that of equivalence, but the pattern is not completely clear. Overall, these results suggest that, within the early school years, these six concepts still need more time to develop and that some concepts will need more time than others. A longitudinal study that follows the same children from even before they start their formal schooling until they are well into their school years is needed to be able to ascertain this.

Current Knowledge About Children's Understanding of Arithmetic Concepts in the Early School Years

In the absence of a longitudinal study that investigates the long-term development of multiple concepts across many years, what can we conclude about what is known about children's knowledge of arithmetic concepts? First, arithmetic concepts matter. Children with strong conceptual knowledge of arithmetic are more likely to have stronger mathematical skills overall. Even when they are presented with novel problems that they have never seen before, children with good conceptual knowledge are able to apply what they know to come up with their own ways to solve these problems (Siegler & Stern, 1998) and so are not completely reliant on formal instruction.

Second, strong conceptual knowledge of basic arithmetic may be the foundation for success with more complex mathematics. For example, children with a strong understanding of the meaning of the equal sign at the beginning of Grade 5 achieved higher scores on an algebra test at the end of the school year (Fyfe, Matthews, Amsel, McEldoon, & McNeil, 2017).

Third, children often have a markedly weak understanding of arithmetic concepts. The results of the study with Grade 1–3 children presented in this chapter highlight that arithmetic concepts are often difficult and take time to develop. This leads to the fourth point that more instruction is needed to promote the development of the conceptual knowledge of arithmetic. Classroom mathematics tends to focus on procedural and factual knowledge but, as many recent mathematics curricula highlight, there is a need to spend instructional time on assessing and enhancing conceptual knowledge of arithmetic (Nunes et al., 2008). Even though some children seem to acquire conceptual knowledge without the need for prolonged formal instruction (Siegler & Stern, 1998), many children need help or scaffolding in order to achieve a good understanding of arithmetic operations and the relations among them.

Promoting Conceptual Knowledge

What are some methods for parents or teachers to increase children's understanding of arithmetic and increasing a child's chance of success with later, more complex mathematical skills? Four possible approaches are proposed here. First, the most obvious approach is to teach children about the arithmetic concepts. For example, inversion, associativity, and equivalence are often not taught to children. In the USA, equivalence has been specifically included in mathematics curricula, but in Canada, many provincial mathematics curricula do not specifically address the concept of equivalence. From a researcher perspective, that inversion, associativity, and sometimes equivalence problems are novel problems to children is a major advantage. If children have never seen these problems before, it means that they have never been taught a problem-solving procedure to solve them. Children who then make use of the inverse or associative relation between addition and subtraction or make both sides of an equation equal are doing so because they know the underlying arithmetic concept instead of having been taught that "when you see X problem, apply Y procedure" without understanding the meaning or concept underlying the procedure. However, procedures can be taught with meaning and getting students to reflect on why a problem-solving procedure works increases conceptual knowledge (NMAP, 2008).

A second approach to increasing conceptual knowledge is encouraging flexibility in children's problem-solving procedures (Newton, Star, & Lynch, 2010). For most arithmetic problems, there are several ways to solve a problem and get the correct answer (Robinson, 2001; Siegler, 1987). However, children often get used to solving problems a certain way that becomes increasingly familiar and practiced over time. Just as arithmetic facts become encoded into long-term memory and are quickly and easily accessed, so too can problem-solving procedures. As discussed near the beginning of the chapter, some problem-solving procedures are more appropriate than others, depending on the task demands. Much of the time, children solve problems in the same way that they read from left to right. Most arithmetic problems in textbooks and worksheets are, when presented in a horizontal format, presented from left to right. For example, in few instances would children encounter problems in the following format: $? = 3 + 8$ (McNeil et al., 2006). Instead, they see the math problem on the left and the answer belongs on the right.

After several years of seeing problems in that left-to-right format, children become used to solving them using a left-to-right approach. On an inversion problem such as $4 + 29 - 29$ or an associativity problem such as $43 + 867 - 865$, applying a left-to-right approach is going to result in a slow, cumbersome, and error-prone process. When children are given demonstrations of how they could solve these problems by dealing with the right side of the problem first (i.e., $29 - 29$ or $867 - 865$), some children are astounded and delighted by the cleverness of this approach while other children reject this approach as a way of "cheating." (Robinson & Dubé, 2012). These children often follow up this allegation of cheating by asserting that arithmetic problems are always supposed to be solved from left to right. So, some children are flexible and some children are inflexible when solving arithmetic problems. Teachers and parents can encourage problem-solving flexibility by encouraging children to come up with several different ways of correctly solving a problem and also by not encouraging children to think that there is only one "proper" way to solve a problem. This is an approach that has been successfully used in Asian classrooms for many years (Stevenson et al., 1990) and may be at least part of the reason for why Asian children tend to rank among the best in the world in mathematics (OECD, 2016). This approach also is the foundation of the book *Children's mathematics: Cognitively based instruction* (Carpenter, Fennema, Loef Frank, Levi, & Empson, 2014) which is based on many years of mathematical education research and supported by online videos developed for teachers and parents who are trying to improve children's mathematical knowledge and understanding.

Third, and related to both teaching problem-solving procedures and encouraging flexibility, is the idea of mixing problem formats. McNeil (2014) proposed that one of the reasons that children have such difficulties with equivalence problems is that children become entrenched in their thinking. This change-resistance account posits that over time, children become so used to seeing two-term problems presented in the same format or pattern which they then solve by using the same left-to-right algorithmic approach, that they form a misconception of the equal sign. The equal sign becomes associated with something needs to be done, i.e., as an operation to be performed. So, children who suddenly encounter problems such as 5 + 8 + 7 = 5 + ? think that the equal sign means that they need to add all of the numbers and therefore the answer should be 5 + 7 + 8 +5. Indeed, this misconception is so strong that even attempts to correct this misunderstanding of the equal sign have often been unsuccessful (e.g., Alibali, Philips, & Fischer, 2009) as children are unable to transfer the instruction they receive to new or different problems. So, even if children are taught how to solve the problem and also encouraged to come up with different problem-solving procedures, they are often so used to seeing problems in a certain format that they are unable to overcome a strongly entrenched way of solving problems. Instead, if children are regularly presented with problems in non-traditional formats such as 3 + ? = 8, ? + 3 = 8, 8 = 3 + ?, or 8 = ? + 3?, this may prevent them from becoming change-resistant as they will not have had the chance to form a misconception such as what an equal sign means.

Another way of mixing up formats is to also present children with problems that have more than two terms or have more than one operation. They will need to be able to solve problems with these characteristics when they are introduced to algebra, and seeing these "different" problems earlier may remove some of the novelty of algebra problems when they are exposed to them. In many of my studies, children are startled when they see an inversion, associativity, or equivalence problem, and the basis for their surprise appears to be due to either the problem having more than two numbers and/or the inversion and associativity problems including more than one operation. Disappointingly, it is not uncommon for children to tell me that they do not know how to solve that kind of problem because they have never seen them before and then choose to skip the question and move on to the next problem. Overall, increasing children's flexibility in their problem-solving strategies may not only be about teaching children to consider more than one problem-solving strategy, but also being presented with flexibly formatted problems so that they have even less opportunity to develop a "this is the right way and the only way" attitude to problem solving.

Finally, and also related to the above points, is the notion of attention. Conceptual knowledge is more likely to be applied to inversion problems such as 85 + 23 − 23 than it is to associativity problems such as 17 + 98 − 95 (Gilmore & Bryant, 2006). The inversion concept is probably easier for children to understand as it builds logically on the identity and negation concepts, but the associativity concept does not have similarly easy concepts to build on. However, the inversion concept may also be easier to apply during problem solving as even a quick glance will show that there are two numbers that are the same. Indeed, when inversion problems are mixed up and presented as $23 + 8 - 23$, application of the inversion concept decreases. This finding supports the ideas of flexibility and entrenchment, but it also supports the idea of encouraging children to pay attention to the whole problem before applying a problem-solving procedure. This lack of attention, combined with entrenchment, may be the reason why children and adults do not apply the associativity concept when solving associativity problems because they simply do not notice that there is an easier way to solve the problem (Robinson & LeFevre, 2012). In studies where children are given a demonstration of how they can apply the associativity concept in problem solving, many children are amazed and delighted when their attention is drawn to the right side of the problem. Dubé and Robinson (2010) found that when adults' attention was drawn to the right side of the problem by having the problem appear on a computer screen one symbol at a time going from right to left versus left to right, participants were more likely to apply both the inversion and associativity concepts in their problem-solving procedures. Thus, getting children to look at the whole problem before choosing and implementing a problem-solving procedure may promote conceptual thinking during problem solving.

Arithmetic concepts are an integral part of children's mathematical knowledge, and increasing this understanding is considered critical both for current and future success in mathematics. As the current research shows, there is significant room for improvement about what children know about numbers, operations, and the relations among them. There is also still a need for more research on arithmetic concepts in order to determine which concepts develop when, which concepts are dependent on each other, and which concepts might need the most instruction. In the meantime, there are several straightforward ways in which teachers and parents can easily enhance the development of conceptual knowledge of arithmetic in their students and children. Such efforts in the early school years have the potential to have long-lasting impacts on children's performance in the domain of mathematics.

References

Alibali, M. W., Knuth, E. J., Hattikudur, S., McNeil, N. M., & Stephens, A. C. (2007). A longitudinal examination of middle school students' understanding of the equal sign and equivalent equations. *Mathematical Thinking and Learning, 9,* 221–247. https://doi.org/10.1080/10986060701360902.

Alibali, M. W., Phillips, K. M. O., & Fischer, A. D. (2009). Learning new problem-solving strategies leads to changes in problem representation. *Cognitive Development, 24,* 89–101. https://doi.org/10.1016/j.cogdev.2008.12.005.

Asquith, P., Stephens, A. C., Knuth, E. J., & Alibali, M. W. (2007). Middle school mathematics teachers' knowledge of students' understanding of core algebraic concepts: Equal sign and variable. *Mathematical Thinking and Learning, 9,* 249–272. https://doi.org/10.1080/10986060701360910.

Baroody, A. J., & Gannon, K. E. (1984). The development of commutativity principle and economical addition strategies. *Cognition and Instruction, 1,* 321–329.

Baroody, A. J., Lai, M., Li, X., & Baroody, A. E. (2009). Preschoolers' understanding of subtraction-related principles. *Mathematical Thinking and Learning, 11,* 41–60. https://doi.org/10.1080/10986060802583956.

Bisanz, J., & LeFevre, J. (1990). Strategic and nonstrategic processing in the development of mathematical cognition. In D. F. Bjorklund (Ed.), *Children's strategies: Contemporary views of cognitive development* (pp. 213–244). Hillsdale, NJ: Erlbaum.

Canobi, K. H., Reeve, R. A., & Pattison, P. E. (1998). The role of conceptual understanding in children's addition problem solving. *Developmental Psychology, 34,* 882–891. https://doi.org/10.1037/0012-1649.34.5.882.

Canobi, K. H., Reeve, R. A., & Pattison, P. E. (2002). Young children's understanding of addition concepts. *Educational Psychology, 22,* 513–532. https://doi.org/10.1080/01443410220000023608.

Canobi, K. H., Reeve, R. A., & Pattison, P. E. (2003). Patterns of knowledge in children's addition. *Developmental Psychology, 39,* 521–534. https://doi.org/10.1037/0012-1649.39.3.521.

Carpenter, T. P., Fennema, E., Loef Franke, M., Levi, L., & Empson, S. B. (2014). *Children's mathematics: Cognitively guided instruction* (2nd ed.). Portsmouth, NH: Heinemann.

Crooks, N. M., & Alibali, M. W. (2014). Defining and measuring conceptual knowledge in mathematics. *Developmental Review, 34,* 344–377. https://doi.org/10.1016/j.dr.2014.10.001.

Dubé, A. K, & Robinson, K. M. (2010). The relationship between adults' conceptual understanding of inversion and associativity. *Canadian Journal of Experimental Psychology, 64,* 60–66. https://doi.org/10.1037/a0017756.

Falkner, K. P., Levi, L., & Carpenter, T. P. (1999). Children's understanding of equality: A foundation for algebra. *Teaching Children Mathematics, 6,* 232–236.

Fyfe, E. R., Matthews, P. G., Amsel, E., McEldoon, K. L., & McNeil, N. M. (2017). Assessing formal knowledge of math equivalence among algebra and pre-algebra students. *Journal of Educational Psychology.* https://doi.org/10.1037/edu0000208.

Gelman, S., & Gallistel, C. R. (1978). *The child's understanding of number.* Cambridge, MA: Harvard University Press.

Gilmore, C. K., & Bryant, P. (2006). Individual differences in children's understanding of inversion and arithmetical skill. *British Journal of Educational Psychology, 76,* 309–331. https://doi.org/10.1348/000709905X39125.

Gilmore, C., & Bryant, P. (2008). Can children construct inverse relations in arithmetic? Evidence for individual differences in the development of conceptual understanding and computational skills. *British Journal of Developmental Psychology, 26,* 301–316. https://doi.org/10.1348/026151007X236007.

Ginsburg, H. P., Cannon, J., Eisenband, J., & Pappas, S. (2006). Mathematical thinking and learning. In K. McCartney & D. Philips (Eds.), *The handbook of early child development* (pp. 208–229). Malden, MA: Blackwell.

Groen, G. J., & Resnick, L. B. (1977). Can preschool children invent addition algorithms? *Journal of Educational Psychology, 69,* 645–652.

Hiebert, J., & Lefevre, P. (1986). Procedural and conceptual knowledge. In J. Hiebert (Ed.), *Conceptual and procedural knowledge: The case of mathematics* (pp. 1–27). Hillsdale, NJ: Lawrence.

Hudson, C., Price, D., & Gross, J. (2009). *The long term costs of numeracy difficulties.* London, UK: Every Child a Chance Trust.

Kilpatrick, J., Swafford, J., & Findell, B. (2001). *Adding it up: Helping children learn mathematics.* Washington, DC: National Academy Press.

Klein, J. S., & Bisanz, J. (2000). Preschoolers doing arithmetic: The concepts are willing but the working memory is weak. *Canadian Journal of Experimental Psychology, 54,* 105–116. https://doi.org/10.1037/h0087333.

LeFevre, J., Sadesky, G. S., & Bisanz, J. (1996). Selection of procedures in mental addition: Reassessing the problem size effect in adults. *Journal of Experimental Psychology. Learning, Memory, and Cognition, 22,* 216–230.

McNeil, N. M. (2014). A change-resistance account of children's difficulties understanding mathematical equivalence. *Child Development Perspectives, 8,* 42–47. https://doi.org/10.1111/cdep.12062.

McNeil, N. M., & Alibali, M. W. (2005). Why won't you change your mind: Knowledge of operational patterns hinders learning and performance on equations. *Child Development, 76,* 883–889. https://doi.org/10.1111/j/1467-8624.2005.00884.x.

McNeil, N. M., Grandau, L., Knuth, E. J., Alibali, M. W., Stephens, A. C., Hattikudur, S., et al. (2006). Middle-school students' understanding of the equal sign: The books they read can't help. *Cognition and Instruction, 24,* 367–385. https://doi.org/10.1207/s1532690xci2403_3.

National Council of Teachers of Mathematics. (2000). *Principles and standards for school mathematics.* Reston, VA: NCTM.

National Council of Teachers of Mathematics, & National Association for the Education of Young Children. (1999). *Mathematics in the early years.* Reston, VA: NCTM.

National Governors' Association Center for Best Practices, & Council of Chief State School Officers. (2010). *Common core state standards for mathematics.* www.corestandards.org/wp-content/uploads/Math_Standards1.pdf.

National Mathematics Advisory Panel. (2008). *Foundations for success: The final report of the National Mathematics Advisory Panel.* Washington, D.C.: U.S. Department of Education.

Newton, K. J., Star, J. R., & Lynch, K. (2010). Understanding the flexibility of struggling algebra students. *Mathematical Thinking and Learning, 12,* 282–305. https://doi.org/10.1080/10986065.2010.482150.

Nunes, T., & Bryant, P. (1998). *Children doing mathematics.* Cambridge, MA: Blackwell.

Nunes, T., Bryant, P., Burman, D., Bell, D., Evans, D., Hallett, D., et al. (2008). Deaf children's understanding of inverse relationships. In M. Marschark & P. C. Hauser (Eds.), *Deaf cognition: Foundations and outcomes* (pp. 201–225). Oxford: New York, NY.

OECD. (2016). *PISA 2015 results (Volume 1): Excellence and equity in education.* Paris: PISA, OECD Publishing. https://doi.org/10.1787/9789264266490-en.

Perry, M., Church, R. B., & Goldin-Meadow, S. (1988). Transitional knowledge in the acquisition of concepts. *Cognitive Development, 3,* 359–400.

Piaget, J. (1952). *The child's conception of number.* London, UK: Routledge & Kegan Paul.

Prather, R. W., & Alibali, M. W. (2009). The development of arithmetic principle knowledge: How do we know what learners know? *Developmental Review, 29,* 221–248. https://doi.org/10.1016/j.dr.2009.09.001.

Rasmussen, C., Ho, E., & Bisanz, J. (2003). Use of the mathematical principle of inversion in young children. *Journal of Experimental Child Psychology, 85,* 89–102. https://doi.org/10.1016/S0022-0965(03)0031-6.

Rittle-Johnson, B., & Alibali, M. W. (1999). Conceptual and procedural knowledge of mathematics: Does one lead to the other? *Journal of Educational Psychology, 91,* 175–189. https://doi.org/10.1037/0022-0663.91.1.175.

Rittle-Johnson, B., Schneider, M., & Star, J. R. (2015). Not a one-way street: Bidirectional relations between procedural and conceptual knowledge in mathematics. *Educational Psychology Review, 27,* 587–597. https://doi.org/10.1007/s10648-015-9302-x.

Rittle-Johnson, B., & Siegler, R. S. (1998). The relation between conceptual and procedural knowledge in learning mathematics: A review. In C. Donlan (Ed.), *The development of mathematical skills: Studies in developmental psychology* (pp. 75–101). Hove, UK: Psychology Press.

Rittle-Johnson, B., Siegler, R. S., & Alibali, M. W. (2001). Developing conceptual understanding and procedural skill in mathematics: An iterative process. *Journal of Educational Psychology, 93,* 346–362. https://doi.org/10.1037/0022-0663.93.2.346.

Robinson, K. M. (2001). The validity of verbal reports in children's subtraction. *Journal of Educational Psychology, 93,* 211–222. https://doi.org/10.1037/0022-0663.93.1.211.

Robinson, K. M., & Dubé, A. K. (2009). Children's understanding of addition and subtraction concepts. *Journal of Experimental Child Psychology, 103,* 532–545. https://doi.org/10.1016/j.jecp.2008.12.002.

Robinson, K. M., & Dubé, A. K. (2012). Children's use of arithmetic shortcuts: The role of attitudes in strategy choice. *Child Development Research,* 10. https://doi.org/10.1155/2012/459385.

Robinson, K. M., & Dubé, A. K. (2013). Children's additive concepts: Promoting understanding and the role of inhibition. *Learning and Individual Differences, 23,* 101–107. https://doi.org/10.1016/j.lindif.2012.07.016.

Robinson, K. M., Dubé, A. K., & Beatch, J.-A. (2017). Children's understanding of additive concepts. *Journal of Experimental Child Psychology, 156,* 16–28. https://doi.org/10.1016/j.jecp.2016.11.009.

Robinson, K. M., & LeFevre, J. (2012). The inverse relation between multiplication and division: Concepts, procedures, and a cognitive framework. *Educational Studies in Mathematics, 79,* 409–428. https://doi.org/10.1007/s10649-011-9330-5.

Robinson, K. M., Ninowski, J. E., & Gray, M. L. (2006). Children's understanding of the arithmetic concepts of inversion and associativity. *Journal of Experimental Child Psychology, 94,* 349–362. https://doi.org/10.1016/j.jecp.2006.03.004.

Robinson, K. M., Price, J. A. B., & Demyen, B. (2018). *Understanding of arithmetic concepts: Does operation matter?* Manuscript under review.

Schneider, M., Rittle-Johnson, B., & Star, J. R. (2011). Relations among conceptual knowledge, procedural knowledge, and procedural flexibility in two samples differing in prior knowledge. *Developmental Psychology, 47,* 1525–1538. https://doi.org/10.1037/a0024997.

Schneider, M., & Stern, E. (2009). The inverse relation of addition and subtraction: A knowledge integration perspective. *Mathematical Thinking and Learning, 11*(1), 92–101. https://doi.org/10.1080/10986060802584012.

Sherman, J. (2007, April). *From failure to success on equivalence problems and how teachers perceive the process.* Paper presented at the biennial conference for the Society for Research in Child Development, Boston, MA.

Sherman, J., & Bisanz, J. (2007). Evidence for use of mathematical inversion by three-year-old children. *Journal of Cognition and Development, 8,* 334–344. https://doi.org/10.1080/15248370701446798.

Sherman, J., & Bisanz, J. (2009). Equivalence in symbolic and nonsymbolic contexts: Benefits of solving problems with manipulatives. *Journal of Educational Psychology, 101,* 88–100. https://doi.org/10.1037/a0013156.

Siegler, R. S. (1987). The perils of averaging data over strategies: An example from children's addition. *Journal of Experimental Psychology: General, 116,* 250–264. https://doi.org/10.1037/0096-3445.116.3.250.

Siegler, R. S. (1996). *Emerging minds: The process of change in children's thinking.* New York: Oxford University Press.

Siegler, R. S., & Araya, R. (2005). A computational model of conscious and unconscious strategy discovery. In R. V. Kail (Ed.), *Advances in child development and behaviour* (Vol. 33, pp. 1–42). New York, NY: Elsevier.

Siegler, R. S., & Jenkins, E. (1989). *How children discover new strategies.* Hillsdale, NJ: Erlbaum.

Siegler, R. S., & Shipley, C. (1995). Variation, selection, and cognitive change. In T. Simon & G. Halford (Eds.), *Advances in child development and behaviour* (Vol. 16). New York: Academic Press.

Siegler, R. S., & Shrager, J. (1984). Strategy choice in addition: How do children know what to do? In C. Sophian (Ed.), *Origins of cognitive skills* (pp. 27–46). Hove, UK: Psychology Press.

Siegler, R. S., & Stern, E. (1998). Conscious and unconscious strategy discoveries: A microgenetic analysis. *Journal of Experimental Psychology: General, 127,* 377–397. https://doi.org/10.1037/0096-3445.127.4.377.

Starkey, P., & Gelman, R. (1982). The development of addition and subtraction abilities prior to formal schooling in arithmetic. In T. P. Carpenter, J. M. Moser, & T. A. Romberg (Eds.), *Addition and subtraction: A cognitive perspective* (pp. 99–116). Hillsdale, NJ: Erlbaum.

Stevenson, H. W., Lee, S., Chen, C., Stigler, J. W., Hsu, C.-C, Kitamura, S., & Hatano, G. (1990). Contexts of achievement: A study of American, Chinese, and Japanese children. *Monographs of the Society for Research in Child Development, 55*(1–2, 221), 123. https://doi.org/10.2307/1166090.

Vilette, B. (2002). Do young children grasp the inversion relationship between addition and subtraction? Evidence against early arithmetic. *Cognitive Development, 17,* 1365–1383. https://doi.org/10.1016/S0885-2014%2802%2900125-9.

Watchorn, R. P. D., Bisanz, J., Fast, L., LeFevre, J., Skwarchuk, S.-L., & Smith-Chant, B. L. (2014). Development of mathematical knowledge in young children: Attentional skill and the use of inversion. *Journal of Cognition and Development, 15,* 161–180. https://doi.org/10.1080/15248372.2012.742899.

An Integrated Approach to Mathematics and Language Theory and Pedagogy

José Manuel Martínez

To explore the relationship between mathematics and language in bilingual classrooms, I begin with a vignette from a third-grade Spanish immersion class in the Midwest region of the USA. I use *bilingual* as an umbrella term to refer to contexts where the teacher and the students communicate in more than one language. All students' native language was English, and the language of instruction was Spanish. During a whole-class discussion conducted in Spanish, the class was determining the number of squares in a 25×8 grid. After a few students had suggested different strategies that involved counting all the squares, one student suggested counting the squares in the top row and adding that number eight times. This student went to the board and, counting one by one, stated that there were 25^2 in each row and that now they needed to add.

To elicit the usefulness of multiplication in this strategy, the teacher asked what would be an efficient way to figure out the total number of squares if they know there are eight rows. Daniel (all names are pseudonyms) responded by saying "Podemos dividir" (*we can divide*). His suggestion immediately received a loud rejection from many of his classmates, several of whom talked simultaneously telling Daniel they needed to multiply. Modeling the mathematical practice of considering others' thinking, the teacher asked Daniel to show what he meant. Daniel went to the board where the teacher had projected the grid. Moving his finger along the vertical line that had 20 columns to the left and five to the right, he explained: "Podemos dividir esto [the grid] en veinte y cinco, y ocho por cinco es cuarenta, y sumar ocho por vente. Esto es más fácil que ocho por veinticinco" (*We can divide this [the grid] in twenty and five, and eight times five is forty, and add eight times twenty. That's easier than eight times twenty-five*). Daniel was using the word "divide" in the informal everyday sense of breaking into smaller components. He thought of splitting the 25×8 problem into what he thought was the easier $5 \times 8 + 20 \times 8$ problem.

J. M. Martínez (✉)
Michigan State University, 300F Erickson Hall, 620 Farm Lane,
East Lansing, MI 48824, USA
e-mail: mart1580@msu.edu

K. M. Robinson et al. (eds.), *Mathematical Learning and Cognition in Early Childhood*,
https://doi.org/10.1007/978-3-030-12895-1_11

One interpretation of the relationship between mathematics and language in this vignette is that mathematics and language are two separate components of the class discussion. Following this interpretation, it would be reasonable to say that Daniel needed more or different language learning opportunities before being able to communicate the mathematical idea he wanted to share. This approach could focus, for example, on opportunities to develop Daniel's understanding of the word *divide* in mathematics as referring to the arithmetic operation of division. In this case, Daniel would need to know the problem called for multiplication and that he could use the distributive property to solve the problem. He may avoid using the informal, out-of-school connotation of the word divide and instead would use the formal mathematical connotation.

This interpretation has problematic implications for teaching and learning. This is the typical challenge of "pull-out" programs (Honigsfeld, 2009) in which non-native speakers of the language of instruction miss regular content classes to take language classes. This format continues until students are supposed to have developed sufficient skills in the language of instruction. Assuming that students need to learn the language first to then learn mathematics focuses students' schooling on language learning, delaying mathematics learning opportunities (Moschkovich, 2013). Many of the so-called English language learners in the USA, students whose native language is not English and who attend schools where English is the language of instruction, are segregated in these pull-out programs (García & Kleifgen, 2010).

An alternative and perhaps more compelling interpretation is that mathematics and language are interrelated and inseparable and, thus, using language to communicate about mathematical ideas (re)shapes ideas and vice versa (Brown, 2002). Following this interpretation, it would be reasonable to say that there were integrated elements of language use and mathematical reasoning in Daniel's contributions. This interpretation would argue that through communication Daniel simultaneously engaged in the second language use and in mathematics, thus avoiding the drawbacks and problematic assumptions discussed above. This integrated perspective implies providing bilingual students the same mathematics learning opportunities their monolingual peers have, while providing opportunities to learn the language.

The purpose of this chapter is to articulate a theoretical framework from psychology about mathematics education (situated learning) and one from sociolinguistics about the second language education (communicative language teaching) to understand mathematics–language integration. I draw on examples from research in an enrichment language immersion classroom. I follow Brisk's(2011) definition of language immersion classroom as one where students are native users of the community's dominant language and the language of instruction (in this case Spanish) is a second or additional language. I illustrate how pedagogical practices consistent with a situated perspective on mathematics education provide opportunities to engage with the second language. Simultaneously, pedagogical practices consistent with a communicative perspective on the second language education provide opportunities to engage in mathematical activity.

The chapter is organized into two main parts. First, I discuss points of convergence between research that has examined mathematics teaching in bilingual classrooms and research that has examined the second language teaching in mathematics classrooms. Based on this discussion, I propose a unified theoretical framework of content–language integration that is specific to mathematics. In the second part, I draw on examples from a classroom to illustrate the types of analysis that the integrated framework facilitates.

Conceptualizing the Mathematics–Language Integrated Phenomenon

Mathematics and Language as Related but Separated

The notion that articulating an idea helps further develop the idea and helps further develop language (Halliday, 1978) has informed research in mathematics education in general. For example, Pimm (1987) has argued that contrary to popular opinion, mathematics is not a language-free discipline and language use in mathematics is different from general language use. Similarly, Brown (2002) has drawn on a post-structuralist view to challenge dichotomizations of language and mathematics as separate. This line of research has highlighted that language plays a crucial role in mathematical activity. These views have informed research in bilingual classrooms. Previous research has made visible the challenges and tensions that teachers and students face in bilingual classrooms (Adler, 1998; Barwell, 2005; Moschkovich, 2007; Planas & Civil, 2013; Planas & Setati-Phakeng, 2014). This body of literature has called for pedagogies that meet bilingual mathematics learners' specific needs.

Despite these important contributions, recommendations for bilingual mathematics classrooms have tended to conceptualize mathematics and language as separated (Nikula, Dalton-Puffer, Llinares & Lorenzo, 2016). Specifically, the relationship between the two has been portrayed as linear, arguing for students to master specific language structures before engaging in meaningful mathematical activity. For example, Cambridge ESOL (2010) suggests that teachers identify the vocabulary that students will use in a mathematical unit, so that they can teach those linguistic skills before students engage in mathematics tasks. They give the example of graphs, recommending that the teacher uses crosswords and fill in the blanks exercises for students to learn words such as plot, slope, and axes. Similarly, Smit, Bakker, van Eerde and Kuijpers (2016) argued for identifying the linguistic features of a mathematical concept and teaching them to students. They, too, focused on graphs, suggesting that teachers teach grammar structures such as "I'll plot the line" so that students are not confused when encountering this language in the mathematical unit. This framing implies a separation between language and mathematics and the subordination of language to the ultimate purpose of mathematics teaching and learning.

The search for bilingual pedagogies that integrate the content and second language has been housed outside of mathematics education research. This research has proposed models such as Content and Language Integrated Learning (CLIL) and the Sheltered Instruction Observation Protocol (SIOP) (Echevarria, Vogt, & Short, 2004; Kareva, 2013; Mehisto, Frigols, & Marsh, 2008; Snow, Met, & Genesee, 1989). These models are motivated by the need to structure teachers' efforts to simultaneously teach a language and a subject matter. Rather than being content-specific, however, these programs are generic. Proponents argue that these models are flexible enough to be applicable to the different content areas that sometimes one single teacher covers. The lack of content-specific recommendations, however, tasks teachers with figuring out the under-researched connections between the content and second language. Very few studies have examined these models of integrated content and language in mathematics education (for exceptions see Barwell, 2005; Chval, Pinnow, & Thomas, 2015; Novotna & Hofmannova, 2000).

To summarize, research on mathematics and language education in bilingual classrooms rarely maintains the interdisciplinary and interconnected nature of this phenomenon. Instead, research has tended to focus on language learning as a prerequisite for mathematics learning. Moreover, each of these two bodies of work is disseminated in its corresponding field, with few opportunities for interdisciplinary analysis. Seeking to address this issue, I simultaneously draw on a theory of mathematics education from psychology (situated learning) and a theory of the second language education from sociolinguistics (communicative language teaching).

An Integrated Theoretical Framework

Situated learning has been extensively used in mathematics education (Boaler, 2000; Greeno, 1991; Moschkovich, 2002; Sfard, 1998). This theory emerged as an alternative to frameworks that overemphasize cognition and the individual (Cobb & Bowers, 1999; Sfard, 1998). Other theories such as cognitivism regard learning as an individual cognitive process of constructing knowledge. Instead, situated learning theory considers learning both individual and social, as learners interact with others to be apprenticed into the ways of participating and using knowledge in a specific community of practice (Greeno, 1998; Lave & Wenger, 1991).

Similarly, communicative language teaching emerged as both a theoretical and pedagogical approach that offered an alternative to frameworks that overemphasize knowledge about language (Brumfit, 1988; Littlewood, 2014). Views of language as an existing system of grammar structures, vocabulary, and phonemic rules focus on individuals' identification of these structures and rules. Accompanying teaching approaches through which individuals got repetitive practice of the structures explicitly shown includes the audiolingual and the direct translation methods (Larsen Freeman, 2011). The purpose of repetition is for the learner to gain automaticity in recognizing and reproducing these structures. Conversely, communicative language teaching is concerned with the need to use language to communicate (Richards, 2006). The focus is on language use naturally occurring in interactions, paying attention to language functions, that is, what people do with language in daily situations.

I conceptualize situated learning and communicative language teaching as both compatible and complementary. They are compatible in that both prioritize interaction to use knowledge in tasks and situations relevant to a community. At the same time, these two theories have complementary foci. Situated learning focuses on how individuals interact around mathematical activity in a particular context. Communicative language teaching focuses on how individuals interact in a second language during meaningful tasks. Thus, the integration of situated learning and communicative language teaching helps attend to mathematical activity and language use simultaneously.

I focus on three practices that are common among pedagogies consistent with situated learning perspectives and with communicative language teaching: (1) students' autonomous exploration, (2) authentic communication, and (3) balance between fluency and accuracy. First, both frameworks highlight the importance of *students' autonomous exploration* as a means to engage in relevant disciplinary practices and in meaningful situations. Autonomous exploration involves interactions among students to communicate ideas around a particular situation. Through this communication, mathematical ideas emerge and develop. For example, from the point of view of situated learning, Greeno (1998) has described the role that tasks contextualized in situations relevant to students play in eliciting learners' autonomous exploration of mathematical ideas. Similarly, from the point of view of communicative language teaching, Larsen Freeman (2011) has provided examples of language development through student independent engagement with a situation that requires purposeful communication. Therefore, an integrated framework to analyze the inseparable mathematics–language phenomenon pays attention to students' autonomous engagement with tasks in which relevant mathematical and linguistic practices emerge and develop.

A second pedagogical practice common to both frameworks is the focus on *authentic communication*. Students' autonomous exploration entails naturally occurring communication that typically involves students, the teacher, and textbooks. From the point of view of situated learning, communication is inherent in the process of learning to participate in a community of practice (Lave & Wenger, 1991). Some of the practices regarded as important in mathematical activity include making claims, hypothesizing, justifying, making arguments, and considering others' thinking (Greeno, 1991; Sfard, 1998; Sfard, Nesher, Streefland, Cobb, & Mason, 1998). Students have opportunities to learn how these practices are enacted in mathematical activity when they engage in communication. From the point of view of communicative language teaching, communication is at the center of the second language learning. In this approach, language functions drive the learning process (Larsen Freeman, 2011). Language functions refer to the purposes for which people use language, such as to greet, to apologize, or to make arrangements. The mathematical practices mentioned above correspond to language functions. Accordingly, an integrated framework considers authentic communication as constitutive of both mathematics and language teaching and learning.

Finally, a common consideration of both situated learning and communicative language teaching is the *balance between fluency and accuracy*. A situated learning perspective analyzes the practices in which students engage that are valued by the community of practice (Greeno, 1991). Currently, mathematics educators regard both fluency and accuracy as important. The National Council of Teachers of Mathematics (2000) states that "computational fluency refers to having efficient and accurate methods for computing" (p. 152). That is, the ultimate goal of mathematics education is not accuracy exclusively, but the flexible and appropriate use of diverse methods in mathematical activity.

From the perspective of communicative language teaching, fluency refers to language use in a way that allows the smooth flow of communication, while accuracy refers to correct word use and production of grammatical structures (Lightbown & Spada, 2006). Effective communication requires *sufficient* fluency and accuracy. That is, communication is not effective when it is interrupted and unnaturally slow for the sake of accuracy, or when it happens at a natural pace, but it is incomprehensible due to frequent or major errors. For the purpose of devising an integrated framework, accuracy and fluency in language use are accuracy and fluency in mathematical activity and vice versa. Teaching and learning involve a balance between fluency and accuracy in the co-constitutive phenomenon of mathematics–language.

These three pedagogical practices of an integrated approach to mathematics and language have implications for teachers. Under this framework, teachers' roles are in opposition to views of teachers as transmitters of knowledge. Both situated learning (Sfard, 2001) and communicative language teaching (Richards, 2006) acknowledge the importance of these roles. Specifically, in a classroom where teachers are aware of the interaction between language and mathematics, teachers act as facilitators of discussions among students. To foster these discussions, teachers also take on the role of designing tasks that promote autonomous exploration of mathematically and linguistically rich situations that are relevant to students. For this exploration to take place, teachers foster a class environment in which students are comfortable taking risks, making mistakes, and exploring ideas. In the second part of the chapter, I exemplify how the integrated framework can guide analyses of the interrelationship between mathematics and language.

Classroom Example

This third-grade classroom, located in the Midwest region of the USA, followed a language enrichment full immersion model (Brisk, 2011) with Spanish as the language of instruction. The teacher was Señora Abad, a US-born Latina who considered both English and Spanish her native languages. Her background was in Spanish–English bilingual education, and she had had four years of teaching experience, all at this school. There were 23 students (ages eight to nine) in this classroom, including 9 female students (1 Latina, 1 African American, 7 White) and 14 male students (2 African American, 12 White). All students had attended this language immersion program since kindergarten, except for one student (White female) who had started during first grade. All students were English native speakers and were comfortable communicating in Spanish.

I had worked with Señora Abad for two years as participant–researcher (Dreyer, 1998, 2016), using interpretive ethnographic data generation methods such as classroom observations, lesson video recording, and field note writing (Erickson, 2006, 2012; Spradley, 2016). My involvement included co-planning and co-teaching mathematics lessons with Señora Abad once a week for most of the school year (October 2015 to April of 2016) and every day during a teaching unit on geometry (May 2016), and on number sense and problem solving (June 2016).

The Lesson

The examples I present come from the first lesson of a geometry unit. The teacher and I adapted Mack's (2007) task to elicit students' understanding of two-dimensional shapes and to explore the relationship between a shape's name and its attributes. There were five parts in this lesson. The first part was a whole-class discussion that I facilitated and in which students answered the question "What do you know about geometry?" During the second part, each student received the image of a two-dimensional shape (i.e. a triangle, square, pentagon, hexagon, or circle). We asked students to find classmates whose shapes were in the same category, without telling students what we meant by "same category."

In the third part of the lesson, we asked students to work within their category groups deciding what the name of the category was and preparing a description to share with the rest of the class. In the fourth part, each group described their shape to the whole class. In the final part of the lesson, each student drew and named one shape they made up. In their small groups, they shared their creation and what the relationship was between the shape's attributes and its name. I first describe the integrated mathematical practices and language use that the tasks fostered and then I focus on one student, analyzing a student's interaction with the concept of hexagons throughout the lesson.

Mathematics and Language Throughout the Lesson

Students had several mathematics and language learning opportunities through the pedagogical practices highlighted in the integrated framework. These three pedagogical practices seem overlapping and mutually constitutive. The lesson fostered *students' autonomous exploration* of ideas as learners helped determine the direction of whole-class discussions based on their contributions. Additionally, during the second and third part of the lesson (making groups according to shapes, and naming and describing the shapes), students worked independently to figure out shapes' relevant attributes that made them belong to a category. This autonomous exploration required *authentic communication* for students to contribute and expand their ideas. Through unrehearsed conversation, students showed what they knew about shapes and showed possible directions to further their learning. One of those directions stemmed from the *balance between fluency and accuracy*. For example, students freely expressed some ideas, without the teacher or the researcher objecting, and at the same time, those ideas helped us see the need to develop more accurate definitions of particular shapes and categories.

One example of students' contributions guiding the lesson took place during the first whole-class discussion. As a geometry example, one student mentioned Rubik's Cubes which is a puzzle in the shape of a cube where each face is made up of nine individually rotatable squares of the same color. This idea resonated with several students who mentioned they had one or that they had a Pyramix, a puzzle similar to a Rubik's Cube but in the shape of a tetrahedron. The teacher asked what shapes students could find in a Rubik's Cube and in a Pyramix. Students mentioned cubes, squares, triangles, and what a student called "mixed-up shapes." When one of the students, Bill, said that Rubik's Cubes had hexagons, the following conversation unfolded. Italicized text in parenthesis is my translation to English of speech originally uttered in Spanish. Text in brackets describes non-verbal communication. Bold letters show speakers' emphasis.

Researcher	¿Quién puede dibujar un hexágono? (*Who can draw a hexagon?*)
Bill	¡Yo puedo! (*I can!*). [Goes to the board and draws an octagon.]
Researcher	¿Cómo sabes que es un hexágono? (*How do you know that's a hexagon?*)
Bill	[Erases the octagon and draws another octagon.]
Duke	**Esto** es un hexágono. (*This is a hexagon*). [Goes to the board and draws a hexagon.]
Researcher	¿Y cómo sabes que es un hexágono? (*And how do you know that's a hexagon?*)
Daniel	¿Porque tiene seis partes? (*Because it has six parts?*)
Researcher	¿Seis partes? (*Six parts?*)
Daniel	¡Puntos! ¡Hay seis puntos! (*Points! There are six points!*)
Josh	Lados. (*Sides.*)
Daniel	¡Lados! No. ¡Lados, lados! (*Sides! No. Sides, sides!*) [Gestures with his arm and hand stretched out as if representing a straight line.]

Analyses of this excerpt from the perspective of the independent theories, situated learning and communicative language teaching, could foreground either mathematics or language. Following a situated learning perspective would elicit an analysis of important mathematical practices that emerged during the lesson. In this example, those practices included naming geometric objects, using multiple representations to make thinking public and attending to others' reasoning. Following a communicative language teaching perspective would elicit an analysis of important language use. In the example, language use included naming and describing objects and images, repairing miscommunication and understanding who, why, and what type questions which require more than a yes or a no as a response. Focusing on situated learning or on communicative language could yield interpretations of which phenomenon, mathematics practices or language use, was driving the lesson. These interpretations, however, respond to the particular focus of each framework, artificially separating mathematics and language.

The simultaneous consideration that the integrated framework facilitated seems closer to the events as the students, the teacher, and the participant–researcher experienced them. In this example, participation in mathematical practices constituted language use and language use constituted participation in mathematical practices. Through participation in mathematical practices, we explored ideas about what a hexagon is and what sides are. This exploration involved language skills such as using vocabulary appropriately and accurately. Instead of differentiating or separating whether the lesson provided language or mathematical learning opportunities, the juxtaposition of the two merits analysis and constitutes the phenomenon of interest. I further develop this point by expanding on Bill's ideas about hexagons.

A Hexagon According to Bill

In this lesson, collective processes interacted with opportunities for individual sense making. Although situated learning theory and the communicative language teaching approach highlight the importance of the collective, these frameworks do not ignore individual experience. Instead, these frameworks contextualize individual experience in the collective. Accordingly, an integrated framework for mathematics and language should give account of the collective, the individual, and the interaction between the two. In the previous description, I analyzed the lesson from the point of view of the collective. I now focus on one particular student, Bill, and what the consideration of the intersection of his mathematical practices and language use suggests about his conceptualization of hexagons.

As described above, Bill mentioned the hexagon when the class started talking about the Rubik's Cube. In that case, the Rubik's Cube was an artifact that served as a springboard for both mathematical and linguistic exploration. Using artifacts from the community is a mathematics teaching move that helps students participate in multimodal mathematical interaction (O'Halloran, 2015). Using artifacts is also a pedagogical practice in the second language where an artifact acts as a focal tangible object that everybody can see or feel so that there is a common ground for communication (Lantolf, Thorne, & Poehner, 2015). Bill drew an octagon when asked to draw a hexagon. This could be interpreted as suggesting he needed learning opportunities to choose the accurate word in Spanish or to understand the mathematical concept of hexagon. Considering Bill's initial participation around the idea of "hexagon" from the perspective of the integrated framework, however, we know that he recognized "hexagon" as a word that could be used in a conversation about geometry. He also knew that the word "hexagon" had to do with shapes and attempted a representation of a hexagon. Additionally, Bill knew that his contribution fitted within the whole-class discussion, which is a situated practice in school mathematics and is a communicative activity in the second language classrooms. From this integrated perspective, Bill was using the language as part of his exploration of mathematical ideas.

During the second part of the lesson (finding classmates with similar shapes), Bill received a regular pentagon; that is, all angles measured the same, and all sides measured the same. He paired up with one classmate only, Judith, who had received a concave pentagon. The two students proceeded to look for others with shapes in their same category. Judith approached two of their classmates. As all four children looked at their shapes, Bill counted from one to six as he pointed at each of the sides of one of his classmate's shape. He stated: "No. Ellos tienen seis" (*No. They have six*), and he and Judith moved along looking for others. This interaction suggests that he knew that the number of sides is a relevant attribute to categorize a shape. He also knew a shape with five sides was not in the same category as a shape with six sides. He knew to express the relationships between sides and shapes using the verb *to have*.

In the third part of the lesson, Bill and Judith worked on naming and describing their shapes:

Judith	Es un… un… (*It's a… a…*)
Bill	Hexágono. (*Hexagon.*)
Judith	Hexágono. Y tiene cinco part… Cinco… (*Hexagon. And it has five par… Five…*)
Bill	Lados. (*Sides.*)
Judith	Cinco lados. Y es recta. (*Five sides. And it's straight.*)

This interaction reinforces the interpretation that Bill understood sides as relevant attributes of shapes. The support he provided for Judith to participate in this activity went beyond providing language for her to express her mathematical ideas. Their understanding of the concept at the moment was informed by the multi-semiotic juxtaposition of language and image (O'Halloran, 2005). So far, Bill had used the word hexagon to refer to shapes he could see in a Rubik's Cube and also in association with the image of an octagon and of a pentagon. He had also seen a six-sided shape and, without naming it, he had decided it did not belong in the same category as his.

When we transitioned to the fourth part of the lesson (each group sharing the name and description of their shape), Judith suggested starting with the group whose shape had the fewer number of sides:

Judith	Empezamos con los triángulos que tiene tres lados. Luego ellos que tiene cuatro. (*Let's start with the triangles that have three sides. Then them that have four*).
Duke	No. Círculos. Círculos no tiene **nada** de lados. (*No. Circles. Circles doesn't have any sides.*)
Researcher	¿Qué significa lado? (*What does side mean?*)
Grant	Línea. (*Line.*)
Researcher	¿Esto es una línea? (*Is this a line?*) [Drawing a curve on the board.]
Duke	No. Porque es como un círculo. (*No. Because it's like a circle.*)
Bill	Es como *curve*. (*It's like* curve.)

Researcher	¿Es curvo? (*It's curve?*)
Bill	Si. No es como recto. (*Yes. It's not like straight.*)
Duke	Esto no es figura. (*This isn't shape.*)
Bill	Ajá. Los otros son hexágonos y estos son círculos. (*Uh-huh. The other ones are hexagons and these are circles.*) [Pointing at the curve on the board.]

After the group with triangles and quadrilaterals presented their shapes, Judith announced her group was next:

Judith	¿Quién sigue? [Pause] Nosotros. ¿Qué es el nombre? (*Who's next? [Pause] We are. What's the name?*)
Bill	Hexágono. (*Hexagon.*)
Researcher	Un momento… ¿Tenemos aquí alguien que tenga pentágonos? (*Hold on… Do we have anybody with pentagons?*)
Judith	¿Quién tiene cinco? Tenemos cinco, cinco lados. (*Who has five? We have five, five sides.*)
Researcher	A ver, déjame ver. (*Here, let me see.*)
Judith	Cinco lados. Uno, dos, tres, cuatro, cinco. (*Five sides. One, two, three four, five.*) [Holding the shape for the class to see, she points at one side with each number she says]
Researcher	¡Oh! ¿Como se llama la figura de ustedes? (*Oh! What's the name of your shape?*)
Judith	[Shrugs]
Bill	Hexágono. (*Hexagon.*)
Ariel	Pentágono. (*Pentagon.*)
Judith	Hexágono. (*Hexagon.*)
Mike	¡Penta! ¡Penta! (*Penta! Penta!*)
Researcher	[Addressing the whole class] Tenemos una duda. Esta figura, unas personas opinan que es un **pentágono** y otras personas opinan que es un **hexágono**. ¿Qué piensas? (*We're not sure about something. This shape, some people think it's a pentagon and others think it's a hexagon. What do you think?*)
Leila	Pentágono. (*Pentagon.*)
Researcher	¿Por qué? (*Why?*)
Leila	Porque el pentágono tiene cinco lados. (*Because the pentagon has five sides.*)
Researcher	¿El pentágono tienen cinco lados? ¿Cómo sabes? (*The pentagon has five sides? How do you know?*)
Leila	**Pen**tagrama y **pen**tágono. Cinco líneas y cinco lados. (*Staff and pentagon. Five lines and five sides.*)
Researcher	¿Entonces el nombre de la figura y la figura tienen que ver? (*So, the name of the shape and the shape have something to do?*)
Milo	Si. **Tri** ángulo. **Tres** ángulo. **Pen** tagono. Tienen cinco lados y tienen cinco esquinas. (*Yes. Tri angle. Three angle. Pen tagon. They have five sides and they have five corners.*)

Bill Ohhh!

First, Bill made a distinction between circles and hexagons. After that, it is clear that Bill knew there are three-sided, four-sided, and five-sided shapes. His classmates' naming their shapes as triangles and quadrilaterals did not surprise him. When the time came for his group to share, I heard him saying their shape was a hexagon before I could see their shape. Thinking we had skipped a shape, I asked for pentagons. Judith responded expressing doubt, but Bill insisted that their shape was a hexagon. The remaining of the interaction made explicit connections between shapes' names and their number of sides, to which Bill expressed surprised. After Bill's expression of realization at the end of this vignette, he stated their shape was a pentagon and Judith agreed, counting the five sides.

Based on these interactions, it is not possible to be completely sure whether Bill's conceptualizations of a hexagon had to do with his language or his mathematical understanding. Instead, the integrated framework I use here draws attention to both at the same time, considering this an issue of the mathematics–language phenomenon. The *autonomous exploration*, *authentic communication,* and *favoring of fluency* allowed Bill to use language in ways he thought appropriate in the context of this mathematical discussion. These practices made Bill's thinking about hexagons public. In turn, the search for accuracy was grounded in the interaction between mathematics and language, as students explained the relationship between a shape's name and its attributes.

Consider the alternative: having students follow a short lecture on the names of geometric shapes and their number of sides, and a drilling exercise for them to produce specific sentences and match them with shapes, such as "This is a hexagon. The hexagon has six sides." Language production would be accurate. There would be, however, insufficient evidence regarding whether accurate language use reflected mathematical understanding or just the recitation of memorized sentences. The accuracy (or lack thereof) of language use relates to Bill's understanding of what a hexagon is. Similarly, his understanding of "hexagon" as a distinct concept comes to be because of language use. Students cannot communicate what they do not know and they do not know what they cannot communicate.

Conclusion

From a theoretical point of view, this chapter contributes and operationalizes a framework that considers mathematics and language as integrated. In doing so, it initiates a conversation between educational research in mathematics and in the second language. Although the examples come from an enrichment language immersion classroom, the analyses presented here could inform research in other contexts. The tenets of mathematics–language activity described here could illuminate our understanding of programs for English language learners. Specifically, an integrated framework can guide analyses that are close to classroom experience, where teachers and students engage in mathematical activity and language use at the same time.

From an analytical point of view, each of the examples provided could have been analyzed from one perspective only: situated learning in mathematics or a communicative approach to language teaching. As discussed in the conceptualization of the mathematics–language phenomenon, previous attempts to relate the two usually draw on linearity and causality to show how language knowledge allows mathematics activity and vice versa. These analytical options are limited when the boundaries between exploring mathematical ideas and using language are blurred. Instead, the proposed integrated framework focuses on the phenomenon of mathematics–language as it realistically happens in the classroom. Teachers and students do not set aside specific formal language to use it in the learning of mathematics, as models such as CLIL or SIOP suggest. Instead, teachers and students deal with the integrated phenomenon of mathematics–language at once, and learning one involves learning the other.

From a pedagogical point of view, this chapter has implications for teachers. A productive alternative to the current compartmentalization of language and mathematics in some of the models that teachers use is the integration of the two. The pedagogical practices of fostering students' autonomous exploration, authentic communication, and a balance between fluency and accuracy could become part of the principles that teachers consider when designing tasks. Since both the teacher and the researcher in this study have expertise in the second language teaching, future research could explore how to extend this integrated approach to other teachers with different expertise.

This chapter has suggestions for parents supporting their children's learning process. The integrated framework invites parents to shift away from overemphasizing children's accurate use of the second language and of mathematical procedures. Instead, parents' expectations could continue to place value in children's development of accuracy and precision while simultaneously recognizing the mathematically and communicative skills that students use in language immersion classrooms. Language immersion classrooms depend on parents' support and commitment to this type of context that are not the well-known mainstream monolingual classroom. Acknowledging children's developing skills can further parents' appreciation of and patience with the complex task that teachers and children in these classrooms undertake.

This chapter used the integrated theoretical framework to make sense of third-graders working on geometry. From research on the second language, we know young students have different ways of learning languages (Pinter, 2017), and from research on mathematics education, we can infer that the interrelationship between language and content such as numbers and operations is different from the interrelationship between language and geometry. Therefore, more research is necessary to operationalize this framework, acknowledging learners' characteristics—including their age—and specific mathematical content in diverse contexts.

References

Adler, J. (1998). A language of teaching dilemmas: Unlocking the complex multilingual secondary mathematics classroom. *For the Learning of Mathematics, 18*(1), 24–33.

Barwell, R. (2005). Integrating language and content: Issues from the mathematics classroom. *Linguistics and Education, 16*(2), 205–218. https://doi.org/10.1016/j.linged.2006.01.002.

Boaler, J. (2000). Exploring situated insights into research and learning. *Journal for Research in Mathematics Education, 31*(1), 113–119.

Brisk, M. E. (2011). Bilingual education. In E. Hinkel (Ed.), *Handbook of research in second language teaching and learning* (Vol. II, pp. 7–24). New York, NY: Routledge.

Brown, T. (2002). *Mathematics education and language: Interpreting hermeneutics and post-structuralism* (pp. 1–281). Dordrecht, Netherlands: Kluwer Academic Publishers.

Brumfit, C. (1988). Applied linguistics and communicative language teaching. *Annual Review of Applied Linguistics, 8,* 3–13. https://doi.org/10.1017/S0267190500000982.

Cambridge, E. S. O. L. (2010). *Teaching maths through English—A CLIL approach.* Cambridge, UK: University of Cambridge.

Chval, K. B., Pinnow, R. J., & Thomas, A. (2015). Learning how to focus on language while teaching mathematics to English language learners: A case study of Courtney. *Mathematics Education Research Journal, 27,* 103–127. https://doi.org/10.1007/s13394-013-0101-8.

Cobb, P., & Bowers, J. (1999). Cognitive and situated learning perspectives in theory and practice. *Educational Researcher, 28*(2), 4–15.

Dreyer, J. S. (1998). The researcher: Engaged participant or detached observer? *Journal of Empirical Theology, 11*(2), 5–22.

Dreyer, J. S. (2016). Knowledge, subjectivity, (de)coloniality, and the conundrum of reflexivity. In J. A. Mercer & B. J. McLemore (Eds.), *Conundrums in practical theology* (pp. 90–109). Leiden, The Netherlands: Brill Academic Publishers.

Echevarria, J., Vogt, M., & Short, D. (2004). *Making content comprehensible for English learners: The SIOP model* (2nd ed.). Boston: Allyn and Bacon.

Erickson, F. (2006). Definition and analysis of data from videotape: Some research procedures and their rationales. In J. L. Green, G. Camilli, & P. B. Elmore (Eds.), *Handbook of complementary methods in education research* (pp. 177–191). Hillsdale, NJ: Erlbaum.

Erickson, F. (2012). Qualitative research methods for science education. In B. J. Fraser & K. G. Tobin (Eds.), *International handbook of science education* (pp. 1451–1469). Dordrecht, The Netherlands: Springer.

García, O., & Kleifgen, J. A. (2010). *Educating emergent bilinguals: Policies, programs, and practices for English language learners*. New York, NY: Teachers College Press.

Greeno, J. G. (1991). Number sense as situated knowing in a conceptual domain. *Journal for Research in Mathematics Education, 22*(3), 170–218.

Greeno, J. G. (1998). The situativity of knowing, learning, and research. *American Psychologist, 53*(1), 5–26.

Halliday, M. (1978). *Language as social semiotic: The social interpretation of language and meaning*. Baltimore, MD: University Park Press.

Honigsfeld, A. (2009). ELL programs: Not 'one size fits all'. *Kappa Delta Pi Record, 45*(4), 166–171.

Kareva, V. (2013). Using the SIOP model for effective content teaching with second and foreign language learners. *Journal of Education and Training Studies, 1*(2), 239–248. http://doi.org/10.11114/jets.v1i2.173.

Lantolf, J., Thorne, S. L., & Poehner, M. (2015). Sociocultural theory and second language development. In B. van Patten & J. Williams (Eds.), *Theories in second language acquisition* (pp. 207–226). New York, NY: Routledge.

Larsen Freeman, D. (2011). *Techniques and principles in language teaching* (3rd ed.). Oxford, U.K.: Oxford University Press.

Lave, J., & Wenger, E. (1991). *Situated learning: Legitimate peripheral participation*. Cambridge, England: Cambridge University Press.

Lightbown, P., & Spada, N. M. (2006). *How languages are learned* (3rd ed.). Oxford: Oxford University Press.

Littlewood, W. (2014). Communication-oriented language teaching: Where are we now? Where do we go from here? *Language Teaching, 47*(3), 349–362. https://doi.org/10.1017/S0261444812000134.

Mack, N. K. (2007). Gaining insights into children's geometric knowledge. *Teaching Children Mathematics, 14*(4), 238–245.

Mehisto, P., Frigols, M. J., & Marsh, D. (2008). *Uncovering CLIL content and language integrated learning in bilingual and multilingual education*. Oxford, U.K.: Macmillan Education.

Moschkovich, J. (2002). A situated and sociocultural perspective on bilingual mathematics learners. *Mathematical Thinking and Learning, 4*(2–3), 189–212.

Moschkovich, J. (2007). Bilingual mathematics learners: How views of language, bilingual learners, and mathematical communication affect instruction. In N. S. Nasir & P. Cobb (Eds.), *Improving access to mathematics: Diversity and equity in the classroom* (pp. 89–104). New York, NY: Teachers College Press.

Moschkovich, J. (2013). Principles and guidelines for equitable mathematics teaching practices and materials for English language learners. *Journal of Urban Mathematics Education, 6*(1), 45–57.

National Council of Teachers of Mathematics. (2000). *Principles and standards for school mathematics*. Reston, VA: National Council of Teachers of Mathematics.

Nikula, T., Dalton-Puffer, C., Llinares, A., & Lorenzo, F. (2016). More than content and language: The complexity of integration in CLIL and bilingual education. In T. Nikula, E. Dafouz, P. Moore, & U. Smit (Eds.), *Conceptualising integration in CLIL and multilingual education* (pp. 73–100). Bristol, UK: Multilingual Matters.

Novotna, J., & Hofmannova, M. (2000). CLIL and mathematics education. In A. Rogerson (Ed.), *Proceedings of the International Conference on Mathematics Education into the 21st Century: Mathematics for living* (pp. 226–230). Aman, Jordan: Third World Forum.

O'Halloran, K. L. (2005). *Mathematical discourse: Language, symbolism and visual images*. New York, NY: Continuum.

O'Halloran, K. L. (2015). The language of learning mathematics: A multimodal perspective. *Journal of Mathematical Behavior, 40*(Part A), 63–74. http://doi.org/10.1016/j.jmathb.2014.09.002.

Pimm, D. (1987). *Speaking mathematically: Communication in mathematics classrooms*. New York, NY: Routledge.

Pinter, A. (2017). *Teaching young language learners*. Oxford, U.K.: Oxford University Press.

Planas, N., & Civil, M. (2013). Language-as-resource and language-as-political: Tensions in the bilingual mathematics classroom. *Mathematics Education Research Journal, 25*(3), 361–378.

Planas, N., & Setati-Phakeng, M. (2014). On the process of gaining language as a resource in mathematics education. *ZDM Mathematics Education, 46*(6), 883–893.

Richards, J. C. (2006). *Communicative language teaching today*. New York, NY: Cambridge University Press.

Sfard, A. (1998). On two metaphors for learning and the dangers of choosing just one. *Educational Researcher, 27*(2), 4–13.

Sfard, A. (2001). There is more to discourse than meets the ears: Looking at thinking as communicating to learn more about mathematical learning. *Educational Studies in Mathematics, 46*, 13–57.

Sfard, A., Nesher, P., Streefland, L., Cobb, P., & Mason, J. (1998). Learning mathematics through conversation: Is it as good as they say? *For the Learning of Mathematics, 18*(1), 41–51.

Smit, J., Bakker, A., van Eerde, D., & Kuijpers, M. (2016). Using genre pedagogy to promote student proficiency in the language required for interpreting line graphs. *Mathematics Education Research Journal, 28*(3), 457–478. https://doi.org/10.1007/s13394-016-0174-2.

Snow, M. A., Met, M., & Genesee, F. (1989). A conceptual framework for the integration of language and content in second/foreign language instruction. *TESOL Quarterly, 23*(2), 201–217.
Spradley, J. P. (2016). *Participant observation*. Long Grove, IL: Waveland Press.

Schema-Based Instruction: Supporting Children with Learning Difficulties and Intellectual Disabilities

Kim Desmarais, Helena P. Osana and Anne Lafay

Children who have learning difficulties or intellectual disabilities have similar challenges when solving mathematical word problems, including creating an internal representation of the problem structure and organizing the information to generate a solution strategy. Students with learning difficulties in mathematics and those with intellectual disabilities benefit from mathematics instruction that incorporates visual aids and repetition, and promotes strategy flexibility to help develop conceptual understanding. With regard to mathematical word problem solving, one approach has shown promise for individuals with learning difficulties and typically developing youth. Schema-Based Instruction (SBI) uses visual representations to teach students the mathematical structure of word problems. In this paper, we draw on existing literature to outline some of the cognitive deficits that have been observed in children with learning difficulties in mathematics and in those with intellectual disabilities and describe the ways in which those deficits can manifest themselves in the context of mathematical problem solving. We then describe the data we collected from our own delivery of SBI to a group of students with intellectual disabilities and compared their performance to students with and without learning difficulties. We focus on instances of meaningful problem solving after the intervention, with a focus on how the students may have circumvented or compensated for specific cognitive deficiencies. We conclude the chapter with a discussion about the elements of the instruction that may account for the students' performance after the intervention.

Classrooms are made up of many different types of learners, ranging from students who excel academically to those who struggle to learn the material being taught. Some students may have learning difficulties, intellectual disabilities, or other challenges, such as autism spectrum disorders. Indeed, it is estimated that in Canada, 57.6% of children aged 5–14 with intellectual disabilities are within the mild to moderate

K. Desmarais
Department of Psychology, Concordia University, Montreal, QC, Canada

H. P. Osana (✉) · A. Lafay
Department of Education, Concordia University, Montreal, QC, Canada
e-mail: helena.osana@concordia.ca

K. M. Robinson et al. (eds.), *Mathematical Learning and Cognition in Early Childhood*,
https://doi.org/10.1007/978-3-030-12895-1_12

range, indicating that more than half are likely placed in inclusive school settings (Human Resources and Skills Development Canada (HRSDC), 2011).

Existing educational policies in North America stipulate that children with learning difficulties or intellectual disabilities should have access to equal opportunities for high-quality education that meet their needs (e.g., Education Act of Ontario, Ontario Ministry of Education, 1990; No Child Left Behind Act (NCLB), 2001; Ministère de l'éducation et de l'enseignement supérieur, 1999). Further, policies advocate for access to general education in the regular classroom to help these children meet the developmental goals established for all students (Individuals with Disabilities Education Improvement Act (IDEA), 2004).

Typical school mathematics curricula and pedagogies are rarely tailored to children with special needs (in our case, children with learning difficulties or intellectual disabilities), despite the fact that they are often integrated with other students in the classroom (Rose & Rose, 2007). Given the potential number of students who need support in mainstream classrooms, it is of utmost importance that appropriate programs are put into place to help support all children's mathematical development.

Although research on effective mathematics instruction is often focused on the typically developing student population, there is less research on effective practices specifically for children with learning difficulties. Even less research attention has been paid to mathematics instruction for children with intellectual disabilities and the pedagogies that can support their learning. It would appear that, for the most part, the mathematics instruction provided to children with intellectual disabilities has by and large emphasized rote, procedural instruction, with little focus on the development of conceptual understanding (Baroody, 1999; Cawley, Parmar, Yan, & Miller, 1998). More recently, Powell, Fuchs, and Fuchs (2013) argued that for students with mathematics difficulties, instruction should increase its focus on fluency with basic arithmetic skills, which are needed to solve problems that require conceptual knowledge. Traditional views of children with mathematics difficulties and intellectual disabilities are that they are "passive learners," capable of learning lower-level skills, but unable to devise or learn new strategies or transfer the skills previously learned. Research has shown, however, that children with intellectual disabilities are indeed capable of learning a number of mathematical skills and concepts with proper instruction (Baroody, 1999; Fletcher, Lyon, Fuchs, & Barnes, 2006; Ginsburg, 1997).

In this chapter, we describe an instructional intervention that we delivered to a small group of students with intellectual disabilities and compare their performance to that of children with and without learning difficulties. For purposes of clarity, we use the term "learning difficulty" to describe students who are either (a) diagnosed with a specific mathematics disability according the researcher's criteria, or (b) students who are underperforming in mathematics relative to their peers in the classroom, but who have no known diagnoses. When we use the term "mathematics learning disability," we are referring only to those students with specific diagnoses reported by the researchers. The intervention we delivered was a modified version of schema-based instruction (SBI; Jitendra & Star, 2011), an empirically validated teaching approach designed to support students' understanding of the mathematical structure of word problems.

We begin the chapter by referring to the existing literature to outline some of the cognitive deficits that have been observed in children with learning difficulties and intellectual disabilities and the ways in which those deficits can manifest themselves in the context of mathematical problem solving. We will then turn our attention to the student data we collected to describe particular instances of meaningful problem solving after the intervention, with an eye toward how the students may have circumvented or compensated for specific cognitive deficiencies. We conclude the chapter with a discussion about the elements of the instruction that may account for the students' performance after the intervention. We note that the conclusions that we can draw are limited because of the small number of participants in our study. Because of this limitation, we do not claim that our data generated any robust effects, causal, or other. Because of the dearth of research on the problem solving of children with intellectual disabilities, however, our findings nevertheless make an important contribution to this literature.

Mental Representation in Problem Solving

De Corte, Verschaffel, and De Win (1985) proposed a theoretical model of the processes involved in solving mathematical word problems. The model consists of five stages. In the first stage, the child processes the verbal text and creates a mental representation of the word problem structure. In the second stage, the child selects the appropriate arithmetic operation for finding the unknown. This selection is in large part dependent on the mental representation constructed in the first stage. In the third stage of the problem-solving model, the child executes the operation he or she has chosen. During the fourth stage, the child reactivates the mental representation, inserting the answer that was calculated. In the last stage, the child verifies if the answer is correct.

A study by Boonen, de Koning, Jolles, and van de Schoot (2016) nicely illustrates how successful problem solving is contingent on a correct mental representation of the problem structure. The authors found that children tended to write numbers sentences with operations that were consistent with the relational terms (i.e., "more than," "less than") used in the problem. In other words, the children were more likely to write number sentences with "+" when "more than" was used in the word problems than when "less than" was used, even when subtraction may have been a legitimate operation for the problem. This indicates that for children to be successful on inconsistent problems (when the relational terms do not match the required operation), they need to rely on a mental representation of the problem structure and cannot get by with superficial aspects of the problem text (e.g., the words "more" or "less").

Students with learning difficulties are often challenged when creating mental representations of problems and identifying the relevant information for solving them (Xin, Jitendra, and Deatline-Buchman, 2005). This has been shown to negatively impact problem solving in a number of ways, including reduced accuracy, difficulty generating number sentences, and applying inappropriate strategies for solving the problems (Hutchinson, 1993; Montague & Applegate, 1993). Some of the difficulties they have in creating mental representations of mathematics problems can be explained by executive functioning deficits that have been identified in the literature for children with intellectual disabilities (Oznoff & Schetter, 2007). Children with learning difficulties are challenged when they solve problems that require visualization and working memory capacity (Stein & Krishnan, 2007).

SBI specifically targets the first stage of the model—students' internalization of an appropriate problem structure. Without an appropriate mental representation, students will be hindered in choosing an appropriate operation, which in turn, will affect subsequent computations. Appropriate mental representations of word problem structures help students see the relationships among the quantities in the problem, which then supports the identification of suitable strategies for solving it (Lucangeli, Tressoldi, and Cendron, 1998).

Domain General Predictors of Word Problem Solving

Daroczy, Wolska, Meurers, and Nuerk (2015) proposed a model of the cognitive factors that are predictive of successful word problem solving. Their model includes domain general abilities as well as linguistic and numerical capabilities, and the authors describe the ways in which these factors account for student performance. In this section, we focus specifically on domain general abilities, such as executive functioning skills (e.g., working memory, shifting, and inhibition). The research has identified two specific executive functions, namely working memory and cognitive flexibility, as being especially important in mathematical problem solving (e.g., Geary, 2004; Geary, Hoard, Byrd-Craven, Nugent, and Numtee, 2007). Below, we present a brief overview of the literature describing the impact of working memory and flexibility on specific aspects of word problem solving.

Working memory. Working memory is the ability to hold a mental representation of information in one's mind while simultaneously using other mental processes to complete a task (Geary et al., 2007). Working memory plays a major role in predicting problem-solving accuracy (Andersson, 2007; Swanson & Beebe-Frankenberger, 2004; Zheng, Swanson, and Marcoulides, 2011); it is involved in all aspects of the problem-solving process because the students need to keep a number of pieces of information in mind during text comprehension, all while selecting an appropriate operation, executing the operation, and verifying the response.

Several researchers have demonstrated that students with learning difficulties and those with intellectual disabilities have significant working memory deficits, which in part explain their difficulty solving mathematics problems (Geary, 2004; Henry, Messer, and Poloczek, 2018). These deficits are manifested in various ways when children solve problems. For example, children with mathematics learning difficulties and with intellectual disabilities have trouble executing the required operation, and as such will use less mature strategies and make more errors than children without difficulties. For instance, children with learning difficulties often resort to finger-counting because it reduces the demands on working memory (Geary, 2004).

In addition, although little is known about the role of working memory in the creation of a mental representation, it has been shown that working memory is implicated in the construction of the mental number line (Geary, Hoard, Nugent, and Byrd-Craven, 2008). It is therefore reasonable to attribute children's struggles in constructing a useful mental model for a word problem to their working memory challenges. Furthermore, students with intellectual disabilities and learning difficulties likely struggle to keep a mental representation in mind throughout the problem-solving process (Lee, Ng, & Ng, 2009; Swanson & Sachse-Lee, 2001).

Judd and Bilsky (1989) attempted to alleviate cognitive load by providing a visual aid (i.e., dots that represented the quantities in the word problem) to students with and without intellectual disabilities while they were solving addition and subtraction word problems. The authors observed that students in both groups who were provided the visual representations were better able to retain the relevant information in the problem. They also showed that of all the students who were provided the visual aids, those who employed counting strategies were more likely to be successful (i.e., accurate) relative to those who were not provided the visual aids. Finally, Judd and Bilsky found that the errors made by children with intellectual disabilities were often characterized by overcounting or undercounting during the execution of the solution strategy.

Strategy flexibility. Flexibility in strategy use is defined as adapting one's strategies to the characteristics of the task at hand (Van der Heijden, 1993, cited in Verschaffel, Torbeyns, De Smedt, Luvwel, and Van Dooren, 2007). Ostad (1997) examined the addition strategies of children with mathematics learning difficulties and found that the children in the primary and upper elementary grades relied more on "back-up" strategies than retrieval strategies relative to their typically developing peers, which appears to be related to working memory deficits (Geary, Hoard, Byrd-Craven, and DeSoto, 2004). Back-up strategies are overt strategies that are visible or audible, such as counting on one's fingers. Retrieval strategies are those in which the answers are retrieved from long-term memory and can support performance on complex tasks because they require fewer demands on working memory (Powell et al., 2013; Raghubar, Barnes, and Hecht, 2010). In addition, Ostad found that children with mathematics learning difficulties exhibited considerably less flexibility in their strategy use compared to those students without difficulties. In particular, they tended to use one strategy repeatedly, as opposed to their typically developing peers, who used a range of strategies when solving problems. Ostad used the term "strategy rigidity" to describe those children who repeatedly use a smaller number of primitive back-up strategies when solving word problems.

Children with intellectual disabilities also struggle with strategy flexibility. Children with autism spectrum disorders, for example, have significant limitations with regard to cognitive flexibility and planning (Oznoff & Schetter, 2007), specifically at the conceptual level, making it difficult for these individuals to shift from one concept to the next. This is especially problematic when they engage in mathematics activities that require frequent shifting between various strategies or operations. It also appears that there is a relationship between the construction of an accurate representation of a problem's structure and flexibility in strategy use. Indeed, children who lack conceptual understanding of the quantitative relationships in a problem—that is, who lack an accurate mental representation of the problem structure—tend to be rigid in terms of the solution strategies they generate (Baroody, 1999; Lee et al., 2009).

Word Problem-Solving Instruction and SBI

There is evidence that children with learning difficulties and children with intellectual disabilities can acquire the same mathematical knowledge as their typically developing peers if additional and appropriate instruction is provided (e.g., Clements & Sarama, 2009; Fletcher et al., 2006). For example, young children with intellectual disabilities can learn oral counting, one-to-one correspondence, and cardinality (Baroody, 1999). In addition, they can learn basic numeracy skills (e.g., counting and subitizing) to the same level as their typically developing peers, as long as instruction is explicit and provides opportunities for practice (Bird & Buckley, 2001).

Three reviews of the literature focused on the effects of mathematics interventions on the learning of children with intellectual disabilities (i.e., Browder, Spooner, Ahlgrim-Delzell, Harris, and Wakemanxya, 2008; Butler, Miller, Lee, and Pierce, 2001; Mastropieri, Bakken, and Scruggs, 1991) and of children with autism spectrum disorders (Browder et al., 2008). Together, these reviews showed that interventions targeting word problem solving with children and teenagers presenting with intellectual disabilities and autism were effective when they focused on the training of cognitive self-control strategies (e.g., checklists), the analysis of problem statements, and the use of concrete objects during the execution of solution strategies.

Bissonnette, Richard, Gauthier, and Bouchard (2010) conducted an overview of several reviews of the literature on mathematics interventions in students with learning difficulties. They demonstrated that explicit instruction is more effective than pedagogical methods based on constructivism for teaching word problem solving in children with learning difficulties. More recently, Jitendra, Nelson, Pulles, Kiss, and Houseworth (2016) reviewed the literature on instructional interventions centered specifically on teaching students the structure of mathematical problems with either visual representations (e.g., schematic drawings of part-whole and compare word problems) or with concrete representations, such as manipulatives. The 25 studies reviewed by the authors targeted students with learning difficulties and those at risk for mathematics learning disabilities. The findings showed that visual representations, whether on their own or in combination with concrete objects, positively impacted students' problem-solving performance.

SBI is an instructional approach that uses visual representations of problem structures to teach students how to solve a variety of word problems (Jitendra & Star, 2011). The findings from several studies have shown that SBI supports youth with learning difficulties, those at risk for mathematics learning disabilities, and typically developing students in their efforts to solve different types of word problems (Fuchs, Fuchs, Finelli, Courey, and Hamlett, 2004a; Fuchs et al., 2004b; Jitendra, DiPipi, and Perron-Jones, 2002; Jitendra & Star, 2011). A number of studies have also shown evidence of conceptual understanding following SBI as evidenced by performance on transfer problems (e.g., Fuchs et al., 2004a, 2004b; Jitendra et al., 2002), and Rockwell, Griffin, and Jones (2011) found maintenance effects for up to six weeks.

SBI's framework is based on schema theory. Schemata are knowledge structures that organize information in the learner's long-term memory (Bransford & Johnson, 1972; Griffin & Jitendra, 2009). In problem solving, schemata assist the learner in categorizing information, identifying the relationships between the quantities in a problem, and determining the best strategy for solving the problem (Chen, 1999). Chen (1999) found that when students are able to internalize what he termed "general schemata," defined as abstract representations of a problem's structure, their performance on transfer problems is enhanced. In addition, a general schema is one that is not linked to a specific procedure (Chen, 1999). When teachers provide students with a multitude of problems and diverse solution strategies, children can abstract a general schema which can then be used to solve novel problems, offering more flexibility across a range of contexts. The use of general schemata allows children to understand the semantic relations between the sets in the problem, which in turn supports a conceptual understanding about increases, decreases, and combinations involving sets (Cummins, 1991).

Xin and Jitendra (1999) argued that one of the reasons for the success of SBI is that it emphasizes conceptual understanding by creating representational links between the various aspects of word problems, thus enhancing students' ability to successfully solve them. SBI has been said to address the working memory and attention deficits of children with learning difficulties, and greatly differs from traditional mathematics teaching for children with intellectual disabilities, which tends to emphasize rote, procedural instruction (Cawley et al., 1998). Another possible reason for the success of SBI, particularly for children with learning difficulties, is that the creation of visual representations of the problem structure helps children solve problems by reducing cognitive load.

A Modified Version of SBI

Given that SBI has been found to be effective with students with a wide range of mathematical and cognitive abilities, we delivered a version of SBI to support the problem solving of three groups of first- and second-grade students: (a) a group of children with comorbid intellectual disabilities and autism spectrum disorders, (b) a group of children with learning difficulties (i.e., who were identified by their teacher as performing below the level of their peers), and (c) a group of children who were not struggling in mathematics in school.

Our delivery of SBI was a slightly modified version of the SBI protocol published by its designers (i.e., Fuchs et al., 2004a, 2004b; Jitendra & Star, 2011). We modified the typical SBI protocol by breaking the problem-solving process into smaller units for instruction. In addition, we also supplemented SBI with instructional features that were recommended by the National Mathematics Advisory Panel (2008), namely collaborative activity during problem solving and the sharing of solution strategies. We also encouraged students to solve the problems in whichever ways they found meaningful, which is also a departure from the typical SBI protocol. Finally, the instructor asked follow-up questions to encourage students to explain their strategies to their peers as clearly as possible, and also to encourage the students to identify any errors.

Participants

Our study had three groups of children, with three students in each group. All students were between 7 and 8 years old. The first group consisted of three students with comorbid intellectual disabilities and autism spectrum disorders who were finishing the first grade (ID group). The second group consisted of three second-grade students with learning difficulties, all from the same classroom (LD group). The final group consisted of three second-grade students, from the same class as the LD group, who exhibited average mathematics performance in school (AM group).

SBI and Student Data

All nine children received three instructional hours of SBI over four sessions (45 min each) that were delivered in small groups by the first author, a trained graduate student in educational research. During instruction, children were specifically taught "Action" problems (also known as join and separate problems; Carpenter, Fennema, Franke, Levi, and Empson, 2014; but also called "Group" or "Change" problems in the SBI literature; Jitendra & Star, 2011). Action problems describe an action where a given set is either increased or decreased, resulting in a different final quantity. In these problems, the unknown can be the initial set, the change set, or the end set. Consider the following word problem, "There are 9 apples in the bin. Five apples fall out of the bin. How many apples are left?" Described is a decreasing (or separating) action of apples falling from the bin, which changes the initial amount in the problem (9 apples). Two other problem types (part-whole and compare; Carpenter et al., 2014) were not used during our instruction, but were used as transfer problems on the assessments before and after the intervention.

Traditionally, SBI consists of two separate phases that we refer to here as the Problem Learning Phase and the Solution Generation Phase. In our study, each phase was completed in two instructional sessions spanning 45-min each. We used the same schematic representation, or schema,[1] for Action problems as those used in previous SBI studies (e.g., Fuchs et al., 2004a, 2004b), which can be found in Fig. 1. In the Problem Learning Phase, we provided explicit explanations of the different components of the schema, as is standard to SBI. Because the goal of this phase was for the children to learn the different components of the schema and where the numbers should be placed in it, we used story scenarios instead of word problems, which is also standard SBI practice. Story scenarios are word problems in which there is no unknown—that is, all of the numbers in the problem are provided. To illustrate using the word problem involving apples above, the corresponding story problem would be, "There are 9 apples in the bin. Five apples fall out of the bin. Now there are 4 apples left."

In the second phase, the Solution Generation Phase, the children were encouraged to generate their own strategies to solve a set of word problems, this time with unknowns. The students were provided with manipulatives (plastic chips) and paper and pencil to solve the problems in any way that was meaningful to them. The instructor also encouraged the students to use more than one strategy to solve a given problem and to share their strategies with the other students in their group. After the children had solved a given problem using a strategy of their choice, the instructor asked them if there was any other way the problem could be solved.

Before and after the modified SBI, we administered two tasks to the students to assess their mental representations of different problem types and their problem solving. Mental representations were assessed using the Problem Structure Test (PST), a multiple-choice test that we constructed specifically for this study. The PST contained six items, each of which required the student to read a word problem and choose one visual representation among three that best matched the structure of the problem (the fourth choice was "none of these"). A sample item from the PST is presented in Fig. 2. Four of the problems on the PST were action problems, and to assess transfer, the PST included one compare and one a part-whole problem. Correct answers were assigned 1 point and incorrect answers 0 points. The points were summed and converted to percent.

The students were also given word problems to solve before and after the intervention. Before the intervention, they solved six problems (i.e., four action, one compare, and one part-whole), and after the intervention, they solved eight problems, which

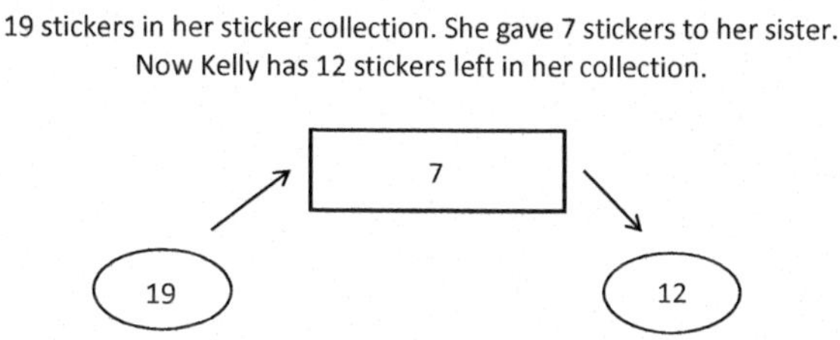

Fig. 1 Action schema illustrated on a workbook page used during the Problem Learning Phase

[1]For the remainder of the chapter, we use the term "schema" to refer to the schematic drawing that represents the structure of the word problem.

Fig. 2 Choices for the following problem on the Problem Structure Test: "Jesse's mom made cookies for the bake sale. She sold 19 cookies and she has 5 left over. How many cookies did Jesse's mom bake for the bake sale?"

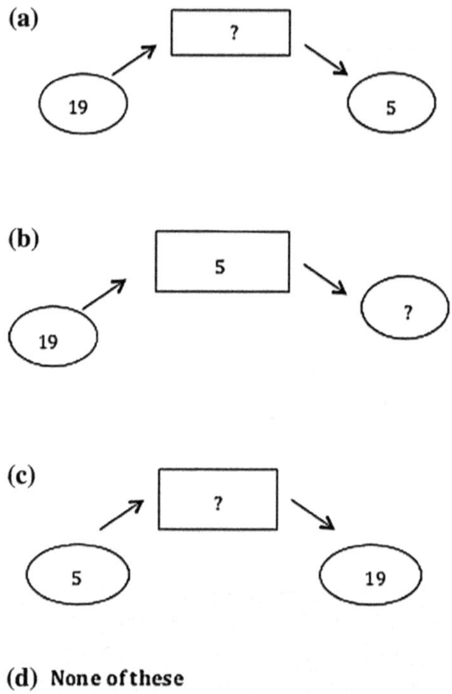

consisted of isomorphic versions of the problems given before the intervention, and two additional action problems in which the quantities were not presented in the typical start–action–end order. The compare, part-whole, and atypical-sequence action problems were used to assess transfer.

We examined two aspects of the students' problem solving. First, regardless of counting or computation errors, we assessed whether the strategy used by the student reflected the structure of the problem. Those that were aligned with the problem structure were coded as "appropriate" and those that were not aligned were coded as "not appropriate." In addition, we coded the type of strategies the children used on the same task. We used Carpenter et al.'s (2014) taxonomy of problem solving strategies as our coding scheme. Direct modeling strategies were characterized by physical representations of the objects and actions in the problem. Counting strategies were those where the child was able to first represent one quantity abstractly (i.e., without representing it physically) and used some tools, such as fingers or tallies, to keep track of counts to find the solution. Derived fact strategies were those where the student used a known fact (e.g., single-digit addition facts) to derive a solution. For example, for 6 + 7, a child using a derived fact strategy may explain that the answer is 13 because she knows that 6 and 6 are 12, and then one more is 13. Strategies based on known facts only were coded as recall. Strategy type allowed us to examine whether students were able to use more than one strategy type after the intervention.

Aspects of Problem Solving Before and After the Intervention

Identification of Word Problem Structure

As reviewed earlier in this chapter, one of the biggest challenges experienced by children with learning difficulties and children with intellectual disabilities is to identify the underlying word problem structure. This is consistent with what we observed prior to SBI instruction. Students in the ID group and in the LD group were unable to identify the word problem structures before the intervention. Specifically, the mean PST percent score for the ID group was 5% (SD = 9.2) and for the LD group was 27% (SD = 9.8) before the intervention. This suggests that these students were either unable to extract an appropriate mental representation of the word problems on the test, or had difficulty interpreting the visual representations provided in the choices on each item. If they did identify the correct structure, it was almost always for the action problems. The AM group performed better than the other two groups before the intervention on the problem structure test with an average PST score of 44% (SD = 25.5) before the intervention. Although their performance left room for improvement, they identified the correct structure for action problems more frequently and were also better able to choose the appropriate structure for the part-whole problem on the test.

Following the intervention, the mean PST scores for the students in the ID and LD groups improved. Mean scores on the PST were at 45% (SD = 14.4) for the ID group and 62% (SD = 25) for the LD group, with the AM group's performance remaining unchanged. After the intervention, the students in both the ID and LD groups were more consistently able to identify the structure for action problems and most of the students (5 of 6) across both groups were also able to correctly identify the structure for part-whole problems. This was an interesting finding, as the students were not exposed to part-whole problems during the intervention. This, paired with the fact that students in the AM group were able to correctly identify the structure of part-whole problems before the intervention, suggests that there may be something intuitive about the representation we created for part-whole problems, which was a different representation of the one originally created for SBI. Further, it could be that learning the structure of action problems facilitates transfer to problems with different mathematical structures.

Strategy Use

The students in the ID and LD groups were observed to more often use appropriate problem-solving strategies after the intervention compared to before. It is possible that having a concrete way of organizing the information presented in the problem—that is, the schema—and to see the relevant information in the schema decreased working memory load, thereby allowing for the allocation of cognitive resources to finding an appropriate solution strategy. It is also possible that the availability of manipulatives used during problem solving alleviated cognitive load. They were able to physically represent the mathematical actions required to solve the problems (e.g., removing tokens or joining them). Because the children could represent the quantities in the problem using tokens, they did not have to keep the numbers in their minds, thus offloading valuable working memory resources.

The use of the schemas appeared to support the problem solving of the children in all three groups, but the ID and LD groups seemed to benefit the most from the visual representations. The majority of the students in these two groups combined (i.e., 5 of 6) continued to use the schema after the intervention to help organize the information in the problem. We observed one student in the LD group rely on the schema when she reached an impasse while attempting to solve the problems. When she was unsure of what solution strategy to use, she would draw the schema and insert the numbers into it. Organizing the information in this way seemed to help clarify critical aspects of the problem and she was then able to determine an appropriate strategy for solving it.

Strategy Flexibility

Most students demonstrated little change in the ability to use more than one type of strategy for the same problem after the intervention. This is consistent with the literature on strategy rigidity describing children with difficulties in mathematics using one type of strategy across problems (Ostad, 1997). We observed that when the researcher asked for an alternative strategy, most children gave a response in the same category as their first strategy, both before and after the SBI instruction. This was seen for most students in all three groups for all problems. For example, one child in the LD group used direct modeling (physically representing the quantities and actions in the problem) using tokens for his first strategy, but when asked if there was another way to solve the problem, he drew the objects in the problem on a piece of paper and acted on those representations by circling and crossing out objects. Although both strategies looked different on the surface, they were both coded as direct modeling, and as such, did not constitute strategy flexibility.

We also observed three students (two from the AM group and one from the LD group) change the operation for the second strategy, both before and after the intervention. That is, they used the standard written algorithm for their first strategy, and for the second, used the standard algorithm for a different operation. For example, on his first strategy, one child in the AM group performed the standard written algorithm for $198 - 116$ and correctly solved the problem. When asked for a second strategy, he used the standard algorithm for the inverse operation and computed $198 + 116$. Thus, we saw some evidence for strategy rigidity as most students would use the same type of strategy to solve a given problem.

Only one student in the LD group was able to successfully use more than one type of strategy on the problems after the intervention. On one problem, he used the standard written algorithm as his first problem-solving strategy for one of the end unknown problems. When asked for a different strategy, he explained what he would do to directly model the problem. He did not physically act it out by counting out tokens or drawing tallies, but clearly explained his direct modeling strategy verbally.

Pedagogical Implications

Despite the tentative nature of the conclusions that can be drawn from our small data set, the results are still promising with regard to intervention planning for inclusive classrooms. All students benefited from the instruction in one way or another. In the following section, we will describe some of our observations of the students' problem solving during the instruction that may account for their performance after the intervention.

Overall, the instruction appeared beneficial for children on most of the assessments we administered. Explicitly teaching the children the different components of the schema helped them prepare for the problem solving that followed. Indeed, we noticed that about half of the children continued to use the schemas during problem solving following the intervention, which presumably helped them to monitor their work and verify their responses. This may have supported the construction of appropriate mental representations of the mathematical structure of the problems, which could in part account for the greater use of appropriate strategies after the intervention.

A practice that we found to be especially helpful was when the instructor asked questions to encourage the children to reflect on their thinking so they could themselves correct any errors and find a more promising avenue for the solution. Instructor questions, together with the feedback of the other members of the group, encouraged the students to "talk out" the problem and change course during problem solving if necessary. This may have increased students' reflections about the relative appropriateness of a number of different solution strategies. By asking questions, the instructor was also provided with information about how the children were thinking about the problems. This, in turn, let the instructor modify her instruction to address the children's specific difficulties or misconceptions.

Another aspect that seemed to benefit the students was when the instructor focused on the structural and conceptual aspects of the problems when teaching the different problem types. For example, for action problems, the instructor would describe the action as something that would lead to a decrease or increase in the start number. When children understood this, their ability to monitor and correct their own work appeared to improve, thereby needing less prompting from the instructor. To illustrate, one child completed an action problem but made an error. To help the child see his error, the instructor asked questions about the structure of the problem. She reread the problem with the child. She asked him what would happen to the start number (the initial number of apples in the problem) if apples fell out of the bin. The child answered that the number would go down. The instructor asked him to look at the end number (the final number of apples after the action) and compare it to the start. In doing so, the child said, "Wait, that's not right. The number can't be bigger!"

When the children used the schemas during instruction, the visual representations appeared to help them organize the information provided in the problem. Prior to instruction, most children in the ID group struggled to make sense of how the numbers related to one another. At times, we observed some children in this group having difficulty using the correct numbers, despite having the information in front of them. The use of the schemas during and after instruction may have had a positive effect on students' ability to manipulate several pieces of information at once; the schemas presumably allowed the students to offload the numerical information in the problem so they could better focus on the problem structure and select an appropriate strategy. This is consistent with research showing that children with learning difficulties have working memory deficits and tend to rely on immature problem-solving strategies like finger counting as a result (e.g., Geary, 2004). The organizational support provided by the schemas appeared to provide easier access to key parts of the problem, thereby freeing up their working memory capacity.

While our instruction brought forth positive change in students' problem solving, some modifications are required to further enhance performance in students with learning difficulties as well as those with intellectual disabilities. First, the Action problems presented during the instruction all described the start, action, and end sets in that order (e.g., Lisa had 324 pennies (start). She found some more pennies on the sidewalk (action). Now she has 434 pennies (end). How many pennies did Lisa find on the sidewalk?). This led to difficulties in successfully solving the atypical-sequence action transfer problems, especially for the ID group. These students began to enter the numbers into the schema in a rote manner. That is, they would automatically put the first number in the problem in the first part of the schema, the second number in the problem in the second component, and the third number in the last component, disregarding what each component meant within the structure of the problem. This was evident as most of the students in the ID and LD groups failed to correctly solve problems that presented information out of sequence at posttest. This is consistent with research on "psychological sets" (Duncker, 1945): Strategies can become automatic after repetitive use (i.e., in this case, putting the numbers into the schema in a certain order), so that it becomes rote. In these situations, the students did not appear to stop to think about the problem, but rather engaged in a behavior that had served them well in the past.

Another issue centered on the use of multiple strategies. Across the groups, both before and after the intervention, most students used only one strategy. Sometimes, they offered two different variations of the same strategy type category (e.g., direct modeling on fingers and direct modeling by drawing tallies on paper). There are several possible explanations for the lack of flexibility, which are important to consider in the design of future implementations of SBI with children with learning difficulties and children with intellectual disabilities. For one, the length of the intervention may have impacted the children's ability to demonstrate flexibility. As previously mentioned, all students received a total of three hours of instruction over a two-week period. It is possible that three hours are not enough time for children to become more flexible in their strategy use. Increased opportunities for practice are especially important for children who exhibit strategy rigidity (Baroody, 1996; Ostad, 1997). Perhaps with more practice and increased exposure to different types of strategies, children may have been more flexible following the intervention.

Although we believe more time would have been beneficial, we also suggest that flexibility gains were not observed because the strategies that were shared during the intervention may have all been of the same type. That is, as opposed to seeing a direct modeling strategy followed by a counting strategy, for example, children may have seen two different ways to directly model the same problem or two different ways to use counting strategies. In fact, evidence that this occurred was observed following instruction. Furthermore, previous mathematics instruction could also have played a role in children's choice to use the same strategy. It is possible that during the instruction they had received in school, they were not given the opportunity to explore different solution strategies.

Another possibility is that perhaps students with learning difficulties and children with intellectual disabilities need explicit instruction on how to use a variety of strategies to solve a given problem. In fact, there is a debate in the literature as to whether mathematics instructors should explicitly teach strategy flexibility to students who have learning difficulties. Specifically, the controversy centers on whether children should be taught to use a variety of strategies flexibly, or only a small handful of strategies for solving problems (Verschaffel et al., 2007). One argument is that children who struggle could benefit from using a small number of strategies repeatedly, which would alleviate pressures on working memory (Baxter, Woodward, and Olson, 2001). On the other hand, others have argued that flexibility should be explicitly targeted from the beginning (e.g., Butler et al., 2001; Verschaffel et al., 2007).

Conclusion

In this chapter, we described an instructional intervention based on SBI that we delivered to three small groups of children, one group with intellectual disabilities and autism spectrum disorders, a second group who were identified by their teacher because they were performing poorly relative to their peers, and a third group of students who were average performers. We described elements of improved problem solving after the intervention and speculated about how our instantiation of SBI may have supported students' performance. Our conclusions are necessarily tentative because our sample was small; nevertheless, the speculations we draw offer an existence proof that students with learning difficulties and intellectual disabilities can indeed learn and use key mathematical concepts in the context of appropriate instruction. Our approach deviates from existing work in the field in that we shifted attention away from students' deficits toward their strengths. In this way, our observations can inspire new lines of research on the mathematical potential of students with difficulties and intellectual disabilities.

Children with intellectual disabilities need to learn adaptive mathematical skills (such as problem solving) to help them become as autonomous as possible in their daily lives. In this chapter, we provided some evidence that students with intellectual disabilities and autism spectrum disorders, as well students presenting with mathematical learning difficulties, are able to develop both conceptual and procedural knowledge. Our observations are encouraging teachers and other practitioners who work with this population. Further, given the positive effects of instruction that was delivered in small-group settings, it appears that SBI could be a promising approach for teaching mathematical word problem solving in inclusive classrooms.

References

Andersson, U. (2007). The contribution of working memory to children's mathematical word problem solving. *Applied Cognitive Psychology, 21*(9), 1201–1216. https://doi.org/10.1002/acp.1317.

Baroody, A. J. (1996). Self-invented addition strategies by children with mental retardation. *American Journal on Mental Retardation, 101,* 72–89.

Baroody, A. J. (1999). The development of basic counting, number, and arithmetic knowledge among children classified as mentally handicapped. *International Review of Research in Mental Retardation, 22,* 51–103. https://doi.org/10.1016/s0074-7750(08)60131-7.

Baxter, J. A., Woodward, J., & Olson, D. (2001). Effects of reform-based mathematics instruction on low achievers in five third-grade classrooms. *The Elementary School Journal, 101*(5), 529–547. https://doi.org/10.1086/2F499686.

Bird, G., & Buckley, S. (2001). Number skills for individuals with down syndrome—An overview. In *Down Syndrome Issues and Information.* https://doi.org/10.3104/9781903806135.

Bissonnette, S., Richard, M., Gauthier, C., & Bouchard, C. (2010). Quelles sont les stratégies d'enseignement efficaces favorisant les apprentissages fondamentaux auprès des élèves en difficulté de niveau élémentaire? Résultats d'une méga-analyse. *Revue de Recherche Appliquée sur l'Apprentissage, 3,* art. 1.

Boonen, A. J., de Koning, B. B., Jolles, J., & van der Schoot, M. (2016). Word problem solving in contemporary math education: A plea for reading comprehension skills training. *Frontiers in Psychology, 7,* 191. https://doi.org/10.3389/fpsyg.2016.00191.

Bransford, J. D., & Johnson, M. K. (1972). Contextual prerequisites for understanding: Some investigations of comprehension and recall. *Journal of Verbal Learning and Verbal Behavior, 11*(6), 717–726. https://doi.org/10.1016/s0022-5371(72)80006-9.

Browder, D. M., Spooner, F., Ahlgrim-Delzell, L., Harris, A. A., & Wakemanxya, S. (2008). A meta-analysis on teaching mathematics to students with significant cognitive disabilities. *Exceptional Children, 74*(4), 407–432. https://doi.org/10.1177/001440290807400401.

Butler, F. M., Miller, S. P., Lee, K. H., & Pierce, T. (2001). Teaching mathematics to students with mild-to-moderate mental retardation: A review of the literature. *Mental Retardation, 39*(1), 20–31. https://doi.org/10.1352/0047-6765(2001)039%3c0020:tmtswm%3e2.0.co;2.

Carpenter, T. P., Fennema, E., Franke, M. L., Levi, L., & Empson, S. B. (2014). *Children's mathematics: Cognitively guided instruction* (2nd ed.). Portsmouth, NH: Heinemann.

Cawley, J. F., Parmar, R. S., Yan, W., & Miller, J. H. (1998). Arithmetic computation performance of students with learning disabilities: Implications for curriculum. *Learning Disabilities Research & Practice, 13*(2), 68–74.

Chen, Z. (1999). Schema induction on children's analogical problem solving. *Journal of Educational Psychology, 91*(4), 703–715. https://doi.org/10.1037/0022-0663.91.4.703.

Clements, D. H., & Sarama, J. (2009). *Learning and teaching early math: The learning trajectories approach.* New York: NY: Routledge. https://doi.org/10.4324/9780203883389.

Cummins, D. D. (1991). Children's interpretations of arithmetic word problems. *Cognition and Instruction, 8*(3), 261–289. https://doi.org/10.1207/s1532690xci0803_2.

Daroczy, G., Wolska, M., Meurers, W. D., & Nuerk, H. C. (2015). Word problems: A review of linguistic and numerical factors contributing to their difficulty. *Frontiers in Psychology, 6,* 348. https://doi.org/10.3389/fpsyg.2015.00348.

De Corte, E., Verschaffel, L., & De Win, L. (1985). Influence of rewording verbal problems on children's problem representations and solutions. *Journal of Educational Psychology, 77*(4), 460–470. https://doi.org/10.1037/0022-0663.77.4.460.

Duncker, K. (1945). On problem solving. *Psychological Monographs, 58*(5), 270.

Fletcher, J. M., Lyon, G. R., Fuchs, L. S., & Barnes, M. A. (2006). *Learning disabilities: From identification to intervention.* Guilford Press. 10.1080%2F09297040701455171.

Fuchs, L. S., Fuchs, D., Finelli, R., Courey, S. J., & Hamlett, C. L. (2004a). Expanding schema-based transfer instruction to help third graders solve real-life mathematical problems. *The American Educational Research Journal, 41*(2), 419–445. https://doi.org/10.3102/00028312041002419.

Fuchs, L. S., Fuchs, D., Prentice, K., Hamlett, C. L., Finelli, R., & Courey, S. J. (2004b). Enhancing mathematical problem solving among third-grade students with schema-based instruction. *Journal of Educational Psychology, 96*(4), 635–647. https://doi.org/10.1037/0022-0663.96.4.635.

Geary, D. C. (2004). Mathematics and learning disabilities. *Journal of Learning Disabilities, 37*(4), 4–15. https://doi.org/10.1177/00222194040370010201.

Geary, D. C., Hoard, M. K., Byrd-Craven, J., & DeSoto, M. C. (2004). Strategy choices in simple and complex addition: Contributions of working memory and counting knowledge for children with mathematical disability. *Journal of Experimental Child Psychology, 88*(2), 121–151. https://doi.org/10.1016/2Fj.jecp.2004.03.002.

Geary, D. C., Hoard, M. K., Byrd-Craven, J., Nugent, L., & Numtee, C. (2007). Cognitive mechanisms underlying achievement deficits in children with mathematical learning disability. *Child Development, 78*(4), 1343–1359. https://doi.org/10.1111/j.1467-8624.2007.01069.x.

Geary, D. C., Hoard, M. K., Nugent, L., & Byrd-Craven, J. (2008). Development of number line representations in children with mathematical learning disability. *Developmental Neuropsychology, 33*(3), 277–299. https://doi.org/10.1080/87565640801982361.

Ginsburg, H. P. (1997). Mathematics learning disabilities: A view from developmental psychology. *Journal of Learning Disabilities, 30*(1), 20–33. https://doi.org/10.1177/002221949703000102.

Griffin, C. G., & Jitendra, A. K. (2009). Word problem-solving instruction in inclusive third-grade mathematics classrooms. *The Journal of Educational Research, 12*(3), 187–201. https://doi.org/10.3200/joer.102.3.187-202.

Henry, L. A., Messer, D. J., & Poloczek, S. (2018). Working memory and intellectual disabilities. In T. P. Alloway (Ed.), *Working memory and clinical developmental disorders: Theories, debates and interventions* (pp. 9–21). Abingdon-on-Thames, UK: Routledge. https://doi.org/10.4324/9781315302072.

Human Resources and Skills Developmental Canada. (2011). *Disability in Canada: A 2006 profile* (Cat. No. HS64/2010). Retrieved from http://www.hrsdc.gc.ca/eng/disability_issues/reports/disability_profile/2011/index.shtml.

Hutchinson, N. L. (1993). Effects of cognitive strategy instruction on algebra problem solving of adolescents with learning disabilities. *Learning Disability Quarterly, 16*(1), 34–63. https://doi.org/10.2307/1511158.

Individuals with Disabilities Education Improvement Act. (2004). 20 U.S.C. § 1400.

Jitendra, A. K., & Star, J. R. (2011). Meeting the needs of students with learning disabilities in inclusive mathematics classrooms: The role of schema-based instruction on mathematical problem-solving. *Theory into Practice, 50*(1), 12–19. https://doi.org/10.1080/00405841.2011.534912.

Jitendra, A. K., DePipi, C. M., & Perron-Jones, N. (2002). An exploratory study of schema-based-word-problem-solving instruction for middle school students with learning disabilities: An emphasis on conceptual and procedural understanding. *The Journal of Special Education, 36*(1), 23–38. https://doi.org/10.1177/00224669020360010301.

Jitendra, A. K., Nelson, G., Pulles, S. M., Kiss, A. J., & Houseworth, J. (2016). Is mathematical representation of problems an evidence-based strategy for students with mathematics difficulties? *Exceptional Children, 83*(1), 8–25. https://doi.org/10.1177/0014402915625062.

Judd, T. P., & Bilsky, L. H. (1989). Comprehension and memory in the solution of verbal arithmetic problems by mentally retarded and nonretarded individuals. *Journal of Educational Psychology, 81*(4), 541–546. https://doi.org/10.1037/0022-0663.81.4.541.

Lee, K., Ng, E. L., & Ng, S. F. (2009). The contributions of working memory and executive functioning to problem representation and solution generation in algebraic word problems. *Journal of Educational Psychology, 101*(2), 373–387. https://doi.org/10.1037/a0013843.

Lucangeli, D., Tressoldi, P. E., & Cendron, M. (1998). Cognitive and metacognitive abilities involved in the solution of mathematical word problems: Validation of a comprehensive model. *Contemporary Educational Psychology, 23*(3), 257–275. https://doi.org/10.1006/ceps.1997.0962.

Mastropieri, M. A., Bakken, J. P., & Scruggs, T. E. (1991). Mathematics instruction for individuals with mental retardation: A perspective and research synthesis. *Education and Training in Mental Retardation, 26*, 115–129.

Ministère de l'éducation et de l'enseignement supérieur (1999). *Policy on special education: Adapting our schools to the needs of all students.* Retrieved from: http://www.education.gouv.qc.ca/fileadmin/site_web/documents/dpse/adaptation_serv_compl/polite00A.pdf.

Montague, M., & Applegate, B. (1993). Mathematical problem-solving characteristics of middle school students with learning disabilities. *Journal of Special Education, 27*, 175–201. https://doi.org/10.1177/002246699302700203.

National Mathematics Advisory Panel. (2008). *Foundations for success: The final report of the National Mathematics Advisory Panel.* Washington, D.C..: U.S. Department of Education.

No Child Left Behind Act of 2001. (2002). P.L. 107–110, 20 U.S.C. § 6319.

Ontario Ministry of Education. (1990). The Education Act, Bill 82, § 7 et seq.

Ostad, S. A. (1997). Developmental differences in addition strategies: A comparison of mathematically disabled and mathematically normal children. *British Journal of Educational Psychology, 67*(3), 345–357. https://doi.org/10.1111/j.2044-8279.1997.tb01249.x.

Ozonoff, S., & Schetter, P. L. (2007). Executive dysfunction in Autism spectrum disorders: From research to practice. In L. Meltzer (Ed.), *Executive function in education: From theory to practice* (pp. 287–308). New York, NY: Guilford. https://doi.org/10.5860/choice.45-1588.

Powell, S. R., Fuchs, L. S., & Fuchs, D. (2013). Reaching the mountaintop: Addressing the common core standards in mathematics for students with mathematics difficulties. *Learning Disabilities Research & Practice, 28*(1), 38–48. https://doi.org/10.1111/ldrp.12001.

Raghubar, K. P., Barnes, M. A., & Hecht, S. A. (2010). Working memory and mathematics: A review of developmental, individual difference, and cognitive approaches. *Learning and Individual Differences, 20*(2), 110–122. https://doi.org/10.1016/j.lindif.2009.10.005.

Rockwell, S. B., Griffin, C. C., & Jones, H. A. (2011). Schema-based strategy instruction in mathematics and the word problem-solving performance of a student with. *Focus on Autism and Other Developmental Disabilities, 26*(2), 87–95. https://doi.org/10.1177/1088357611405039.

Rose, D., & Rose, K. (2007). Deficits in executive function processes. In L. Meltzer (Ed.), *Executive function in education: From theory to practice* (pp. 287–308). New York, NY: Guilford. https://doi.org/10.5860/choice.45-1588.

Stein, J. A., & Krishnan, K. (2007). Nonverbal learning disabilities: The challenges of effective assessment and teaching. In L. Meltzer (Ed.), *Executive function in education: From theory to practice* (pp. 287–308). New York, NY: Guilford. https://doi.org/10.5860/choice.45-1588.

Swanson, H. L., & Beebe-Frankenberger, M. (2004). The relationship between working memory and mathematical problem solving in children at risk and not at risk for serious math difficulties. *Journal of Educational Psychology, 96*(3), 471–491. https://doi.org/10.1037/0022-0663.96.3.471.

Swanson, H. L., & Sachse-Lee, C. (2001). Mathematical problem solving and working memory in children with learning disabilities: Both executive and phonological processes are important. *Journal of Experimental Child Psychology, 79*(3), 294–321. https://doi.org/10.1006/jecp.2000.2587.

Verschaffel, L., Torbeyns, J., De Smedt, B., Luwel, K., & Van Dooren, W. (2007). Strategy flexibility in children with low achievement in mathematics. *Educational and Child Psychology, 24*(2), 16–27.

Xin, Y. P., & Jitendra, A. K. (1999). The effects of instruction in solving mathematical word problems for students with learning problems: A meta-analysis. *The Journal of Special Education, 32*(4), 207–225. https://doi.org/10.1177/002246699903200402.

Xin, Y. P., Jitendra, A. K., & Deatline-Buchman, A. (2005). Effects of mathematical word problem-solving instruction on middle school students with learning problems. *The Journal of Special Education, 39*(2), 181–192. https://doi.org/10.1177/00224669050390030501.

Zheng, X., Swanson, H. L., & Marcoulides, G. A. (2011). Working memory components as predictors of children's mathematical word problem solving. *Journal of Experimental Child Psychology, 110*(4), 481–498. https://doi.org/10.1016/j.jecp.2011.06.001.

Tablets as Elementary Mathematics Education Tools: Are They Effective and Why

Adam K. Dubé, Sabrina Shajeen Alam, Chu Xu, Run Wen and Gulsah Kacmaz

Descriptors such as 'easy to use', 'accessible', and 'fun' often accompany the promotion of tablets in education (Connell, Lauricella, & Wartella, 2015). Tablets are perceived this way because they are controlled by simple gestures that appear to require a low degree of formal instruction (Aziz, 2013) and because many of the educational apps are seen either as interactive hands-on approaches to traditional academic tasks or are framed as educational games (McEwen & Dubé, 2017). According to a survey of the educational applications (apps) category on iTunes, four of the top five apps are for STEM subject areas (Science, Technology, Engineering, and Mathematics) and mathematics apps account for the greatest number of subject-specific content (Shuler, 2012). It is possible that the reason tablets are used as mathematics learning tools and the reason mathematics apps are so popular is because the proclaimed 'easy to use', 'accessible', and 'fun' aspects of mathematics apps offset the attitude that the subject is 'difficult', 'inaccessible', and 'boring' (Larkin & Jorgensen, 2016).

Indeed, an optimistic attitude towards both tablets generally and tablet mathematics apps specifically seems to be held by children, educators, and parents alike but the reasons behind the optimism vary. For children, a study of kindergarten to grade 2 children's use of mathematics tablet apps found that they were far more engaged

A. K. Dubé (✉) · S. S. Alam · C. Xu · R. Wen · G. Kacmaz
Department of Educational and Counselling Psychology, McGill University, Montreal, QC, Canada
e-mail: adam.dube@mcgill.ca

S. S. Alam
e-mail: sabrina.alam@mail.mcgill.ca

C. Xu
e-mail: chu.zu2@mail.mcgill.va

R. Wen
e-mail: run.wen@mail.mcgill.ca

G. Kacmaz
e-mail: gulsah.kacmaz@mail.mcgill.ca

K. M. Robinson et al. (eds.), *Mathematical Learning and Cognition in Early Childhood*,
https://doi.org/10.1007/978-3-030-12895-1_13

(i.e. played twice as long) with apps that incorporated interactive hands-on activities or gaming than they were with apps that mimicked more traditional academic tasks (i.e. mad minutes, flash cards; McEwen & Dubé, 2016). This suggests that children enjoy mathematics apps of a certain kind and do not simply like them because they are on a tablet. A survey of 1234 teachers on their use of educational technology found that teachers who report more student-centred teaching pedagogies were more likely to use educational technology in the classroom and that tablets were among the top four most used devices in the classroom—alongside digital cameras, video games, and computers (Blackwell, Lauricella, & Wartella, 2014), which are all technologies now subsumed by tablets. Perhaps the intimate (i.e. individualized) nature of tablets espoused in the original iPad marketing (Apple, 2010) contributed to teachers both viewing tablets as student-centred and subsequently using them in their classrooms. For parents, Wood et al. (2016) found that over 94% of parents allowed their children to access mobile technologies like tablets and believed that their children were both familiar with and held favourable attitudes towards iPads. In fact, one study found that 61% of parents' believed children should be introduced to iPads prior to 2.5 years of age and fewer than 10% of parents believed that introduction to iPads should wait until school age (Wood et al., 2016). Further, Kosko and Ferdig (2016) found that parents specifically singled out the ability of tablets to increase their child's interest in and understanding of mathematics. Clearly tablets are popular with children and teachers, but whether the efficacy of tablets as mathematics learning tools warrants this level of optimism and interest is a more important question to consider.

Are Tablets Effective Mathematics Education Tools?

Since the introduction of the iPad in 2010, there has been a surge in research on the use of tablets in mathematics education. Fabian, Topping, and Barron (2016) conducted a meta-analysis on the effectiveness of mobile devices (i.e. cell phones, tablets, iPods) as mathematics learning tools and identified over 60 studies. One issue highlighted by Fabian and colleagues is the problem of how to measure effectiveness; is a mathematical app effective if it simply gets children to use it (behavioural engagement), does it have to make children better at math (achievement), or does it even have to make them like math (attitudes)? Previous research and theory have argued that each of these outcomes makes an equally important contribution to understanding the use of information communication technologies (ICT) for learning (Mouza & Lavigne, 2012). However, it is not common for all three aspects to be measured in one study and previous research on other related ICTs (e.g. educational video games) has been criticized for focusing on one aspect (engagement) to the detriment or neglect of others (i.e. attitudes and achievement, Mayer, 2014). Thus, a review on the effectiveness of tablet mathematics apps requires consideration of all possible outcome metrics.

While Fabian et al's. (2016) review included multiple outcome metrics, drawing conclusions from this study about the potential of tablets as mathematics learning tools for elementary education is limited due to three issues. First, all of the studies included in the review were conducted in 2012 or earlier and a substantial proportion of them were published in 2011 or sooner (e.g. 78% of the studies investigating achievement outcomes). This means that they were conducted prior to the tablet surge caused by the release of the iPad in 2010 and, resultantly, were conducted with technology markedly different than modern tablets. Second, the studies include all types of mobile devices and not just tablets, which further limit the generalizability of the review to modern tablet computers. Third, the reviewed studies covered mathematics education from kindergarten to grade 12 and therefore groups together very different areas of mathematics (e.g. counting, arithmetic, calculus). The present review addresses these issues and focuses on studies published since 2012 in which tablet computers are used as mathematics learning tools by early elementary students (i.e. kindergarten to grade 5).

Review Methodology and Goals

The present review involved a search of math-related tablet learning studies published in the years of 2012–2017. The search process involved four researchers using indexing databases (Scopus, Web of Knowledge, Google Scholar, ProQuest, WorldCat, OAlster) and academic hosting websites (ResearchGate, Academia.edu) to search for relevant studies. The keywords used in the search were tablets, touch screens, learning, mathematics, arithmetic, and variations of these terms (e.g. 'tablet' included tablets, tablet computer, iPad, Android tablet, LeapPad). This search resulted in 204 articles, which were saved into a Mendeley shared database and then reviewed to ensure the bibliographic information and abstracts were accurate. The abstracts of the articles were then exported and screened to remove any non-tablet studies (e.g. cell phone) leaving 47 studies in total and then further screened to remove studies with participants outside of kindergarten to grade 5 (e.g. adult or infant studies). The final 25 studies were then coded according to their outcome metrics (i.e. engagement, math attitudes, achievement; see Table 1) and were used to inform the present literature review. Some of the studies that did not meet the strict inclusion criteria were also used when they reinforce or extend the conclusions from the selected 25 studies.

The goal of this review is to discuss whether tablet computers are useful mathematics learning tools and why. To this end, the studies will be compared to determine whether tablet computers are effective at one of three tasks: (1) engaging children with mathematics, (2) improving children's attitudes towards mathematics, and (3) improving children's mathematics achievement. Addressing why tablets are effective is critical because even though several studies may show a significant effect of tablets, the cause for the effect can differ from study to study and this has important implications for how to use tablets effectively. Simply introducing tablets into a classroom may not improve engagement, attitudes, and achievements, but smartly introducing them into classrooms may.

Table 1 Recent studies of tablets in elementary mathematics education

Studies	Engagement	Achievement	Attitudes
Bebell and Pedulla (2015)		X	
Bray and Tangney (2016)	X		
Carr (2012)		X	
Clarke and Abbott (2016)	X		
Desoete, Praet, Velde, Craene, and Hantson (2016)	X		
Dubé and McEwen (2016)	X		
Fabian et al. (2016)	X	X	X
Falloon (2013)	X		
Hwang, Shadiev, Tseng, and Huang (2015)		X	
Ingram, Williamson-Leadley, and Pratt 2016	X	X	
Jackson, Brummel, Pollet, and Greer (2013)	X		
Jong, Hong, and Yen (2013)		X	
Kiili, Ketamo, Koivisto, and Finn (2014)	X		
Kyriakides et al. (2016)	X		
Larkin and Jorgensen (2016)			X
McEwen and Dubé (2016)	X		
McEwen and Dubé (2015)	X		
Musti-Rao and Plati (2015)		X	
Riconscente (2013)		X	X
Sinclair and Heyd-Metzuyanim (2014)	X		
Sinclair, Chorney, and Rodney (2016)	X		
Strouse et al. (2017)	X	X	X
Tucker, Moyer-Packerman, Westenskow, and Jordan (2016)	X	X	
Yang, Chang, Cheng, and Chan (2016)		X	
Zhang, Trussell, Gallegos, and Asam (2015)		X	

In the discussion on engagement, the review will outline how most studies focus on behavioural engagement (i.e. keeping children playing) whereas relatively few studies investigate cognitive engagement and how it is facilitated through visual design, gesture-based interactions, gamification, and discursive collaboration (i.e. children working together). The discussion on math attitudes uses Pekrun's (2006) theory of achievement emotions to argue that the interactive nature of tablets gives children a sense of control over their math learning, which fosters feelings of mastery and improves math attitudes. Finally, the review of math achievement research will demonstrate how the learning theory used by the researcher determines if and how tablets improve children's mathematics ability.

Can Tablets Engage Children with Math?

The belief that new technologies can engage learners is at the heart of why they have been introduced into education since audio-visual materials (i.e. photographs) were considered a new educational innovation (Dale, 1954). In the context of tablet mathematics apps, there are three levels of engagement, cognitive, emotional, and behavioural (Annetta, Minogue, Holmes, & Cheng, 2009). Cognitive engagement means that tablet users are mentally invested in the learning activity and willing to exert effort to complete the academic task (Fredricks, Blumenfeld, & Paris, 2004). Emotional engagement refers to how the use of the mathematical app can elicit positive (i.e. interest) and negative (i.e. boredom) feelings that serve as motivators to either maintain or cease users' continued cognitive or behavioural engagement with the app (Chang, Evans, Kim, Deater-Deckard, & Norton, 2014). Behavioural engagement includes attentiveness, diligence, and following the rules of the mathematical app (Deater-Deckard, Chang, & Evans, 2013), and it is the most frequently measured aspect of engagement, often indexed by the total amount of tablet use or degree of goal-directed tablet use (e.g. Dubé & McEwen, 2016). While these three aspects of engagement are often measured separately, the goal is for initial behavioural engagement with the app to evoke feelings of interest or joy that translate into prolonged interaction with mathematics content and, consequently, cognitive engagement with the mathematics content itself. However, cognitive engagement or emotional engagement are not a guaranteed result of prolonged behavioural engagement and educational technology research has been criticized for making this assumption (Mayer, 2014). Instead, cognitive and emotional engagement may be somewhat, but not entirely, independent of behavioural engagement.

The most robust finding from research on tablets in mathematics education is that tablets can engage children with mathematics, but the studies differ significantly in terms of why this occurs. Some researchers argue that the gamification of tablet mathematics apps produces this engagement (e.g. Kyriakides, Meletiou-Mavrotheris, & Prodromou, 2016; Strouse et al., 2017); some researchers think that the interaction modalities and visuals of tablets are responsible (Desoete et al., 2016); others argue that including tablet mathematics apps in a classroom increases engagement because it facilitates students' discussions about mathematics (Ingram et al., 2016), and others propose that the gestures used to interact with tablets make them more 'hands-on' and that this is responsible for increased levels of both cognitive engagement and affective engagement (Dubé & McEwen, 2015; Tucker et al., 2016). Basically, most researchers in this area seem to agree that tablet mathematics apps are engaging. However, there is less of a consensus on why tablet mathematics apps are engaging. One source of disagreement stems from the type of mathematical app experience researchers choose to study. Some researchers evaluate apps currently available to children, while others design their own custom apps. Not surprisingly, how app developers, who need to make a profit, attempt to engage children with math seems to differ from how researchers attempt to engage children with math.

Why Mathematics Apps Available 'in Stores' Engage Children

A recent tablet study by Strouse et al. (2017) frames the behavioural engagement produced by tablets as a means to foster and promote math practice, an often-ignored topic, and proposes that gamification is the reason why tablets are engaging. The authors noticed that math practice is not supported outside of topic-specific homework assignments (cf., reading practice provided in libraries) and that parents have difficulty constructing and providing feedback with mathematics tasks. In response, they propose that tablet mathematics apps are particularly well suited to solving these issues because they keep children engaged and contain ready-made sets of math problems.

In their study, children from kindergarten to grade 6 ($n = 181$) individually used the iPad app IXL math either in a school setting or in a summer programme to practice basic addition, subtraction, multiplication, division, or fractions for 1–20 h (Strouse et al., 2017). IXL math is based on the US common core curriculum and reportedly contains over 24 billion unique math questions presented in a gamified manner with feedback and reward systems (i.e. badges). Not only did the IXL app engage children, with organizers reporting no behavioural problems during the practice sessions and being able to 'hear the drop of a needle' during the practice sessions (Strouse et al., 2017, p. 179), but also that students found the math practice fun, that they reported feeling less nervous about math during practice, and that they reported believing that the practice helped them become better at math. Thus, tablets' ability to keep children engaged not only makes them good vehicles for practising math outside of school, a difficult and often-ignored challenge, but this engagement may go beyond mere continued exposure to math and can result in improved attitudes towards both practice and mathematics itself. The question remains, however, how are mathematics apps like IXL math able to produce these results?

How commercially available tablet mathematics apps go beyond mere behavioural engagement requires a look at the apps themselves. Desoete et al. (2016) identified the 80 most downloaded children's mathematics apps available on the iTunes app store and categorized them according to how they cognitively engage users through their interaction types and visuals. For interaction types, Desoete et al. (2016) categorized commercial mathematics apps as offering either modifiable, manipulable, or passive interaction, with each of these containing varying levels of cognitive engagement (Goodwin & Highfield, 2013). Apps with modifiable interactions allow learners to construct new representations of mathematics by having the user modify or change content on the screen and then reflect on this change. Slice Fractions is an example of just such an app, in which learners segment whole objects into fractional amounts through swipes/slices of their fingers (a la Fruit Ninja) to solve novel fraction problems (Shapiro, 2017). Apps with manipulative interactions allow for discovery and experimentation with existing mathematics representations but do not focus on building new understanding. The app Park Math HD is a good example of this type of gesture, in which learners practise their understanding of the equal sign by dragging numbers to either side of a scale to balance an equation (McEwen &

Dubé, 2017). Apps with passive interactions focus on instruction and the practice of existing knowledge (i.e. mad minutes) and tend to produce engagement by giving the learner control over the pace of their mathematics learning/practice (Kiili et al., 2014).

For visuals, Desoete et al. (2016) categorized apps into one of three visual levels. Level 1 apps rely solely on the use of text and numerals. Level 2 apps include visuals and colours as simple 'add-ons' that do not contribute to the mathematics content itself. Level 3 apps use colour, geometric shapes, simulations, graphs, or 3D representations to directly inform the mathematics content. These categories were created based on all the mathematics apps in the review, but the proportion of apps in each category varied substantially. Disappointingly, but not surprisingly, Desoete et al.'s (2016) review found that the majority of existing commercially available mathematics apps are cognitively engaging children at a low level and are not taking full advantage of the interaction or visual available on tablets. Specifically, they found that only 3 of the 78 apps reviewed allowed for modifiable interactions and only 20 of the apps used the interface and visuals at Level 3 (Level 2 = 40 apps). This contextual review of popular mathematics apps supports findings from two previous studies of children's use of commercially available mathematics apps, a behavioural analysis and an eye-tracking study.

McEwen and Dubé (2016) video recorded kindergarten to grade 2 children's use of mathematics apps ($n = 36$) on 1 of 4 different tablet computers (iPad, LeapPad2, Innotab, Acer Tab). Three different mathematics apps were used on each device, to ensure that apps made full use of each tablet's unique features (see McEwen & Dubé, 2016 for a list and description of each app). Children's behavioural interactions with the devices were coded to determine the variety and frequency of gestures used by children and the proportion of time spent passively watching content (i.e. instruction), practising existing math knowledge, or exploring mathematics constructively. While most of the apps offered a variety of gestures (e.g. drag, swipe, tilting) and interaction affordances, children spent 90% of their time engaged in practice of previously learned content using only 1 gesture (i.e. tap) to interact with over 90% of the content on the screen. McEwen and Dubé's (2015) eye-tracking study compared how grade 2 children's gaze data differed between visually simple and complex tablet mathematics apps. The visually simple apps in this study fall under Level 3 of Desoete et al.'s (2016) classification system, as they used colour, geometric shapes, and 3D representations to directly inform the mathematics content. The visually complex apps fall under Level 2, as they included colour and images but ones that did not inform the mathematics content. The labels simple and complex also refer to the overall amount of visual information present on the screen, with simple apps containing fewer visual elements than complex app. The study further coded the visual content as either extraneous, germane or intrinsic, according to cognitive load theory (Sweller, 1994). Extraneous content is unrelated to the math content and adds unnecessary cognitive load, while germane and intrinsic content either inform or are central to the math content and add necessary cognitive load (Sweller, Van Merriënboer, & Paas, 1998). McEwen and Dubé (2015) found that the more visually complex the apps, the more children paid attention to the extraneous, unimportant content in those apps. Further, difficulty ignoring unrelated content was exacerbated for children with low attentional control (i.e. those that need the most help).

What is particularly worrisome is that the visually complex apps in the eye tracker study were the ones children played longer and reported liking more in the behavioural observation study (McEwen & Dubé, 2016). Considering the results from Desoete et al. (2016) and McEwen and Dubé (2015, 2016) together, it seems that the majority of commercially available mathematics apps attempt to engage children using visuals that are simple add-ons that do not inform the mathematics content and, unfortunately, that these Level 2 apps engage children the most, for the wrong reasons, and with the wrong kind of content. So, commercial tablet apps are engaging children but why they are engaging children is troublesome. The majority of the apps available to children, parents, and teachers rely on relatively simplistic cognitive engagement approaches that are drawing more attention to the entertainment aspect of the apps than the math itself (Falloon, 2014).

Why Mathematics Apps Made or Chosen by Researchers Engage Children

The potential of tablet mathematics apps to engage children in more complex ways, as seen in a handful of commercially available apps, has spurred researchers into either studying these small handful of quality apps or creating custom tablet mathematics apps. The majority of this work has focused on how tablets allow students to physically interact with mathematics through gestures.

Gestures like tap and drag are synonymous with tablets, and this central affordance of the devices has caused many to think of tablets as a 'hands-on' way to learn mathematics (Aziz, 2013; Dubé & McEwen, 2015). For example, commercial apps like Motion Math Zoom go beyond the simple tap to select an answer found in most tablet mathematics apps to using gestures to cognitively engage students with the underlying mathematics content in ways that align and reinforce the conceptual underpinnings of the content. In Motion Math Zoom, learners navigate a virtual number line that scales from 1 to 10, 10 to 100, and 100 to 1000 using the drag gesture to move linearly through the number line from left to right (drag to the right and the numbers increase, i.e. 5, 6, 7, 8) and using the pinch gesture to move exponentially (expand two fingers apart to quickly shift the scale up from 1s to 10s to 100s to 1000s and pinch two fingers together to shift down). How children interact with this app directly aligns with the underlying continuous nature of the mental number line, as dragging spans the distance between numbers while simply tapping a number would treat them as discrete quantities. An alignment between gestures and content cognitively engages children with mathematics concepts in a more meaningful way—partially because gestures constitute an embodied interaction that brings together physical movement and conceptual learning (Abrahamson &

Bakker, 2016; Alibali & Nathan, 2012)—and has been found to improve subsequent performance on related mathematics tasks.

For example, Dubé and McEwen (2015) used two custom number line tablet apps—one where numbers were placed on the number line using a tap gesture and one where numbers where placed with a drag gesture—and found that placement accuracy was greater with the drag gesture and that participants who used the drag gesture performed better on subsequent near and far transfer tasks requiring a continuous understanding of numbers. Further, Moyer-Packenham et al. (2016) found that not only does the gesture-content alignment of an app improve learning of mathematics but that there must also be a gesture-ability alignment as well. In their study, 100 preschool to grade 2 children's interactions with 18 tablet mathematics apps were coded to investigate whether the cognitive ability of the child affected their use of helping and hindering gestures. Interestingly, whether a specific gesture was helping or hindering depended less on the gesture itself and its use in the app than it depended on the child. For some children, the use of a drag gesture to group objects together in an addition task helped them connect the operation of addition to the concept of grouping. For other children, the drag gesture resulted in objects being misplaced or dragged to the incorrect location and a simple tap gesture to place the object would have been better.

This gesture-ability alignment was further explored by Tucker et al. (2016) in a follow-up study involving 45 min observations of 33 grade 2 children using the same mathematics app. They found that children's use of a given gesture changed as a function of time spent with the app. For example, a child might start by grouping objects together in a counting task using the drag gesture but over the course of using the app would switch to using the tap gesture to instantaneously transport the object from one location to the next (Tucker et al., 2016). This change in gesture use suggests that initial interactions with tablet mathematics apps often begin with the use of more concrete and real gesture but, with time, transition towards more abstract ones (Novack et al., 2014).

Sinclair et al. (2016) interdisciplinary approach to the study of tablet mathematics apps goes one step beyond that of Tucker et al. (2016) and Dubé and McEwen (2016). They argue that not only do gestures cognitively engage children with mathematics to varying degrees of success but that the act of physically touching the screen is itself a form of affective engagement that lies at the core of using tablets as a learning tool. Sinclair and colleagues propose that the gestures used to manipulate mathematics content build a 'rhythm' of interaction that constitutes a 'dynamic coupling' of movement and learning that is motivating and engaging. Imagine a child using a tablet app to count objects on the screen through the use of a drag gesture. For each object, the child places their finger on it, drags the object to a given location, and then repeats this motion for every item in the set. This repetition creates a flow like experience, which is said to be intrinsically enjoyable and engaging (Csikszentmihalyi, 1990). In Sinclair and colleagues' observations of grade 1 and 2 children's use of the app Touch Counts, they found that the act of gesturing was somewhat self-perpetuating and could sometimes override the mathematics task itself. For example, one child was observed counting objects using a drag gesture and became so engrossed in the activity that he did not cease counting, or even flinch, when the app accidentally reset the tally. Instead, he just kept executing the gesture again and again, enacting the gesture divorced from its original purpose. This example and the concept of rhythm further highlight how children physically interact with mathematics on tablets (i.e. gestures) may be just as important to engagement as either the mathematics content in the lesson or the visual representations of the content on the screen.

Thus, tablet mathematics apps not only result in children interacting with math for an extended period but, if they are well designed, then the child is experiencing a multi-level interaction that includes behavioural, affective, and cognitive components that may translate into increased interest or competence in math.

Many Kids 'Hate' Math: Can Tablets Improve Children's Attitudes Towards Mathematics?

The idea that tablets can somehow change or shape children's attitudes towards mathematics may be just as important as increasing their competency. Children's negative attitudes towards mathematics start early and intensify across elementary education (Ma & Kishor, 1997; Wigfield & Meece, 1988), are precursors to math anxiety (Ahmed, Minnaert, Kuyper, & Van Den Werf, 2012; Hembree, 1990), and are somewhat responsible for the majority of adults holding a negative attitude towards mathematic, with women holding stronger negative attitudes than men (Dowker, Sarkar, & Looi, 2016). Research suggests that children possess a positive opinion of tablets in education (Dündar and Akçayir 2014) and the possibility that tablets might ameliorate children's attitudes towards mathematics should not be undervalued.

There is a relative dearth of research on the effect of tablets on children's attitudes towards mathematics (Fabian et al., 2016). The preponderance of work in this area has focused on children's attitudes towards the technology (e.g. Jackson et al., 2013), on the effect of other related technologies (e.g. Shin, Sutherland, Norris, & Soloway, 2012), or on the attitudes of older students (e.g. Ross, Sibbald, & Bruce, 2009). Fabian et al. (2016) also found there was a lack of research on math attitudes (5 studies in total), most of the research reviewed was conducted before the iPad was released, and that the results from these studies were mixed (2 positive: Main & O'Rourke, 2011; Wu, Hsiao, Chang, & Sung, 2006, 3 neutral: Jaciw, Toby, & Ma 2012; Miller & Robertson 2010, 2011). The following review suggests that tablets can improve children's attitudes towards mathematics and Pekrun's control-value theory of achievement emotions (2006) will be used to understand why.

One study by Larkin and Jorgensen (2016) used the video feature of iPads to collect the math attitudes of children in grades 3 and 6 ($n = 105$) over a 10-week period in a sort of 'Big Brother' confession room format. Even though this study did not look at the effect of tablets on math attitudes, it reinforced previous findings that math attitudes become increasingly negative in the later school years (e.g. Wigfield & Meece, 1988) and provides guidance as to how tablets could improve students' attitudes towards math. One year 6 student reported that, 'It absolutely sucks. And it sucks because it's hard and you have to do it every single day' (Larkin & Jorgensen, 2016, p. 938). The most commonly reported negative attitudes towards math included the idea that math is boring, that is useless/not applicable to everyday life, that it involves too many paper and pencil practice problems that students must copy off the board by 'writ[ing] them all down' (p. 940), and that the problems are too difficult. These issues could all theoretically be addressed through the use of mathematics apps like Strouse et al. (2017) previously discussed in their study of IXL math, which gamifies learning (boredom), contains 'billions' of problems (writing them all down) that are presented in authentic contexts (useless), and can be adjusted to the learners ability (difficulty). In that study, children who held negative math attitudes used the IXL app to practise math just as frequently as children who held positive attitudes towards math and both groups reported feeling less anxious about practising math after using the app. However, Strouse and colleagues' study was not designed to directly assess math attitudes and these results were anecdotal.

To our knowledge, only one study has systematically investigated the effect of a tablets on children's attitudes towards mathematics, but the results from this single study are compelling. Riconscente (2013) used a repeated measures crossover experimental design in which grade 5 children ($n = 122$) played the fraction-number line game Motion Math (not the same as Motion Math Zoom discussed previously) 20 min a day for 5 consecutive days or received traditional mathematics instruction not on fractions. Given the crossover design of the study, both groups of participants used the app and served as controls. For the first half of the study, one group was the experimental condition (Class 1) and the other was the control condition (Class 2). Then, halfway through the study, the groups switched. Motion Math involves placing fractional amounts on a number line (e.g. place 1/5 on a number line from 0 to 1) and was chosen because of how number line reasoning may elucidate the proportional and continuous nature of fractions (Wu, 2008). Attitudes towards fractions were assessed in terms of fraction self-efficacy (e.g. "I am good at fractions"), fraction liking (e.g. "Fractions are fun"), and fraction-number line knowledge (e.g. "I know where 1/2 goes on the number line") (p. 197).

The change in attitudes reported in Riconscente (2013)'s study is striking (see Fig. 1). Students' sense of self-efficacy for and liking of fractions increased following tablet mathematical app use. More importantly, children's attitudes towards fractions decreased when they received traditional instruction not on fractions and then increased when they did use the app. Specifically, the delayed intervention group reported poorer attitudes towards fractions following traditional instruction (midtest) than they did at pretest, but their attitudes then increased to match or exceed that of the intervention first group. The author argue that the immediate feedback, multiple opportunities to solve each problem (cf., the single submission situation of a classroom assignment), and the self-paced nature of the app contributed to the students' sense of mastery over the content (i.e. control, Pekrun, 2006) and that this increased their self-efficacy and liking. Conversely, the author pointed out that the number line nature of the game was not particularly authentic (i.e. low in value) and that this may account for why fraction knowledge was assessed as being low, as the students

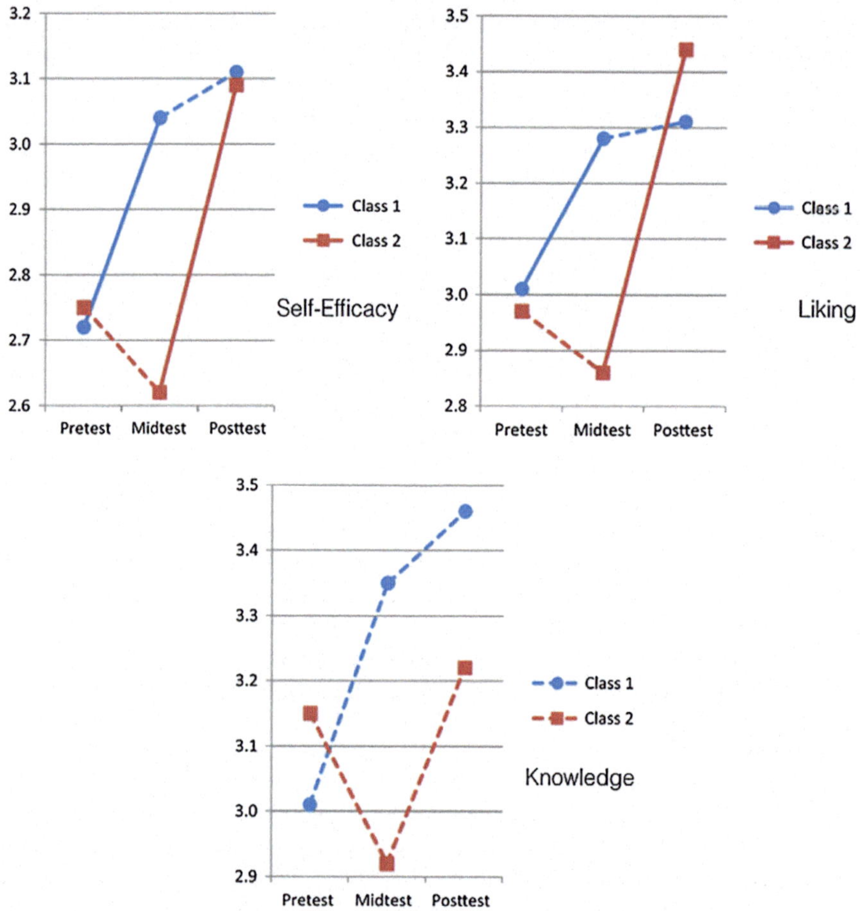

Fig. 1 Change in fraction self-efficacy, liking, and knowledge. Reproduced with Permission from Riconscente (2013)

had difficulty relating the activity in the game to a real-world task. Understanding why mathematics apps may improve children's attitudes requires consideration of how the interactivity and gamification present in tablet mathematics apps increase children's sense of control over and value of mathematics.

Why Tablets Improve Children's Attitudes Towards Math: A Theoretical Approach

Pekrun's control-value theory of achievement emotions (2006) provides a useful framework for understanding how tablet mathematics apps may improve children's attitudes towards mathematics. According to Pekrun, achievement emotions are the result of prospective, retrospective, and activity (i.e. in the moment of learning) appraisals of perceived levels of control and value. Control is the extent to which the learner feels they can influence the outcome or execution of their achievement activity (i.e. agency) while value refers to whether the learner judges the achievement activity to be worthwhile for either extrinsic (grades) or intrinsic (useful in its own right) reasons (Goetz, Frenzel, Pekrun, & Hall, 2007; Pekrun, 2006). Achievement activities judged as providing high levels of control and intrinsic sources of value result in positive emotions such as interest, joy, and pride while activities judged as providing low levels of control and extrinsic sources of value are more likely to produce feelings of boredom, sadness, and shame (Muis, Ranellucci, Trevors, & Duffy, 2015; Pekrun, 2006). Thus, tablet mathematics apps may improve children's attitudes towards mathematics if they are able to provide a sense of control over mathematics learning and convey why mathematics skills are useful.

The interactive and game-like nature of tablet mathematics apps may contribute to children's appraisals of control and value. Interactivity is often identified as a central feature of tablets and cited as a reason why tablets are better than other forms of passive educational technology like video (Dillenbourg & Evans, 2011; Namukasa, Gadanidis, Sarina, Scucuglia, & Aryee, 2016). In terms of achievement emotions, the degree of interactivity (cf., passive learning) may increase children's appraisals of control. For example, in a study of children's tablet mathematical app use, apps with a higher frequency of interaction were judged as more enjoyable and were played longer than apps with a higher proportion of passive consumption (i.e. watching videos or listening to instructions, McEwen & Dubé, 2016). The interactive nature of tablet mathematics apps may result in increased levels of control and this could elicit more positive emotions and improve children's attitudes towards mathematics. However, the extent to which tablet apps enable interactivity varies considerably (Namukasa et al., 2016) and there is more to improving achievement attitudes than just feeling a sense of control over learning.

The gamification of academic tasks that occurs in most tablet mathematics apps may also affect children's attitudes towards mathematics, but the mechanism responsible for this outcome may not be as clear as it first seems. One would think that making mathematics a game results in academic tasks that are inherently fun, but this does not seem to be the case. Dubé and Keenan's (2016) review of educational math game research found that few studies actively and purposefully incorporate fun into the design of math games—instead these studies assume that fun 'just happens' because the mathematics task is labelled as a 'game'. Also, few studies value and measure children's experience of fun (e.g. Lee, Luchini, Michael, Norris, & Soloway, 2004), with those studies that do measure fun finding that children report wanting to play other games, ‘games that are fun’ (Ke, 2008, p. 1614). If the fun aspect of math games does not seem to be responsible for children’s increased appraisals of value, this requires a more complex understanding of what makes an activity a game.

According to Suits (1978, p. 55), ‘Playing a game is the voluntary attempt to overcome unnecessary obstacles’. In many math games, the obstacle to be surmounted is a problem in the game world that can only be solved using specific strategies or facts germane to the mathematical skill being trained (Chorianopoulos, Giannakos, & Chrisochoides, 2014). Solving problems in a math game may increase children’s appraisals of intrinsic value because math knowledge is being applied to achieve a goal (i.e. beat the game) instead of being applied as a form of practice (Boyle et al., 2016).

How well math problems are integrated into gameplay and made meaningful to the player varies considerably from game to game. This is something that is not lost on children, who have exclaimed ‘Oh these are math games …’ while playing games like Candy Factory (Ke, 2008). However, other games like DragonBox Algebra are designed to fully integrate the mathematical skill, in this case algebra, into the gameplay and children playing these games have reported even being unaware that the type of thinking used to play the game involved mathematics (Tucker & Johnson, 2017). Thus, the interactive and game-like interactions found in some tablet mathematics apps may result in increased appraisals of both control and value for mathematical tasks and this could result in increased experiences of positive emotions elicited during mathematics learning and subsequently increased positive attitudes towards mathematics.

There are sound theoretical and practical reasons for investigating the effect of tablets on children’s attitudes towards mathematics. In fact, the one study directly measuring this effect not only suggests considerable promise for this area of research but also aligns with existing theories on the role of achievement emotions. To date, this has been an understudied area and Fabian et al.’s (2016) review of related research on older children and on other ICTs suggests that math attitudes be pursued further by researchers.

Can Tablets Improve Children’s Math Achievement?

Fabian et al. (2016) review of mobile math research paints a hopeful but cautious picture for the use of tablets as mathematics education tools. In Fabian’s review, there was a moderate effect size (i.e. the bigger the effect, the greater the change in math performance) across all of the mobile device interventions but the effect size shrank alongside the length of the intervention and the rigour of the experimental design (i.e. greater effect sizes for pretest–posttest designs without control groups than for longitudinal designs with control groups). Further, interventions targeting elementary education were robustly positive whereas studies targeting high schools were less consistent. These results underline the importance of reviewing methodologically rigourous studies when evaluating the effectiveness of educational technology, as the ability of new educational technology to temporarily engage students can result in a Hawthorne Effect that wanes as technology inexorably transitions from novel tool to common place resource (Bakker, van den Heuvel-Panhuizen, & Robitzsch, 2016; Parsons, 1974).

One of the earliest, methodologically rigourous studies on the use of iPads in mathematics education did not show much promise for tablets as mathematics learning tools. Carr's (2012) study involved one grade five classroom ($n = 56$) using iPads every day during their mathematics course in a 1:1 learning scenario for nine weeks and another comparable grade 5 classroom ($n = 48$) not using iPads but covering the same material (number and number sense; computation and estimation; measurement, geometry, probability and statistics; and patterns, functions, and algebra). A pretest–posttest comparison using a 50-item multiple choice test based on the grade 5 mathematics curricula found no difference between the iPad intervention classroom and the control classroom. In fact, the improvement between pretest and posttest were nearly identical between the iPad and control group (6.74 and 6.67%, respectively). According to the criteria outlined by Fabian et al. (2016), Carr's (2012) investigation was ideal in that it used comparable groups and was conducted over a relatively long period of time. Thus, the results from this more rigourous study would suggest that educators, researchers, and tablet mathematical app designers should be cautious when adopting or developing tablets as mathematics learning tools for achievement purposes. However, Carr's study is not as informative as it first seems because of a methodological practice that is not uncommon among educational technology and educational game studies.

Carr's (2012) study did not specify, control, or investigate how the iPads should be incorporated into the mathematics classroom. Instead, the iPad activities included in the study were open-ended and left up to the discretion of the teachers and students. These activities included playing tablet math games, using iPads to review lessons, watching online video tutorials, or using mathematics apps with manipulatives. Thus, the study lumped together a collection of practices under the umbrella of using tablets as mathematics learning tools and, somewhat unsurprisingly, found no cumulative effect of these different practices. This lumping together of different pedagogical approaches under the umbrella of tablet math use is akin to the problematic practice identified by Dubé and Keenan's (2016) study of educational math games. There, researchers and developers were testing the effectiveness of varying mathematics activities, grouping them all under the category of 'games', and finding inconsistent results. This practice only muddies the waters and makes comparisons between individual studies problematic.

Why Tablets Improve Math Achievement

Reviewing studies to determine if tablets improve mathematics achievement is complicated by each study taking a different approach to why tablets affect learning. In one study, tablets are supposed to support learning because they provide individualized practice (e.g. Musti-Rao & Plati, 2015) while in another study tablets support learning because they provide opportunities for collaboration (Hwang et al., 2015). These differences in how tablet technology results in mathematics achievement is likely due to the different learning theories used in each study. Fabian et al. (2016) identified 'a gap in terms of [the] discussion [of] how mobile technologies support the learning process in mathematics … and that studies should consider linking pedagogical theories to technology' (p. 97). In line with Fabian's critique, no review to date has compared the learning theories researchers use to investigate whether tablet computers aid mathematics achievement. Doing so will not only help answer the question of whether tablets improve mathematics achievement but should also help identify why they improve achievement.

To this end, we had four researchers code 92 of the original 204 studies identified by the search process according to their dominant learning theory. Of the 204 studies originally identified, only 92 focussed on mathematics education. Of the 92 studies on mathematics education only 25 were included in the review as the remaining studies included papers that either did not fit the target age range (i.e. were outside the kindergarten to grade 5 age range) and/or looked at other educational technologies (e.g. smartboards). However, we coded all 92 studies to ensure that the coding system accurately captured the learning theories used to study the effect of technology on mathematics education and therefore could be applied to the 25 studies included in this review.

The 22 learning theories identified by Millwood's (2013) Holistic Approach to Technology Enhanced Learning (HoTEL) were included in the coding scheme, as Millwood's list contains the most commonly used learning theories from educational technology research (see Appendix). An additional 8 theories were added to include approaches not captured by Millwood's classification system (i.e. activity theory, critical making, multimedia learning theory, embodied cognition, TPACK, engagement theory, self-regulated learning, information processing). For a theory to be ascribed to a study, the learning theory had to be specifically mentioned by the authors by name or the theoretical approach had to be explained. For articles where the theory was explained but not named, the descriptions from the article where compared against the definitions provided by HoTEL (Millwood, 2013). To establish the reliability of the coding scheme, one of the researchers randomly selected and recoded 10 of the 25? studies specifically investigating tablets in mathematics education. The resulting inter-rater reliability of the coding scheme was 80%. Disagreements were resolved through a discussion among the original coder, the secondary coder, and the principle investigator.

For the 25 studies focusing on tablets in kindergarten to grade 5 mathematics education, there was considerable diversity in the learning theories deployed. A total of 13 different learning theories were used, with the most common theories being information processing (12%), constructivism (8%), and discovery learning (8%). The majority of studies (40%) used a learning theory not used by another study. In line with research on educational math games (Dubé & Keenan, 2016), 32% of the articles in the review neither named nor explained the learning theory being tested by the study. For the studies that specifically measured the effect of tablets on mathematics achievement, a total of 10 learning theories were used with only discovery learning being used by more than one study. Only 2 studies measuring achievement failed to name or explain their learning theory approach, suggesting that researchers specifically investigating learning outcomes are considering how pedagogical practices translate into technological interventions.

Given the diversity of pedagogical approaches used to study the effect of tablets on mathematics achievement, a comparison between studies is not appropriate but a brief review of the findings provides some common themes. The study by Bebell and Pedulla (2015) further highlights the problem of a theoretically implementing tablets into mathematics classrooms. They set out to establish whether 1 student:1 tablet ratio is an effective means of improving kindergarten to grade 3 children's literacy and mathematics achievement using two experiments. The first experiment involved a randomized pretest–posttest intervention in which kindergarten to grade 3 students ($n = 266$) used iPads or not in a 1:1 ratio for a 9-week period. Following this, a second experiment involved making iPads available to all students in the school district at a 1:1 ratio and comparisons were drawn between students who had access to the iPads and students from previous years who did not. Only the second experiment compared students' math performance (i.e. measurement, numeracy, operations, and patterns) and neither study specifically controlled how the tablets were to be used in the classroom. Rather, they simply compared how access to iPads did or did not affect achievement. The authors argue that the atheoretical nature of the tablet deployment was essential, as it mirrors how tablets are being implemented in schools today with a focus on access over specific pedagogical approaches. Both studies found significant improvements in literacy because of iPad accessibility but this pattern was not found for mathematics. Specifically, no significant differences on any of the subtests were found for any of the kindergarten to grade 3 students and the only non-significant trend was for small improvements in kindergarten.

Bebell and Pedulla's (2015) results further suggest that implementing tablets into mathematics education requires more consideration of the pedagogical approach than simply making the devices available. This outcome could have arisen from teachers not knowing how to effectively use tablets in their mathematics classroom, from students not being engaged by the chosen implementation, or by the mathematics apps available at the time of the study not being of sufficient quality (i.e. Desoete et al., 2016). Obviously, simply putting an iPad in the classroom is not going to improve mathematics performance and determining how tablets can improve mathematics achievement requires consideration of how the tablets are being used.

For example, Zhang et al. (2015) tested the effectiveness of three mathematics apps (Motion Math Zoom, Splash Math, and Long Multiplication) that the researchers considered to be of high quality because they provided feedback, opportunities for practice, and adjusted their difficulty to the learners' ability. Following four sessions of use, grade 4 participants demonstrated improved understanding of the mathematics concepts practised in the apps (i.e. place value, decimals, multiplication) and that using the apps even closed a performance gap between struggling and typical learners. This investigation shows that some tablet mathematics apps can improve mathematics achievement and demonstrates the potential problem of lumping together low-

and high-quality apps to make general claims about tablets as mathematics education tools (i.e. Bebell & Pedulla, 2015). However, Zhang et al. (2015) study did not have a control group and did not identify why or how these high-quality apps improve mathematics achievement, as the specific learning theory was neither named or described, and this makes it difficult to draw conclusions about other tablet apps based on this study. Without a sound theory explaining why Motion Math Zoom and Splash Math are effective, how can educators or developers use the results from this study to identify other effective apps? Failure to connect the investigation to a learning theory limits the utility of the results and places teachers in the same unguided situation as seen in the studies by Bebell and Pedulla (2015) and Carr (2012); the goal of tablet research cannot be to identify high-quality apps three at a time.

When researchers are explicit about the learning theories underlying their investigation, it is easier to make guidelines for finding and using high-quality mathematics apps. For example, Jong et al. (2013) took a learning styles approach and investigated whether the temperaments of 119 kindergartners affected how well tablet counting apps can be implemented in the classroom. Further, the study compared two different types of tablet mathematics apps, gesture (control the app by making a gesture in the air above the screen) or touch-based apps (control the app by touching the screen). The study found that kindergartners performed better with the touch-based apps and that participants with a persistent temperament (i.e. easily on-task/focused) performed particularly well at posttest using the touch-based apps. From this result, the authors recommend that the learning style/temperament of the child be considered when implementing tablets into mathematics.

Similarly, Riconscente (2013) took an embodied cognition approach and found that 5 consecutive days of mathematical app use produced a significant improvement in grade 5 children's fraction knowledge ($d = 1.27$). The authors concluded that the gamification of the number line task helped motivate students to persist long enough (i.e. through all 750 problems) for the app to be effective but that the effectiveness was largely attributable to the embodied nature of the interaction. As was argued by Dubé and McEwen (2015), touching the screen and tilting the device provides proprioceptive feedback that aligns the physical actions of the learner with the underlying mathematics concepts (see Siegler & Ramani, 2009 for similar arguments with mathematics board games). Based on these results, Riconscente recommends that researchers and teachers consider how the learners' physical interactions with the tablet inform their mathematics understanding when choosing among various tablet apps. The results form both Riconscente (2013) and Jong et al. (2013) demonstrate how making recommendations for tablet mathematical app use is easier and more effective when there is a clear link between the learning theory and the investigation. However, evaluating the effectiveness of tablets as mathematics learning tools is more complicated than simply collating the recommendations of individual studies.

Even when multiple studies agree that tablets can improve mathematics achievement, the learning theories employed can produce contradictory explanations and recommendations. Musti-Rao and Plati (2015) used a discovery learning approach to tablets in math education and found that a tablet intervention produced significant improvements in grade 3 children's multiplication fact fluency above and beyond another effective math intervention. In the study, half of the children were assigned to the student-paced iPad app Math Drills, which individually presented 3 sets of 12 multiplication facts (i.e. 36 problems in total) to students and then tested them on those facts with immediate feedback. The other half were assigned to a teacher-paced discovery (complete a set of problems)-practice (students identify errors and practice those items)-repair (timed assessment of all items) intervention (DPR, Poncy, Skinner, & O'Mara, 2006). Not only did both teachers and students prefer the iPad intervention, subsequent performance on paper and pencil 'Mad Minute' worksheets was better for the iPad than the DPR condition and the iPad condition showed greater gains from pretest. The authors concluded that the self-paced nature of the iPad allowed students to practise more items, nearly twice as many as the DPR intervention, but spend less time practising individual items because responses were provided with a tap instead of writing them down. This study concluded that not only are iPads good for practising mathematics, but that this individualized self-paced form of practice is more preferred and more effective than teacher led math interventions.

Interestingly, Musti Rao and Plati's (2015) recommendation contradicts the conclusions from two other studies. Hwang et al. (2015) tested the effectiveness of touchscreen tabletops for improving fourth graders' reasoning about fractions and argued that the devices were effective because they facilitated collaboration (i.e. social constructivism). In contrast, an established PC-based intervention used in the study as a comparison did not allow students to work together and discuss mathematics. Similarly, Yang et al. (2016) found that tablet-supported reciprocal peer tutoring (i.e. social constructivism) was more effective at improving second graders mathematics achievement than 1:1 self-paced practice conducted with traditional materials. Although the learning theory used in each study did not change whether the tablet intervention was effective at improving mathematics achievement, the different theories produced contradictory reasons as to why this type of technology is effective (self-directed vs collaborative learning).

There is a potential for tablets to improve mathematics achievement at the early elementary level. This potential rests on whether tablets are being deployed in mathematics classroom in a purposeful way. If the deployment strategy is structured around a learning theory, then it seems tablets can be useful mathematics learning tools. Unfortunately, not all research takes a theory driven approach and we are not aware of any mass tablet deployments in schools that are theory driven. In fact, the association of community and comprehensive schools in Ireland published a guidebook on the use of tablet devices in schools (Hallissy, Gallagher, Ryan, & Hurley, n.d.) and specifically called for a more pedagogically driven approach to tablets, despite not providing one themselves. Instead, they opted for making general recommendations on how to use tablets as resources in the classroom (i.e. digital textbooks, word processing). The evidence supports the potential of tablets as mathematics education tools in classrooms, but now research needs to go beyond identifying potential and make clear recommendations.

Summary

Whether researchers and educators should consider tablets as tools for elementary mathematics education turns out to be a complex question with many answers. All the research seems to suggest that tablets can engage children with mathematics but the explanations as to why engagement occurs includes the 'cool' factor of new technology, the interactive and visual nature of the apps, the gamification of learning, and the hands-on aspect of touch screens. Little research has been done on whether the engagement students experience with tablets translates into improved math attitudes. The scant existing research suggests that tablet mathematics apps can increase students' liking for specific math subjects (e.g. fractions) by giving the learner a sense of control over their math practice and by increasing their perception of value by using math knowledge to solve meaningful problems. Engaging students and changing their attitudes are both worthy goals but, arguably, the push towards tablets started because of an interest in improving children's math ability.

Will the push to incorporate tablets into the classroom improve children's math skill? If current practices continue and tablets are simply thrust into the classroom without any guidance then the answer seems to be a resounding no (Bebell & Pedulla, 2015; Carr, 2012). This answer reflects the first author's own anecdotal experience studying tablets in elementary settings and teaching educational technology courses to teachers. In both situations, teachers comment that they do not know how to use the iPads provided by the schools, that finding good apps is difficult, and that there is no guidance for figuring out what makes a good educational app. When a good mathematical app is identified then they seem to be effective at improving math achievement (i.e. Hwang et al., 2015; Musti-Rao & Plati, 2015; Yang et al., 2016). Yet, the recommendations for educators and researchers based on these few good apps are not always copasetic and often contradictory. Again, what is needed is for researchers to go beyond studying whether individual apps are effective and start investigating which generalizable tablet-based pedagogical approaches result in learning. Perhaps this could be achieved if more tablet researchers implemented systematic Design Research methodologies (i.e. Sandoval, 2014) in which the design of the study is structured around how the technology is theoretically supposed to result in learning (c.f., studies simply designed to test if learning occurs).

So, where does this leave the state of tablet math research? Are tablets effective math learning tools? Perhaps the answer to this requires a juxtaposition against the idea of Maslow's Hammer (1966), popularized by the phrase, 'I suppose it is tempting, if the only tool you have is a hammer, to treat everything as if it were a nail' (p. 15). Perhaps it is tempting to see tablets as hammers, as a singular-use tool that either does or does not work for mathematics instruction. Testing this hypothesis is easy, introduce tablets into a mathematics classroom and see if engagement, attitudes, or grades improve. However, tablets are not hammers or even tools. They are tool-belts, holsters for the some odd 80,000 + educational apps (Hirsh-Pasek et al., 2015) currently available to researchers and educators, and only research grounded in sound learning theory is going to help us decide which types of mathematics apps should be loaded into the math educators' tool belt.

Appendix

Learning theories (number of the 25 studies included in the review using the theory)

Situated learning	Self-regulated learning
Organizational learning	Activity theory
Experiential learning	Critical making
Learning styles (1)	Multimedia learning theory (1)
De-schooling society	Embodied cognition (1)
Unschooling	TPACK theory
Critical pedagogy	Engagement theory (1)
Montessori education (1)	Instructivism
Experiential education (1)	Theory not clear (8)
Expressive constructivism	
Radical constructivism	
Constructivism (2)	
Constructionism	
Social constructivism (1)	
Connectivism (1)	
Expanse learning (1)	
Discovery learning (2)	
Meaningful learning	
Multiple intelligences (1)	
Mastery learning	
Behaviourism	
Information processing (3)	

References

Abrahamson, D., & Bakker, A. (2016). Making sense of movement in embodied design for mathematics learning. *Cognitive Research: Principles and Implications, 1,* 33. https://doi.org/10.1186/s41235-016-0034-3.

Ahmed, W., Minnaert, A., Kuyper, H., & van der Werf, G. (2012). Reciprocal relationships between math self-concept and math anxiety. *Learning and Individual Differences, 22*(3), 385–389.

Alibali, M. W., & Nathan, M. J. (2012). Embodiment in mathematics teaching and learning: Evidence from learners' and teachers' gestures. *Journal of the Learning Sciences, 21*(2), 247–286.

Annetta, L. A., Minogue, J., Holmes, S. Y., & Cheng, M.-T. (2009). Investigating the impact of video games on high school students' engagement and learning about genetics. *Computers & Education, 53*(1), 74–85. https://doi.org/10.1016/j.compedu.2008.12.020.

Apple. (2010). *Apple launches iPad.* Retrieved May 17, 2017 from https://www.apple.com/pr/library/2010/01/27Apple-Launches-iPad.html.

Aziz, N. (2013). Children's interaction with tablet applications: Gestures and interface design. *International Journal of Computer and Information Technology, 2*(3), 447–450.

Bakker, M., van den Heuvel-Panhuizen, M., & Robitzsch, A. (2016). Effects of mathematics computer games on special education students' multiplicative reasoning ability. *British Journal of Educational Technology, 47*(4), 633–648.

Barack, L. (2011). The kindles are coming: Ereaders and tablets are springing up in schools and librarians are leading the way. *School Library Journal, 57*(3), 58–60.

Bebell, D., & Pedulla, J. (2015). A quantitative investigation into the impacts of 1: 1 iPads on early learner's ELA and math achievement. *Journal of Information Technology Education: Innovations in Practice, 14,* 191–215.

Blackwell, C. K., Lauricella, A. R., & Wartella, E. (2014). Factors influencing digital technology use in early childhood education. *Computers & Education, 77,* 82–90.

Boyle, E. A., Hainey, T., Connolly, T. M., Gray, G., Earp, J., Ott, M., et al. (2016). An update to the systematic literature review of empirical evidence of the impacts and outcomes of computer games and serious games. *Computers & Education, 94*(C), 178–192.

Bray, A., & Tangney, B. (2016). Enhancing student engagement through the affordances of mobile technology: A 21st century learning perspective on realistic mathematics education. *Mathematics Education Research Journal, 28*(1), 173–197.

Carr, J. M. (2012). Does math achievement h'app'en when ipads and game-based learning are incorporated into fifth-grade mathematics instruction? *Journal of Information Technology Education, 11,* 269–286.

Chang, M., Evans, M., Kim, S., Deater-Deckard, K., & Norton, A. (2014). Educational video games and students' game engagement (pp. 1–3). *Presented at the 2014 International Conference on Information Science and Applications (ICISA).* IEEE. https://doi.org/10.1109/icisa.2014.6847390.

Chorianopoulos, K., Giannakos, M. N., & Chrisochoides, N. (2014). Design principles for serious games in mathematics (pp. 1–5). *Presented at the 18th Panhellenic Conference, New York, USA.* ACM Press. https://doi.org/10.1145/2645791.2645843.

Clarke, L., & Abbott, L. (2016). Young pupils', their teacher's and classroom assistants' experiences of iPads in a Northern Ireland school: "Four and five years old, who would have thought they could do that?". *British Journal of Educational Technology, 46*(6), 1051–1064.

Connell, S. L., Lauricella, A. R., & Wartella, E. (2015). Parental co-use of media technology with their young children in the USA. *Journal of Children and Media, 9*(1), 5–21.

Csikszentmihalyi, M. (1990). *Flow: The psychology of optimal performance.* New York, NY: Cambridge University Press.

Dale, E. (1954). *Audio-video methods in teaching* (revised edition). New York: The Dryden Press.

Deater-Deckard, K., Chang, M., & Evans, M. E. (2013). Engagement states and learning from educational games. *New Directions for Child and Adolescent Development, 2013*(139), 21–30. https://doi.org/10.1002/cad.20028.

Desoete, A., Praet, M., Velde, C. Van, De, Craene, De, B., & Hantson, E. (2016). Selection of apps for teaching difficult mathematics topics: An instrument to evaluate touch-screen tablet and smartphone mathematics apps. *International Perspectives on Teaching and Learning Mathematics with Virtual Manipulatives, 7,* 171–187.

Dillenbourg, P., & Evans, M. (2011). Interactive tabletops in education. *International Journal of Computer-Supported Collaborative Learning, 6,* 491–514.

Dowker, A., Sarkar, A., & Looi, C. Y. (2016). Mathematics anxiety: What have we learned in 60 years? *Frontiers in Psychology, 7*(1333), 385–416. https://doi.org/10.3389/fpsyg.2016.00508.

Dubé, A. K., & Keenan, A. (2016). The influence of games on children's mathematical development. In B. Blevins-Knabe, & A. B. Autin (Eds.), *Early childhood mathematics skill development in the home environment* (pp. 165–184). Springer International Publishing.

Dubé, A. K., & McEwen, R. N. (2015). Do gestures matter? The implications of using touchscreen devices in mathematics instruction. *Learning and Instruction, 40,* 89–98.

Dubé, A. K., & McEwen, R. (2016). Abilities and affordances: Factors influencing successful child-tablet interactions. *Educational Technology Research and Development, 65*(4), 889–908. https://doi.org/10.1007/s11423-016-9493-y.

Dündar, H., & Akçayir, M. (2014). Implementing tablet PCs in schools: Students' attitudes and opinions. *Computers in Human Behavior, 32,* 40–46.

Fabian, K., Topping, K. J., & Barron, I. G. (2016). Mobile technology and mathematics: effects on students' attitudes, engagement, and achievement. *Journal of Computers in Education, 3*(1), 77–104.

Falloon, G. (2013). Young students using iPads: App design and content influences on their learning pathways. *Computers & Education, 68,* 505–521.

Falloon, G. (2014). What's going on behind the screens? Researching young students' learning pathways using iPads. *Journal of Computer Assisted Learning, 30*(4), 318–336.

Fredricks, J. A., Blumenfeld, P. C., & Paris, A. H. (2004). School engagement: Potential of the concept, state of the evidence. *Review of Educational Research, 74*(1), 59–109. https://doi.org/10.3102/00346543074001059.

Goetz, T., Frenzel, A. C., Pekrun, R., & Hall, N. C. (2007). Between-and within-domain relations of students' academic emotions. *Journal of Educational Psychology, 99*(4), 715–733. https://doi.org/10.1037/0022-0663.99.4.715.

Goodwin, K., & Highfield, K. (2013). A framework for examining technologies and early mathematics learning. In L. D. English & J. T. Mulligan (Eds.), *reconceptualising early mathematics learning* (pp. 205–226). New York, NY: Springer.

Hallissy, M., Gallagher, A., Ryan, S., & Hurley, J. (n.d.). *The use of tablet devices in ACCS schools.* Retrieved May 31, 2017, from www.accs.ie.

Hembree, R. (1990). The nature, effects, and relief of mathematics anxiety. *Journal of Research in Mathematics Education, 21,* 33–46. https://doi.org/10.2307/749455.

Hirsh-Pasek, K., Zosh, J. M., Golinkoff, R. M., Gray, J. H., Robb, M. B., & Kaufman, J. (2015). Putting education in "educational" apps: lessons from the science of learning. *Psychological Science, 16,* 3–34.

Hwang, W.-Y., Shadiev, R., Tseng, C.-W., & Huang, Y.-M. (2015). Exploring effects of multi-touch tabletop on collaborative fraction learning and the relationship of learning behavior and interaction with learning achievement. *Educational Technology & Society, 18*(4), 459–473.

Ingram, N., Williamson-Leadley, S., & Pratt, K. (2016). Showing and telling: using tablet technology to engage students in mathematics. *Mathematics Education Research Journal, 28,* 123–147. https://doi.org/10.1007/s13394-015-0162-y.

Jaciw, A. P., Toby, M., & Ma, B. (2012). Conditions for the effectiveness of a tablet-based algebra program. In *Proceedings from the Society for Research on Educational Effectiveness.* Washing, DC.

Jackson, A., Brummel, B., Pollet, C., & Greer, D. (2013). An evaluation of interactive tabletops in elementary mathematics education. *Educational Technology Research and Development, 61*(2), 311–332.

Jong, J.-T., Hong, J.-C., & Yen, C.-Y. (2013). Persistence temperament associated with children playing math games between touch panel and embodied interaction. *Journal of Computer Assisted Learning, 29*(6), 569–578.

Ke, F. (2008). A case study of computer gaming for math: Engaged learning from gameplay? *Computers & Education, 51*(4), 1609–1620. https://doi.org/10.1016/j.compedu.2008.03.003.

Kiili, K., Ketamo, H., Koivisto, A., & Finn, E. (2014). Studying the user experience of a tablet based math game. *International Journal of Game-Based Learning, 4*(1), 60–77. https://doi.org/10.4018/IJGBL.2014010104.

Kosko, K., & Ferdig, R. (2016). Effects of a tablet-based mathematics application for pre-school children. *Journal of Computers in Mathematics and Science Teaching, 35*(1), 61–79.

Kyriakides, A. O., Meletiou-Mavrotheris, M., & Prodromou, T. (2016). Mobile technologies in the service of students' learning of mathematics: The example of game application A.L.E.X. in the context of a primary school in Cyprus. *Mathematics Education Research Journal, 28*(1), 53–78. https://doi.org/10.1007/s13394-015-0163-x.

Larkin, K., & Jorgensen, R. (2016). "I hate maths: Why do we need to do maths?" Using iPad video diaries to investigate attitudes and emotions towards mathematics in year 3 and year 6 students. *International Journal of Science and Mathematics Education, 14*(5), 925–944. https://doi.org/10.1007/s10763-015-9621-x.

Lee, J., Luchini, K., Michael, B., Norris, C., & Soloway, E. (2004). More than just fun and games: Assessing the value of educational video games in the classroom. In *CHI'04 extended abstracts on human factors in computing systems* (pp. 1375–1378).

Ludwig, L., & Mayrberger, K. (2012). Next Generation Learning? Learning with tablets as an example for the implementation of digital media in schools. *World Conference on Educational Multimedia, Hypermedia and Telecommunications, 2012*(1), 2179–2187.

Ma, X., & Kishor, N. (1997). Assessing the relationship between attitude toward mathematics and achievement in mathematics: A meta-analysis. *Journal of Research in Mathematics Education, 28,* 26–47. https://doi.org/10.2307/749662.

Main, S., & O'Rourke, J. (2011). New directions for traditional lessons: Can handheld game consoles enhance mental mathematics skills? *Australian Journal of Teacher Education, 36*(2), 43–55.

Maslow, A. H. (1966). *The psychology of science*. New York: Harper & Row.

Mayer, R. (2014). *Computer game for learning: An evidence-based approach*. London: The MIT Press.

McEwen, R., & Dubé, A. K. (2015). Engaging or distracting: children's tablet computer use in education. *Journal of Educational Technology and Society, 18,* 9–24.

McEwen, R., & Dubé, A. K. (2016). Intuitive or idiomatic? An information studies and cognitive psychology study of child-tablet computer interaction. *Journal of the Association for Information Science and Technology, 67,* 1169–1181. https://doi.org/10.1002/asi.23470.

McEwen, R., & Dubé, A. K. (2017). *Understanding tablets from early childhood to adulthood: Encounters with touch technology*. New York: Routledge.

Miller, D. J., & Robertson, D. P. (2010). Using a games console in the primary classroom: Effects of 'Brain Training' programme on computation and self-esteem. *British Journal of Educational Technology, 41*(2), 242–255.

Miller, D. J., & Robertson, D. P. (2011). Educational benefits of using game consoles in a primary classroom: A randomised controlled trial. *British Journal of Educational Technology, 42*(5), 850–864.

Millwood, R. (2013). *Learning theory: What are the established learning theories?* Retrieved May 27, 2017. http://cmapspublic3.ihmc.us/rid=1LNV3H2J9-HWSVMQ-13LH/Learning%20Theory.cmap.

Mouza, C., & Lavigne, N. (2012). *Emerging technologies for the classroom*. New York, NY: Springer Science & Business Media. https://doi.org/10.1007/978-1-4614-4696-5.

Moyer-Packenham, P. S., Bullock, E. K., Shumway, J. F., Tucker, S. I., Watts, C. M., Westenskow, A., … Jordan, K. (2016). The role of affordances in children's learning performance and efficiency when using virtual manipulative mathematics touch-screen apps. *Mathematics Education Research Journal, 28*(1), 79–105.

Muis, K. R., Ranellucci, J., Trevors, G., & Duffy, M. C. (2015). The effects of technology-mediated immediate feedback on kindergarten students' attitudes, emotions, engagement and learning outcomes during literacy skills development. *Learning and Instruction, 38*(C), 1–13. https://doi.org/10.1016/j.learninstruc.2015.02.001.

Musti-Rao, S., & Plati, E. (2015). Comparing two classwide interventions: Implications of using technology for increasing multiplication fact fluency. *Journal of Behavioral Education, 24*(4), 418–437. https://doi.org/10.1007/s10864-015-9228-x.

Namukasa, I. K., Gadanidis, G., Sarina, V., Scucuglia, S., & Aryee, K. (2016). Selection of apps for teaching difficult mathematics topics: An instrument to evaluate touch-screen tablet and smartphone mathematics apps. In P. S. Moyer-Packenham (Ed.), *International perspectives on teaching and learning mathematics with virtual manipulatives* (pp. 275–300). Switzerland: Springer International Publishing.

Novack, M. A., Congdon, E. L., Hemani-Lopez, N., & Goldin-Meadow, S. (2014). From action to abstraction: Using the hands to learn math. *Psychological Science, 25*(4), 903–910.

Parsons, H. M. (1974). What happened at Hawthorne? *Science, 183,* 922–932.

Pekrun, R. (2006). The control-value theory of achievement emotions: Assumptions, corollaries, and implications for educational research and practice. *Educational Psychology Review, 18*(4), 315–341. https://doi.org/10.1007/s10648-006-9029-9.

Poncy, B. C., Skinner, C. H., & O'Mara, T. (2006). Detect, practice, and repair: The effects of a class-wide intervention on elementary students' math-fact fluency. *Journal of Evidence-Based Practices for Schools, 7*(1), 47–68.

Riconscente, M. M. (2013). Results from a controlled study of the iPad fractions game motion math. *Games and Culture, 8*(4), 186–214. https://doi.org/10.1177/1555412013496894.

Ross, J. A., Sibbald, T., & Bruce, C. D. (2009). Characteristics of students assigned to technology-based instruction. *Journal of Computer Assisted Learning, 25,* 562–573. https://doi.org/10.1111/j.1365-2729.2009.00328.x.

Sandoval, W. (2014). Conjecture mapping: An approach to systematic educational design research. *Journal of the Learning Sciences, 23*(1), 18–36. https://doi.org/10.1080/10508406.2013.778204.

Shapiro, J. (2017). *Mind/shift: Guide to digital games + learning*. Retrieved May 31, 2017, from https://a.s.kqed.net/pdf/news/MindShift-GuidetoDigitalGamesandLearning.pdf.

Shin, N., Sutherland, L. M., Norris, C. A., & Soloway, E. (2012). Effects of game technology on elementary student learning in mathematics. *British Journal of Educational Technology, 43*(4), 540–560. https://doi.org/10.1111/j.1467-8535.2011.01197.x.

Shuler, C. (2012). *iLearn II: An analysis of the education category of the itunes app store*. New York: The Joan Ganz Cooney Center at Sesame Workshop.

Siegler, R. S., & Ramani, G. B. (2009). Playing linear number board games—But not circular ones—Improves low-income preschoolers' numerical understanding. *Journal of Educational Psychology, 101,* 545–560.

Sinclair, N., Chorney, S., & Rodney, S. (2016). Rhythm in number: exploring the affective, social and mathematical dimensions of using TouchCounts. *Mathematics Education Research Journal, 28*(1), 31–51.

Sinclair, N., & Heyd-Metzuyanim, E. (2014). Learning number with TouchCounts: The role of emotions and the body in mathematical communication. *Technology, Knowledge and Learning, 19*(1–2), 81–99.

Strouse, G., Hubber, P. J., Kloos, H., Stacy, S. T., Cartwright, M., Arwood, Z., et al. (2017). Addressing the math-practice gap in elementary school: Are tablets a feasible tool for informal math practice? *Frontiers in Psychology, 8,* 179. https://doi.org/10.3389/fpsyg.2017.00179.

Suits, B. (1978). *The grasshopper: Games, life, and utopia*. Toronto, ON: University of Toronto Press.

Sweller, J. (1994). Cognitive load theory, learning difficulty, and instructional design. *Learning and Instruction, 4*(4), 295–312.

Sweller, J., Van Merriënboer, J., & Paas, F. (1998). Cognitive architecture and instructional design. *Educational Psychology Review, 10*(3), 251–296.

Tucker, S. I., & Johnson, T. N. (2017). I thought this was a study on math games: Attribute modification in children's interactions with mathematics apps. *Education Sciences, 7*(2), 50–70.

Tucker, S. I., Moyer-Packenham, P. S., Westenskow, A., & Jordan, K. E. (2016). The Complexity of the affordance–ability relationship when second-grade children interact with mathematics virtual manipulative apps. *Technology, Knowledge and Learning, 21*(3), 341–360.

Wigfield, A., & Meece, J. L. (1988). Math anxiety in elementary and secondary school students. *Journal of Educational Psychology, 80,* 210–216.

Wood, E., Petkovski, M., De Pasquale, D., Gottardo, A., Evans, M. A., & Savage, R. S. (2016). Parent scaffolding of young children when engaged with mobile technology. *Frontiers in Psychology, 7,* 690–701. https://doi.org/10.3389/fpsyg.2016.00690.

Wu, H. (2008). *Fractions, decimals, and rational numbers.* Retrieved May 27, 2017, from https://math.berkeley.edu/~wu/NMPfractions.pdf.

Wu, L. J., Hsiao, H. S., Chang, K. E., & Sung, Y. T. (2006). Construct mathematic path and interactive problem solving discussion system in mobile learning environment. In *Proceeding of the IADIS International Conference on Mobile Learning, Dublin, Ireland* (pp. 288–292).

Yang, E. F. Y., Chang, B., Cheng, H. N. H., & Chan, T. (2016). Improving pupils' mathematical communication abilities through computer-supported reciprocal peer tutoring. *Educational Technology & Society, 19*(3), 157–169.

Zhang, M., Trussell, R. P., Gallegos, B., & Asam, R. R. (2015). Using mathematics apps for improving student learning: An exploratory study in an inclusive fourth grade classroom. *TechTrends, 59*(2), 32–39. https://doi.org/10.1007/s11528-015-0837-y.

Early Understanding of Fractions via Early Understanding of Proportion and Division

Cheryll L. Fitzpatrick and Darcy Hallett

When one thinks about early conceptions of math, fractions do not immediately come to mind. Fractions have been called "without doubt the most problematic area in mathematics education" (Streefland, 1991, p. 6). Not only are they known to be difficult to learn in elementary school (Kerslake, 1986; Mack, 1995; Ni & Zhou, 2005), they are known to be a difficult concept for adolescents as well (Hecht, Vagi, & Torgesen, 2007). At the same time, they are instrumental to later math learning and have been stressed as an area that needs additional focus in our education system (National Mathematics Advisory Panel, 2008).

Nevertheless, just because fractions are inherently difficult to learn does not mean young children lack all understanding of fractions or fraction-related concepts before the start of formal instruction. Researchers have demonstrated that children do indeed have prior conceptions of fractions, and this is often called the "informal" knowledge of fractions (Mack, 1995). For example, many children know, even before learning fractions in school, that if one pizza is shared between three people and another pizza of the same size is shared between four people, three people would get more pizza. At the same time, these same children commonly say that 1/3 is less than 1/4 (Kornilaki & Nuñes, 2005). As this example illustrates, informal knowledge is not complete and can sometimes interfere with fraction understanding (see Ni & Zhou, 2005, for a discussion on the whole number bias). Furthermore, there are also many different kinds of informal knowledge, which relate to the different ways that fractions are used. Previous research in mathematics education has observed that part

This research was supported in part by a Discovery Grant from the Natural Sciences and Engineering Council of Canada awarded to the second author (Grant # 355914-2008).

C. L. Fitzpatrick · D. Hallett (✉)
Department of Psychology, Memorial University of Newfoundland, St. John's, NL A1B 3X9, Canada
e-mail: darcy@mun.ca

C. L. Fitzpatrick
e-mail: cheryllf@mun.ca

K. M. Robinson et al. (eds.), *Mathematical Learning and Cognition in Early Childhood*,
https://doi.org/10.1007/978-3-030-12895-1_14

of the reason that fractions are difficult to learn is that they have different functions, what are sometimes called the subconstructs of fractions (Behr, Harel, Post, & Lesh, 1992; Charalambous & Pitta-Pantazi, 2007; Nuñes, Desli, & Bell, 2003). One of these subconstructs is the part–whole construct, where fractions are used to indicate the relation between a part and the whole to which it belongs. Another subconstruct is the quotient construct, where a fraction represents a division where a certain number of objects are divided into so many pieces. These different subconstructs (and there are others; see Behr et al., 1992; Nuñes et al., 2003) reflect different understandings of fractions and are therefore likely to be based on different kinds of preconceptions.

The purpose of this chapter is to explore the precursors of fraction understanding by examining children's understanding of two related areas: proportional reasoning and division. These topics were chosen for two reasons. First, each of these topics reflects one of the subconstructs of fractions—the part–whole construct is reflected in proportional reasoning, while the quotient construct is reflected in the understanding of division. Second, both of these areas have research literatures investigating children's early conceptions. This chapter reviews and analyzes these literatures with the purpose of better understanding what the early conceptions of these topics are and how they relate to the understanding of fractions. We then draw some conclusions about what these early understandings of proportional reasoning and division can tell us about the early understanding of fractions, as well as the educational implications for fraction learning.

Proportional Reasoning

Proportional reasoning is a complex skill that humans, and non-humans, use daily. This skill has been described as "a pervasive activity that transcends topical barriers in adult life" (Ahl, Moore, & Dixon, 1992, p. 81). Proportional information is vital in dealing with a diverse array of topics such as relational spatial contrasts, temperatures, densities, concentrations, and recipe formulation (Boyer & Levine, 2012; Möhring, Newcombe, & Frick, 2015; Moore, Dixon, & Haines, 1991; Noetling, 1980a, 1980b; Sielger & Vago, 1978). Understanding proportionality is a key concept in mathematics and has been stated by the National Council of Teachers of Mathematics as deserving whatever time can be afforded to its development (as cited in Boyer, Levine, & Huttenlocher, 2008). There are many topics children will encounter in the elementary school mathematics curriculum in which proportional reasoning is central, such as fraction equivalence, long division, and measurement conversion (Lesh, Post, & Behr, 1988, as cited in Boyer & Levine, 2012). One of the interesting things about using proportional reasoning to make decisions in our everyday lives is that judgements can be made quite readily in the absence of numerical scales (Ahl et al., 1992).

The purpose of this section is to explore the research on children's understanding of proportion. Although the focus here is on children's early understanding of proportion, we will start by reviewing the research looking at adolescent understanding of proportional reasoning. This will help us to understand what is different about proportional reasoning in older children compared to that in younger children, which will provide a better picture of what children's early understanding of proportion is really like.

"True" Versus "Intuitive" Proportional Understanding

Like many things, some of the first insights into the development of proportional reasoning start with Piaget and Inhelder (Inhelder & Piaget, 1958; Piaget & Inhelder, 1975). They argued that proportional understanding is beyond the capabilities of young children, as it requires knowledge of formal operations (e.g., reasoning about the "relation between relations"). Piaget and Inhelder based these conclusions on children's responses to questions about the operations of systems, as well as responses to questions about games of chance. One example of a system's task was the balance scale task (Inhelder & Piaget, 1958), where children were given a typical balance scale with which to experiment and asked to explain what happens when weights of varying amounts are placed at different distances from the center on either side, and why it happens that way. They found that only young adolescents (i.e., those who Piaget called formal operational) could explain that the scale would be in equilibrium when the proportional difference in the weights was countered with a proportional difference in the distance of these weights from the center. In this "true" understanding, older children understand the situations to which proportional reasoning applies and can do exact calculations. Younger children (7- to 10-year-olds, i.e., those who Piaget called concrete operational) understand that larger weights have to be closer to the center than smaller weights in order to achieve equilibrium, but do not understand the relation as mathematically proportional. Piaget and Inhelder called this an intuitive understanding of proportion.

Many studies published in the aftermath of the work of Piaget and Inhelder, using different kinds of procedures, supported the assertion that a true understanding of proportion is not achieved until formal operations (e.g., Chapman, 1975; Offenback, Gruen, & Caskey, 1984; Siegler & Vago, 1978). Although some studies seemed to suggest that younger children could solve proportion tasks (e.g., Davies, 1965; Goldberg, 1966; Yost, Siegel, & Andrews, 1962), critics pointed out that these tasks could be solved by choosing the larger number without paying attention to proportion (Chapman, 1975). Studies by Noelting (1980a, 1980b) illustrated this difference. Children were presented with sets consisting of different numbers of glasses with orange juice concentrate and water and had to choose which sets of glasses would result in the strongest (most orange tasting) orange juice mixture. Noelting found that 7-year-olds could solve the task if the amount of water was held constant while the amount of orange juice concentrate changed, as children could just choose the larger number to get the correct answer. However, even 10- and 12-year-olds had difficulty with this task when the amount of water was not held constant, which requires a more complete understanding of proportion.

However, given that our purpose is to look for early conceptions of proportion that can relate to fraction understanding, an intuitive notion of proportion may still be useful. Children might not be able to calculate the proportions, but they may still understand something about how they work, and this may be enough to support fraction learning. Related to this question is the extent to which the study of children's understanding of proportion is tied to the understanding that proportion is needed in the context of a given task. In situations like the balance task, Inhelder and Piaget (1958) are not simply testing children's understanding of proportions, but their understanding that these tasks are governed by a proportional relation. If children can show this understanding, as they do in early adolescence, then it is true that they must also understand proportion. However, it is possible the younger children are able to reason proportionally, but they just do not know that they are supposed to reason proportionally when faced with proportion tasks like the ones described above.

Furthermore, a closer look at some of the studies presented above suggests some potential for proportional understanding at earlier ages. Chapman (1975) concluded that children were not at adult levels in their understanding of proportion, but the Grade 5 children did perform above chance on his task. Similarly, Offenbach et al. (1984) found that just under a third of Grade 6 children were using a proportion strategy. Finally, Siegler and Vago (1978) found that if the children were explicitly instructed on how to do their task, albeit in a very extensive and stepwise way, then 90% of the children (of all ages) could solve it.

The Right Context Can Elicit Proportional Reasoning in Younger Children

As it turns out, more recent research has found that younger children can demonstrate understanding of proportional reasoning if you ask them in the right way. For example, Acredolo, O'Connor, Banks, and Horobin (1989) asked 7-, 9-, and 11-year-olds to guess the probability of drawing a target (jelly beans in one study and ladybugs in another), while they varied the number of targets and the total number in the collection. They found that the children's responses demonstrated an effect of changing the target number (i.e., probability was lower if the number of targets was lower), changing the total number (i.e., probability was lower if the total number was higher), and even the interaction between the two (i.e., the effect of increased probability when the number of targets increased would be larger at lower total item levels). However, the probabilities chosen by the children did not match the actual probabilities. This means that children were able to adapt their answers in a relative way to match changing circumstances, but their absolute judgements of the probabilities were off. These results suggest that children as young as 7 years old are thinking of likelihood in a proportional way, even if they cannot calculate the actual probabilities.

Spinillo and Bryant (1991, 1999) presented children with pictures of boxes filled with blue and white blocks and asked them to choose which of two boxes in front of them matched the picture (i.e., had the same proportion of blue to white blocks). They found that the youngest children, 4- and 5-year-olds, did not succeed on these tasks. However, starting at age 6, children began to correctly identify the proportion and offered proportional justifications for doing so, but they did this mostly when the proportions in the two boxes were on either side of a half. Spinillo and Bryant argued that these trials could be solved by using part–part reasoning (the blue part was more than the white part in one box, while the white part was more than the blue part in the other box), and for this reason, there was something special about the half boundary. For the trials where the proportions were on the same side of half, children would have to use part–whole reasoning, which they found more difficult.

Sophian and Wood (1997) also investigated part–part versus part–whole reasoning in a group of 5- to 7-year-olds. In contrast to Spinillo and Bryant (1991), Sophian and Wood found that 7-year-old children chose the part–whole option above chance both when compared to a non-matching option and when compared to a part–part option, although they argued that their task might elicit a different reasoning than Spinillo and Bryant (1991). Moore, Dixon, and colleagues (Ahl et al., 1992; Moore et al., 1991) had children watch two containers of water, of different temperatures, get mixed together and were then asked what temperature the combined water would be. They found that Grade 5 students demonstrated proportional reasoning when descriptors were used for the temperature (i.e., "very cold," "cold," "medium," "hot," or "very hot"), but not when numbers were used. All these cases provide further evidence that younger children can understand proportions if the task is framed in a particular way.

In the studies reviewed above, children who are about 6 or 7 years old are the ones who are starting to succeed in these proportional tasks, while younger children (if they are included in the study) are still failing them. There are studies, however, that have revealed successful proportional thinking in 3-, 4-, and 5-year-old children. Ng, Heyman, and Barner (2011, Study 2) found that an expectation that resources should be divided equally (i.e., they tell the children that two people worked equally hard to earn some coins to be shared) will help even young children focus on the proportion rather than the absolute amount. Goswami (1989) and Singer-Freeman and Goswami (2001) were able to show that 3-year-olds (Singer-Freeman & Goswami, 2001) and 4-year-olds (Goswami, 1989; Singer-Freeman & Goswami, 2001) were able to match proportions if they were presented as a visual analogy, either as abstract pictures (e.g., half a circle is to half a square, as a quarter of a circle is to blank of a square) or as models of food (e.g., if the experimenter took away 2/8 of their food model, the child had to take away 1/4 of their food model). In both studies, these young children scored above chance on these tasks, although they seemed to do better on 1/2 and 3/4 proportions than they did on the 1/4 proportions.

Going even younger, research has demonstrated that even infants respond to proportional information. Using habituation studies (where infants' looking behavior is observed to infer their ability to notice changes in proportion), 6-month-olds were found to, at the very least, represent proportions and notice large departures from a given proportion, independent of the absolute number of images seen (McCrink

&Wynn, 2007). Similarly, Xu and Denison (2009), and Xu and Garcia (2008) have shown that 8- and 10-month-olds notice when a proportion of colored balls varies significantly from what would be expected. Provided with proportional information, Denison and Xu (2010, 2014) found that, more often than chance, 10- to 14-month-olds chose a cup that was more likely to have a desired object. Interestingly, Girotto, Fontanari, Gonzalez, Vallortigara, and Blaye (2016) found that 3- and 4-year-olds were not above chance on this same forced-choice task (see Xu & Denison, 2009; Xu & Garcia, 2008), although the 5-year-olds were. Nevertheless, this evidence suggests that even infants possess a sensitivity to proportion.

Discrete Versus Continuous Tasks

Whether stimuli are continuous or discrete is another factor that can explain why younger children are succeeding on some proportional tasks while failing on others (Mix, Huttenlocher, & Levine, 2002; Mix, Levine, & Huttenlocher, 1999). For example, Boyer and colleagues (2008) asked kindergarten to Grade 4 students to help Wally Bear mix his juice (similar to the method used by Noelting, 1980a, 1980b). On a computer screen, they saw a proportion of juice to water that Wally Bear wanted to make, and they were then asked to pick which of two alternatives would make the same tasting juice (see Fig. 1). Wally Bear's target exemplar and the choices to match to that exemplar were represented by partially shaded vertical bars; sometimes these bars were continuous, and sometimes they were discrete (i.e., divided into equal sized units using demarcating lines; see Fig. 1). In their first study, second and third graders, but not younger children, were able to do these tasks, but performance was the poorest when the target and the choice alternatives were both presented as discrete quantities. Furthermore, children did worse depending on the wrong choice (or foil) that was presented to them. The juice-matching foil matched the number of units of juice but had a different total, while the total-matching foil matched the total number of units but had a different number of units of juice (all the examples in Fig. 1 are juice-matching foils). The children had worse performance when there was a juice-matching foil than when there was a total-matching foil. These results again suggest that the format of the stimuli may lead children to engage in non-proportional reasoning in tasks such as this.

Jeong, Levine, and Huttenlocher (2007) observed very similar results using donut-shaped spinners with red and blue regions that were sometimes demarcated (discrete) and sometimes not (continuous). Children were presented a choice of two spinners (the diameter of the spinners varied so that the size of the red region would not predict the answer) and asked to pick which one they would like to use to maximize their chance of landing in the red region. They found that all children were above chance on the continuous trials, but the 6-year-olds were not above chance on the discrete trials. The 8- and 10-year-olds, by contrast, were above chance on the discrete trials, but only when the larger number of shaded red regions also corresponded to the larger proportion. Put succinctly, even the youngest children succeeded at the continuous

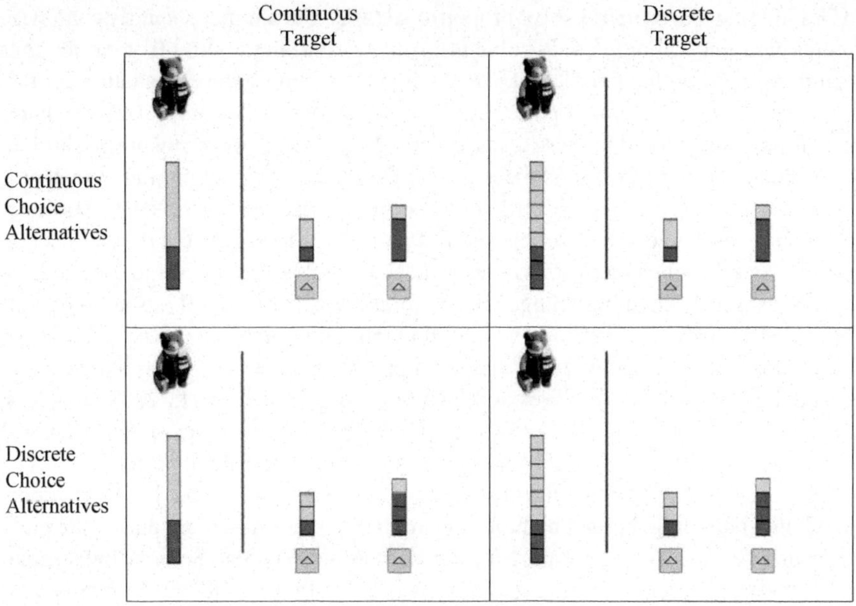

Fig. 1 An illustration of the juice-mixing options where the target could be continuous or discrete and the choice could be continuous or discrete. This is taken from Fig. 1 from Boyer et al. (2008), reproduced with permission

trials, the 8- and 10-year-olds succeeded on the discrete trials if the larger number of units coincided with the larger proportion, and even the oldest children did not succeed on the discrete trials if the larger number of units did not coincide with the larger proportion. All of these results suggest two main points. First, continuous stimuli, more than discrete stimuli, will facilitate children's proportional thinking. Second, the discrete stimuli seemed to lead children, even 8- and 10-year-olds, to use a non-proportional strategy that they do not use with continuous stimuli. The implication is that continuous stimuli might bias children to pay attention to the relative differences between the quantities, while discrete stimuli might bias children to count the units and use absolute number (rather than proportion) to guide their answers.

Reconciling the Evidence for Children's Early Proportional Reasoning

Although children younger than 8 years old may not have a true understanding of proportion, the research reviewed above is abundant and methodologically diverse, and it offers convincing evidence that preadolescent children (indeed, even children in Grade 1) can nevertheless solve proportional tasks. Furthermore, a fairly consistent finding was that children could solve tasks using continuous stimuli at earlier ages than they could with a parallel task using discrete stimuli. One explanation for this difference, tested and supported explicitly by Boyer and colleagues (2008), is that if one presents the stimuli in these tasks as something suggestive of counting, children at these ages are biased to count them and use that count information (rather than proportion) to answer the task. Nevertheless, the studies that demonstrated proportional reasoning with some of the youngest children (i.e., 3- to 5-year-olds) used discrete stimuli. Ng and colleagues (2011) used plates of pennies in their task to demonstrate different sharing situations. Singer-Freeman and Goswami (2001) used models of food where fractions of the food were divided into non-attached pieces (eighths for the experimenter and quarters for the children). Both these sets of stimuli seem as countable as that used by Boyer and his colleagues (Boyer et al., 2008; Boyer & Levine 2012; Jeong et al., 2007), so why did the first set of studies show successful proportional reasoning at very young ages while the second set did not?

Perhaps, it is not simply the countability of discrete stimuli that hinders proportional understanding, but countability in particular situations where one of the parts of the proportion is somehow more focal or seen as more salient. Some of the original studies using discrete stimuli (e.g., Chapman, 1975; Inhelder & Piaget, 1958) were about probability, where children were asked to pick the option that was more likely to get them their target. Perhaps, when you ask a young child to choose which one of two options is more likely to get the target, the child might interpret that as a quantity question, focus on the target, and resort to counting absolute targets. Limiting one's proportional reasoning to a single quantity in this way has been referred to as *univariate reasoning* (Lobato, Hawley Orrill, Druken, & Jacobson, 2011). The orange juice tasks (e.g., Boyer et al., 2008; Noelting, 1980a, 1980b) may be similar, in that children asked to judge the tastiness of orange juice concentrate mixed with water will focus on the amount of the active ingredient, if given an easy way to quantify it. Ng and colleagues (2011), however, structured their task to be about fairness, and when the situation was designed so that it would be socially expected to evenly split the pennies, children paid attention to the proportion and not the absolute amount. Singer-Freeman and Goswami (2001) did not ask children to choose a larger or smaller of two proportions—they asked them to manipulate a model in front of them to match what the experimenter had done with their model. The crucial difference here may be because the child did the manipulation him or herself, or it may be because the situation did not draw attention to an active focal component that, in the child's mind, was the main component in the likelihood of getting a target piece or in the taste of the orange juice. Further research is needed to test this idea, but the research so far suggests that discrete stimuli can still be used in early conceptions of proportion reasoning if the situation is carefully constructed to avoid these pitfalls. At the same time, using continuous stimuli also avoids these pitfalls.

In sum, the research on proportional reasoning suggests that even very young children can attend to proportional reasoning, even if it is also very easy to distract them from attending to it. As such, the early understanding of proportion may be able to support the development of an early understanding of fractions. The next section explores the research regarding the early understanding of division to see if it has a similar potential to support the learning of fractions.

Division

According to the Common Core State Standards in Mathematics, children are not formally taught algorithms to solve division problems until they reach Grade 3 (National Governors Association Center for Best Practices & Council of Chief State School Officers, 2010). Nevertheless, there is a fair amount of the research on children's understanding and performance on division problems occurring before they are taught division at school. This research uses illustrations (e.g., Kornilaki & Nuñes, 2005; Matalliotaki, 2012; Squire & Bryant, 2002b, 2003a) and concrete materials (Correa, Nuñes, & Bryant, 1998; Frydman & Bryant, 1988; Squire & Bryant 2002a), aimed at examining topics like children's understanding of the inverse quotient–divisor relationship (e.g., Correa et al., 1998; Squire & Bryant, 2003b), discrimination of the dividend, divisor, and quotient (e.g., Squire & Bryant 2002b, 2003a), and sharing discrete and continuous quantities (e.g., Kornilaki & Nuñes, 2005; Sophian, Garyantes, & Chang, 1997). Although this research is not as extensive as the studies on proportion, and not as contradictory, there are many features of early division understanding that have the potential to inform our understanding of early fraction knowledge.

Division as Sharing

One of the predominant themes in the research on early division concepts is how it is understood through the concept of sharing, in particular, through a sharing action schema or model of the action (Correa et al., 1998; Fischbein, Deri, Nello, & Marino, 1985; Kornilaki & Nuñes, 2005; Matalliotaki, 2012). The activity of sharing occurs almost daily in the life of a child, and the skill is quite relevant and beneficial to their development. In fact, the concept of sharing is strongly engrained in humans as we have evolved a psychological structure that deals with sharing which has been shaped by our ancestors who engaged in food sharing and resource allocation in order to survive (Cormas, 2014). Naturally, young children's ability to share becomes the basis for their first ideas about division (Bryant & Nuñes, 2002), and because of this notion, division seems a very obvious topic to study in children. However, to determine if young children do have an understanding of division, researchers must examine children's understanding of the principles of division, and that includes the relations between the dividend, divisor, and quotient (e.g., Correa et al., 1998; Kornilaki & Nuñes, 2005; Kouba, 1989; Squire & Bryant, 2002a).

Many of the studies examining children's (e.g., 5- to 8-year-olds) preformal understanding of division compare partitive division tasks to quotitive division tasks. In partitive division tasks, the total number and the number of sets are known, and the solution involves finding the number in each set (e.g., eight children, four tables; how many at each table?). In a quotitive division task, the total number and the number in each set are known, and the solution involves finding the number of sets (e.g., eight children, four at each table; how many tables?). Fischbein and colleagues (1985) liken partitive division tasks to sharing and quotitive division tasks to repeated subtraction (as cited in Matallikotaki, 2012). Fischbein and colleagues (1985) further argue that partitive division is based on action schemas of sharing, but quotitive division must be acquired later through instruction.

If it is true that partitive tasks are solved with sharing, and sharing is something that children easily understand, while quotitive tasks are learned later, then young children should show better performance on partitive questions than quotitive questions. In most studies that use partitive and quotitive division problems, children—at all ages—perform better in the partitive division tasks (e.g., Correa et al., 1998; Matallikotaki, 2012; Squire & Bryant, 2002a, 2002b, 2003b). There also tends to be a developmental jump around 7 years of age, where performance begins to differ between older children and younger children (e.g., 5-year-olds), especially in the quotitive division tasks. Some of these studies, however, have also explored how different ways of pregrouping the items to be divided can affect performance, and what this might say about the nature of the sharing schema. Squire and Bryant (2002a) investigated whether grouping stimuli by the quotient may help students perform better in quotitive division tasks while grouping stimuli by the divisor would work better in partitive tasks. For example, take a partitive version of $6 \div 3$ where 6 children are to sit at 3 tables, and the question is how many children will sit at each table if the same number is seated at each (see Fig. 2). When the children are grouped by divisor, then the number of groups (not the size of the group) is determined by the divisor, so the picture of the problem would group the children together into 3 groups of 2. When children are grouped by quotient, then the number of groups is determined by the quotient (i.e., 2), so the picture of the problem would show 2 groups of 3. A quotitive version of this problem would say that there are 6 children who sit 3 to a table, so how many tables would they need; this problem could also be illustrated with either grouping procedure (see Fig. 2; also see Figs. 2 and 3 in Squire & Bryant, 2002a, e.g., with $12 \div 4$). Children only saw a partitive or quotitive division task and were asked eight questions (four grouping-by-divisor and four grouping-by-quotient). As predicted, they found that performance was better in the partitive division task when the portions were grouped by the divisor and in the quotitive division task when the portions were grouped by the quotient. Looking at only partitive tasks in a different set of three studies, Squire and Bryant (2002b) consistently found an advantage for grouping-by-divisor.

These results suggest partitive and quotitive tasks are easier in grouping-by-divisor and grouping-by-quotient conditions, respectively, because the items are grouped in a way that facilitates a sharing action sequence. Carpenter, Fennema, and Franke (1996) suggest the two types of division problems elicit different strategies to reflect the different actions being described in the problem text. In the example above, having 3 groups of 2 in the partitive task is more easily understood as each table's

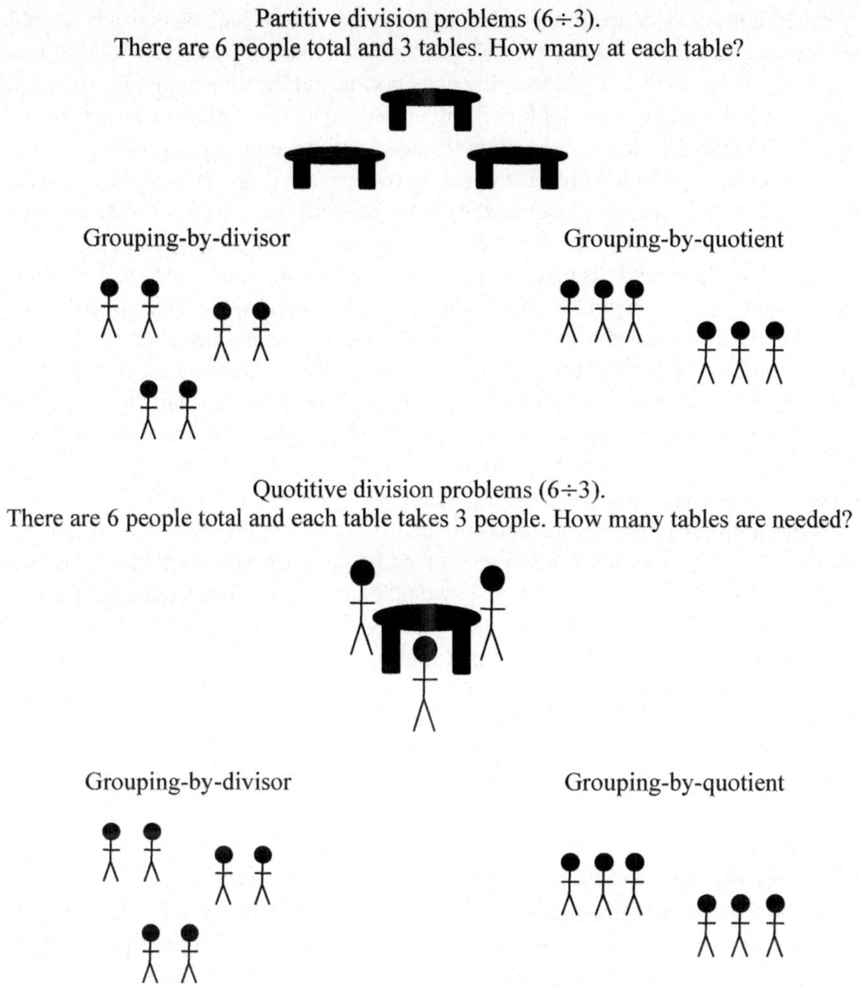

Fig. 2 An illustration of the difference between grouping-by-divisor and grouping-by-quotient for partitive and quotitive division tasks

share, so the size of the group is the answer. If there were 2 groups of 3, children defaulting toward a sharing scheme would have to share out the members of the 3 groups, one group at a time, until each table had 2 at it, and this would be more onerous and error-prone. In the quotitive version of the task, when there are 2 groups of 3 children, the table share is already set as 3, so it is easier to simply count the number of groups. When there are 3 groups of 2, then solving the quotitive problem would involve either regrouping into sets of three or taking one person from each group to make a table of three and realizing you can do that two times, which again would be more error-prone.

Although the evidence reviewed above suggests that 4- and 5-year-olds are able to share out quantities in a one-for-me, one-for-you manner (Desforges & Desforges, 1980; Frydman & Bryant, 1988), this does not mean that they apply all aspects of sharing to their understanding of division. For example, they still have difficulty with understanding equivalence in cardinal values. Frydman and Bryant (1988) asked 4- and 5-year-olds to divide a group of sweets between some dolls. Most children did this using a sharing action schema. Once the children had allotted all the sweets, they were asked how many sweets one of the dolls had. Importantly, they then were asked how many sweets the other dolls had. Not one of the 4-year-olds spontaneously responded and instead began to count the items in front of the other dolls. This suggests they do not understand that shares should be equal and should not need recounting. The experimenter covered the blocks and then asked the children again if they knew how many sweets the doll had without counting, and not even half of the 4-year-old children (10/24) responded correctly. However, Frydman and Bryant, as well as Correa and colleagues (1998), found that this task can be performed successfully by 5-year-olds, which suggests a developmental shift at around this time.

In addition to examining the predominance of the sharing schema in early fraction understanding, many of these same studies also investigated other key aspects of understanding division. One of these key understandings, a core feature of fractions as well as division, is the inverse nature of division—as the divisor increases, the quotient decreases. Younger children still struggle with this concept regardless of how well they can share.

Relations Between Dividend, Divisor, and Quotient

Correa and colleagues (1998) investigated the extent to which young children understand the relation between the three terms that make up a division problem, namely the dividend, divisor, and quotient in a partitive (Study 1) or quotitive (Study 2) division task. This study also looked for evidence of the inverse divisor–quotient relationship. Children were asked to give treats to a group of rabbits under two conditions, the same- and different-divisor conditions, while the dividend remained the same. In the same-divisor condition, there were an equal number of blue and pink rabbits (e.g., two blue rabbits and two pink rabbits); in the different-divisor condition, there were a different number of blue and pink rabbits. In Study 1, the children were asked to determine if the blue rabbits and pink rabbits would receive the same amount of treats, but, in Study 2, the children were asked to determine if the same number of blue and pink rabbits could be invited to the picnic if each rabbit were to receive a specific amount of carrots. They found that understanding of the inverse divisor–quotient relationship was present in about half of the 6-year-olds and only a third of 5-year-olds. In other words, the younger children demonstrated difficulty in reasoning about and comparing the outcome of dividing the same number among different divisors (e.g., sharing 12 carrots among 3 rabbits will result in more carrots per rabbit than when sharing 12 carrots among 4 rabbits). Instead, a common mistake these children made was to erroneously apply a "more is more" rule, thinking the more rabbits there are in a group, the more sweets are needed, and the more sweets means the more the rabbits will get. Older children were also highly susceptible to the more is more rule.

Squire and Bryant (2003a) also examined if 5- to 8-year-old children understood the inverse relation between divisor and quotient. To do so, the authors used stimuli similar to those used by Correa et al. (1998); there were always the same number of carrots for the rabbits, but the number of rabbits varied. The experimenter pointed to one red and one blue rabbit and asked the children if the rabbits would get the same amount of carrots. They found that the children performed better on the same-divisor condition than in the different-divisor condition, and that there were no differences in the different-divisor condition depending on the size of the difference between divisors (small vs. large). As was the case for Correa and colleagues (1998), errors in the different-divisor condition demonstrated that 5-year-olds do not understand the inverse nature of the divisor–quotient relation. At the same time, these studies also demonstrate that most 7- and 8-year-olds do have this understanding, and they have this understanding even before they are taught about division in school.

Discrete Versus Continuous Tasks

Similar to the studies described in the proportional reasoning section, research examining children's understanding of division also differentiates between continuous and discrete quantities. When it comes to proportional reasoning tasks, children tend to perform better when quantities are presented as continuous (e.g., Jeong et al., 2007; Boyer et al., 2008); however, this is not the case when it comes to children's performance in division problems. When division questions are framed as a sharing situation, children historically have more difficulty sharing continuous quantities compared to discrete quantities, usually because children do not know how to properly subdivide the continuous pieces (Kornilaki & Nuñes, 2005). In tasks that examine the relations between the dividend, the divisor, and the quotient, however, the results are a little different.

Kornilaki and Nuñes (2005) used discrete (e.g., small fish) and continuous stimuli (e.g., fish cakes) when examining young children's (5- to 7-year-olds) understanding of division. In the discrete condition, children were presented with two groups of cats, brown and white cats, that were eating fish. There were 12 or 24 fish in a pile for each group of cats. In the continuous condition, the children were told the cats were eating fish cakes (1, 2, or 3). The children were told that the cats were going to share the fish fairly. There were also same- and different-divisor conditions (i.e., either there were the same amount of brown and white cats or there were a different amount of brown and white cats). The experimenter pointed to one brown and one white cat and asked the children if the cats would receive the same amount of fish.

There was no difference in children's performance on the continuous questions versus the discrete questions (Kornilaki & Nuñes, 2005). However, children did have a more difficult time offering justifications for their answers for the continuous questions. This general pattern held true in both partitive and quotitive division tasks. All of the children who performed well on the different-divisor condition also performed well on the same-divisor condition, but the reverse was not always true. The authors suggest children's understanding of the inverse relation between the divisor and the quotient develops later than their understanding of equivalence. They also suggest that children are able to generalize their intuitive understanding of division of discrete items to inform them in their solution process using continuous quantities.

A challenge associated with sharing continuous quantities is that it does not easily facilitate the child's intuitive understanding of sharing, that is, one-for-me and one-for-you. Frydman and Bryant (1988) asked 4- and 5-year-olds to share sweets (12 or 24 blocks) among dolls (2, 3, and 4) so that everyone gets the same amount. However, sometimes the blocks were doubled up (e.g., attached) so that two single blocks were equivalent to a one double block (2:1). To succeed on this task, children would have to consider that giving one doll a double block meant that another doll would have to receive two single blocks to ensure equal sharing. The 4-year-olds ignored the double (and triple) blocks and distributed them among the dolls in a one-to-one fashion, but the 5-year-olds performed significantly better on this task than the 4-year-olds. However, Frydman and Bryant (1988) also trained a separate group of 4-year-olds on how to approach the same task using color cues. During the intervention, children were shown how to group, for example, three single blue blocks as being equivalent to a yellow triple block. Children in the intervention group performed significantly better in the posttest than those in the control group. This is especially important because, during the posttest, the color cues were removed. The 4-year-olds only required minimal exposure showing how the task should be performed, and they were able to generalize that information to a task when color cues were not available to them.

Overall, the evidence supports the idea that children have a sense of the components involved in division long before they receive any formal instruction. While young children do have a sense of division, their early understanding is very much based on their intuitive sense of sharing, more specifically, their sense of sharing on a one-to-one basis. Partitive division tasks are much easier for young children because it corresponds to their natural desire to share by individual rather than by groups of items. It had also been argued that quotitive division problems are more challenging for the same reason; the one-to-one distribution does not equate to the quotient in quotitive division tasks. Despite early understanding of division concepts and the one-to-one allocation method, however, children still struggle with equivalent cardinal items. For example, preschoolers who see a set of objects divided in half, and who are provided with the number of objects in one of the halves, are unable to infer that the other half has the same number of objects (Frydman & Bryant, 1988). However, there are promising signs that young children already have the intuitive knowledge in place to be successful in division and that, prior to any type of formal education, they are demonstrating strategies that are used to instruct formal knowledge of division in the school systems (see Kouba, 1989; Mulligan & Mitchelmore, 1997; Mulligan & Watson, 1998).

Educational Implications for Fractions

In this section, we consider the implications of the research on early understanding of proportion and division on the early understanding of fractions. In doing so, we qualify these suggestions by reconciling these findings with the literature on early understanding of fractions.

Discrete Versus Continuous Stimuli

In the research reviewed above, children's understanding of proportion was found to be facilitated by the use of discrete stimuli compared to continuous stimuli. Studies on the early understanding of fractions also examined this distinction, but these studies do not consistently favor continuous stimuli. Some conclude that discrete stimuli are easier for children in fraction tasks, and there is a debate within the math education field about which stimuli teachers should use for teaching fractions (see Wilkerson et al., 2015). Hunting and Sharpley (1998) investigated 4-year-olds' understanding of a number of different fraction questions and found that children performed the best on the discrete task where they had to share 12 crackers between 3 dolls. Mix et al. (1999) demonstrated that 4- and 5-year-olds could observe a transformation, hidden behind a screen, of separate pieces (e.g., quarters of a circle) being added or subtracted and then match them with a continuous picture of the resulting sum or difference. Wilkerson and colleagues (2015) followed kindergarten and Grade 3 children over a series of lessons that switched between models using discrete and continuous stimuli. The observational evidence suggested that children (and teachers) were more comfortable working with discrete models of fractions. Consistent with this, kindergarten children improved the most on discrete items over the course of the lessons. Grade 3 children, however, improved the most on continuous items. Nevertheless, Wilkerson and colleagues (2015) suggest the bulk of the evidence favors discrete items, largely because children find them easier to share.

It may be that continuous stimuli may best support the understanding of the proportional aspect of fractions (Empson, 1999) while discrete stimuli support the division aspect of fractions (Kornilaki & Nuñes, 2005; Squire & Bryant, 2002a, 2002b). At the same time, young children are still able to succeed on tasks using continuous stimuli (Squire & Bryant, 2002a). The division literature (Kornilaki & Nuñes, 2005) suggests the main difficulty with sharing continuous items is that they are difficult to know how to partition. This parallels findings in the fraction literature as well. In Hunting and Sharpley's (1998) tasks, one of the main difficulties that 3- and 4-year-olds had in dividing materials was to actually make the pieces equal in size. When asked to cut a blanket to share between two dolls, many kids could roughly assess where the halfway point was, but none of them did any kind of checking to see if the pieces were the same size. When asked to divide a sausage or a string between many dolls, the pieces were most often of varying sizes, and quite often there would be an extra piece left over. Given these difficulties, it seems plausible that interventions could be designed that focus on how to subdivide wholes into equal pieces. This extra bit of expertise would build on the intuitions that children already appear to have and may help to improve their fraction understanding.

Instructing children on how to divide into equal pieces can build on the findings that suggest that the notion of a half is special. Spinillo and Bryant (1991; 1999) demonstrated that 6- and 7-year-olds are sensitive to the notion of a half; at least, they were able to distinguish two proportions when they fell on either side of the half boundary but not when they were on the same side of that boundary. Consistent with this idea, Barth, Baron, Spelke, and Carey (2009) demonstrated that 5- and 6-year-olds could mentally halve a component, although Barth and colleagues did not test to see if children were better at halving compared to some other reduction proportions. The notion that half is special is consistent with other research in early fraction understanding. Hunting and Sharpley (1998) tested 3- and 4-year-olds about their conceptions of 1/2, 1/3, and 1/4. To do this, they gave the children scenarios that asked them, for example, to cut a piece of cloth so that it can be shared between two dolls fairly. About a third of these young children could do this task, but almost none of them could do it when they were dividing the cloth between three dolls (although they were not asked to complete this particular task with four dolls). This supports the idea that the notion of half is a privileged reference point, and it has been incorporated through the Common Core Standards for Mathematics. For example, the report mentions that during the teaching of concepts such as measurement and data and geometry, there is a focus on half as being a unique attribute (National Governors Association Center for Best Practices & Council of Chief State School Officers, 2010).

Although fraction learning may be easier with discrete stimuli, Empson (1999) has demonstrated that, with adequate instruction, even first-grade children can succeed on fraction tasks that make use of continuous stimuli. Furthermore, this study demonstrates a way in which fractions other than half can be taught. First-grade students were given a 5-week, 15-lesson instructional intervention that was collectively developed by Empson and the students' teachers. Children were quite skilled at halving and repeated halving to solve fraction problems, even before instruction. However, the training covered ways to build on this intuition, using equal sharing, to partition wholes into pieces other than half. Roughly, half of the first-grade students were able to take their knowledge of equal sharing and apply it to novel fraction problems. This approach not only supports the half as a privileged reference point, it also supports using the notion of sharing to facilitate fraction learning, which we turn to next.

Sharing

One predominant theme in the research on division is that children's early understanding of division is rooted in the notion of sharing. It seems logical, then, that one considers ways in which fractions could be understood as sharing. At one level, this is fairly straightforward, because, in many ways, a fraction is just a division operation. It is the numerator divided by the denominator, and the sharing situation turns that into the number of objects to be shared (the numerator) divided by the number of people to share them between (the denominator). What separates the use of this metaphor in division from its use in fractions, however, is that division problems are usually presented as discrete wholes that do not need to be subdivided to equally distribute them. When in a situation where the wholes cannot be equally distributed, there is a remainder, and as such, there is the potential to delve into the world of fractions. As it turns out, young children are hesitant when dealing with a remainder. When faced with a remainder, children will often ignore it or incorporate it back into their whole number answer (e.g., Guerrero & Rivera, 2001; Lautert & Spinillo, 2004, 2005). Fortunately, there is evidence demonstrating the benefit of intervention training to help children properly incorporate the remainder (Carpenter, Ansell, Franke, Fennema, & Weisbeck, 1993; Guerrero & Rivera, 2001; Lautert, Spinillo, & Correa, 2012; Spinillo & Lautert, 2006). More than simply recognizing the remainder for what it is, however, these situations allow for exploring the notion of fractions by subdividing the remaining pieces to then be shared. This approach is essentially what is used by Streefland (1991, 1997) in his realistic program for teaching fractions. Streefland has devised fraction learning situations through sharing scenarios (often sharing pizzas) that inevitably lead to subdividing the pizza. The research on children's use of sharing to understand division would suggest there is merit to this approach.

Understanding the Relational Aspects of Fractions

Another key aspect to early understanding of division is children's intuitive notions of the relation between the dividend, the divisor, and the quotient. Essentially, this amounts to knowing that the quotient (answer/quantity) will increase as the dividend (numerator) increases, but it will decrease as the divisor (denominator) increases. These exact relations also exist in the understanding of fractions and of proportions. However, even though these notions are the same in both division understanding and proportional reasoning, they each can be framed using fraction subconstructs as either the part–whole situation (proportional reasoning) or a quotient situation (division). Furthermore, research with fractions demonstrates that this can make a difference in performance. Grade 1 students performed better on fraction items (e.g., naming, ordering, and indicating equivalence) when they were presented in a quotient situation than they did in a part–whole situation (Mamede, Nuñes, & Bryant, 2005).

This parallels the findings described above that division problems are easiest for young children when they are presented as partitive tasks (Correa et al., 1998; Squire & Bryant, 2002a, 2002b, 2003a, 2003b).

In sum, these findings suggest that continuous stimuli are used when focusing on the part–whole aspect of fractions and that discrete stimuli are used when in quotient situations. If using discrete stimuli, however, care should be taken not to structure them in a way that encourages counting, and there should be a focus on how to subdivide any remaining pieces. Children also have an intuitive aspect of the relational aspects of fractions, so instruction could also serve to reinforce the parallels between proportion, division, and fractions.

Although the research reviewed here focuses on an early understanding of fractions, what children cannot do is suggestive of what might separate early conceptions of fractions with more mature understandings. The findings from Acredolo and colleagues (1989), for example, suggest that even though young children may understand how proportion changes as numerators and denominators change, they are not able to match them with exact probabilities. Recent research with fraction understanding has found, however, that understanding the quantity that fractions represent is a key predictor of later success on fraction ability (Hamdan & Gunderson, 2017; Siegler, Fazio, Bailey, & Zhou, 2013). The difficulty in properly connecting proportions to quantity may represent a key aspect of fraction understanding that separates early understanding of fractions to a more mature understanding of them.

Building on Children's Intuitions

In the pages above, we have reviewed research findings about children's early conceptions of proportional reasoning and division. Our goal was to identify the informal knowledge that children intuitively knew to support fraction learning. *Cognitively Guided Instruction* (CGI; Carpenter, Fennema, & Franke, 1996; Fennema et al., 1996) is a non-traditional primary school mathematics program developed by researchers that also seeks to identify children's early intuitions about math. More specifically, CGI does not directly instruct teachers on what these intuitions are, but instead trains teachers to listen to children, identify the intuitions themselves, and then build on them.

Analyses of children's solution processes indicate that students model the structure of a problem, first with diagrams or manipulatives and later relying on more efficient counting and fact-derived strategies. Some of the intuitions described by these researchers match quite well with those described above. For example, division problems were often spontaneously solved with an equal sharing strategy, and children would dynamically switch between a grouping-by-divisor and a grouping-by-dividend strategy depending on whether it was a partitive or quotitive division problem (Carpenter et al., 1996). More importantly, CGI researchers have demonstrated the benefits of building on these intuitions. In a multi-year longitudinal study, students in classrooms of teachers trained in CGI showed an improvement in performance and understanding of various mathematical concepts. Those students who received instruction from CGI teachers over more years showed increasingly greater gains. The gains experienced by students in CGI classrooms also had a lasting effect, as non-CGI teachers in later grades reported students coming from CGI classrooms were able to solve more problems, were more enthusiastic about math problems, and were more eager and willing to talk about their problem-solving solutions (Fennema et al., 1996).

These findings reaffirm that children possess intuitive knowledge of mathematical concepts prior to any formal instruction (Kornilaki & Nuñes, 2005; Kouba, 1989; Mulligan & Mitchelmore, 1997; Mulligan & Watson, 1998; Piaget & Inhelder, 1975). If students are entering the classroom already equipped with some of the necessary tools, educational instruction would do well to capitalize on children's existing knowledge. The CGI studies clearly indicate that teacher's and student's mathematical knowledge alike can benefit from intuitive understanding.

Conclusions

This chapter has covered an extensive body of the literature demonstrating the link between early understanding in proportional reasoning and division to the development of fraction understanding. Despite the early findings in the proportional reasoning literature, a reconciled view of this research supports the notion that young children, prior to any type of formal education, have the knowledge to succeed on proportional tasks and these skills can support the development of the early understanding of fractions. The same can be said in regard to the division research: Young children are showing early signs of understanding the relation between the components involved in division, and these concepts are also related to understanding fractions (e.g., part–part and part–whole relations, partitive and quotitive division tasks). If one places the lessons of early proportional reasoning and early division understanding together, recommendations emerge about how to best facilitate early learning of fractions. Building on children's intuitive understanding of proportions and division is a logical place to start (see Empson, 1999). We also have to be mindful of the proper use of discrete and continuous stimuli, as they can help support different aspects of fraction understanding. In particular, proportional reasoning and division understanding can come together when fraction scenarios require children to subdivide pieces, as proportional reasoning (even earlier occurring part–part reasoning) can help children learn to make the pieces equal in size. Many of these recommendations are not new (see Carpenter et al., 1996; Streefland, 1991, 1997), but deriving them from both an understanding of proportion and an understanding of division has the potential to support many different aspects of fraction learning. To the extent that this chapter helps to achieve this end, we can only hope that we have done our part.

References

Acredolo, C., O'Connor, J., Banks, L., & Horobin, K. (1989). Children's ability to make probability estimates: Skills revealed through application of Anderson's Functional Measurement Methodology. *Child Development, 60,* 933–945.

Ahl, V. A., Moore, C. F., & Dixon, J. A. (1992). Development of intuitive and numerical proportional reasoning. *Cognitive Development, 7,* 81–108.

Barth, H., Baron, A., Spelke, E., & Carey, S. (2009). Children's multiplicative transformations of discrete and continuous quantities. *Journal of Experimental Child Psychology, 103,* 441–454. https://doi.org/10.1016/j.jecp.2009.01.014.

Behr, M. J., Harel, G., Post, T. R., & Lesh, R. (1992). Rational number, ratio, and proportion. In D. A. Grouws (Ed.), *Handbook of research on mathematics teaching and learning* (pp. 296–333). New York: Macmillan.

Boyer, T. W., & Levine, S. C. (2012). Child proportional scaling: Is 1/3 = 2/6 = 3/9 = 4/12? *Journal of Experimental Child Psychology, 111,* 516–533. https://doi.org/10.1016/j.jcep.2011.11.001.

Boyer, T. W., Levine, S. C., & Huttenlocher, J. (2008). Development of proportional reasoning: Where young children go wrong. *Developmental Psychology, 44,* 1478–1490. https://doi.org/10.1037/a0013110.

Bryant, P., & Nuñes, T. (2002). Children's understanding of mathematics. In U. Goswami (Ed.), *Blackwell handbook of childhood cognitive development* (pp. 412–439). Malden: Blackwell Publishing.

Carpenter, T., Ansell, E., Franke, M., Fennema, E., & Weisbeck, L. (1993). Models of problem solving: A Study of kindergarten children's problem-solving processes. *Journal for Research in Mathematics Education, 24*(5), 428–441. https://doi.org/10.2307/749152.

Carpenter, T. P., Fennema, E., & Franke, M. L. (1996). Cognitively guided instruction: A knowledge base for reform in primary mathematics instruction. *The Elementary School Journal, 97,* 3–20.

Chapman, R. H. (1975). The development of children's understanding of proportions. *Child Development, 46,* 141–148.

Charalambous, C. Y., & Pitta-Pantzai, D. (2007). Drawing on a theorectical model to study students' understandings of fraction. *Educational Studies in Mathematics, 64,* 293–316. https://doi.org/10.1007/s10649-006-9036-2.

Cormas, P. C. (2014). An evolved psychological structure for dealing, food sharing, and mathematical division. *Psihologijske teme, 23,* 243–263.

Correa, J., Nuñes, T., & Bryant, P. (1998). Young children's understanding of division: The relationship between division terms in a noncomputational task. *Journal of Educational Psychology, 90,* 321–329.

Davies, C. M. (1965). Development of the probability concept in children. *Child Development, 36,* 779–788.

Denison, S., & Xu, F. (2010). Twelve- to 14-month-old infants can predict single-event probability with large set sizes. *Developmental Science, 13,* 798–803. https://doi.org/10.1111/j.1467-7687.2009.00943.x.

Denison, S., & Xu, F. (2014). The origins of probabilistic inference in human infants. *Cognition, 130,* 335–347. https://doi.org/10.1016/j.cognition.2013.12.001.

Desforges, A., & Desforges, C. (1980). Number-based strategies of sharing in young children. *Educational Studies, 6,* 97–109. https://doi.org/10.1080/0305569800060201.

Empson, S. B. (1999). Equal sharing and shared meaning: The development of fraction concepts in a first-grade classroom. *Cognition and Instruction, 17,* 283–342.

Fennema, E., Carpenter, T. P., Franke, M. L., Levi, L., Jacobs, V. R., & Empson, S. B. (1996). A longitudinal study of learning to use children's thinking in mathematics instruction. *Journal for Research in Mathematics Education, 27,* 403–434.

Fischbein, E., Deri, M., Nello, M. S., & Marino, M. S. (1985). The role of implicit models in solving verbal problems in multiplication and division. *Journal for Research in Mathematics Education, 16,* 3–17.

Frydman, O., & Bryant, P. (1988). Sharing and the understanding of the number equivalence by young children. *Cognitive Development, 3,* 323–329.

Girotto, V., Fontanari, L., Gonzales, M., Vallortigara, G., & Blaye, A. (2016). Young children do not succeed in choice tasks that imply evaluating chances. *Cognition, 152,* 32–39. https://doi.org/10.1016/j.cognition.2016.03.010.

Goldberg, S. (1966). Probability judgments by preschool children: Task conditions and performance. *Child Development, 37,* 151–168.

Goswami, U. (1989). Relational complexity and the development of analogical reasoning. *Cognitive Development, 96,* 229–241.

Guerrero, L., & Rivera, A. (2001). Does the acquisition of mathematical knowledge make students better problem solvers? An examination of third-graders' solution of division-with-remainder (DWR) problems. In *Proceedings of the Annual Meeting of the North American Chapter of the International Group for the Psychology of Mathematics Education*. Snowbird, Utah.

Hamdan, N., & Gunderson, E. A. (2017). The number line is a critical spatial-numerical representation: Evidence from a fraction intervention. *Developmental Psychology, 53*, 587–596. https://doi.org/10.10137/dev0000252supp.

Hecht, S. A., Vagi, K. J., & Torgesen, J. K. (2007). Fraction skills and proportional reasoning. In D. B. Berch & M. M. Mazzocco (Eds.), *Why is math so hard for some children? The nature and origins of mathematical learning difficulties and disabilities* (pp. 121–132). Baltimore, MD, US: Paul H Brookes Publishing.

Hunting, R. P., & Sharpley, C. F. (1998). Fraction knowledge in preschool children. *Journal for Research in Mathematics Education, 19,* 175–180.

Inhelder, B., & Piaget, J. (1958). *The growth of logical thinking from childhood to adolescence.* New York: Basic Books.

Jeong, Y., Levine, S. C., & Huttenlocher, J. (2007). The development of proportional reasoning: Effect of continuous versus discrete quantities. *Journal of Cognition and Development, 8,* 237–256.

Kerslake, D. (1986). *Fractions: Children's strategies and errors: A report of the strategies and error in secondary mathematics project.* Winsdon, England: NFER-Nelson.

Kornilaki, E., & Nuñes, T. (2005). Generalising principles in spite of procedural differences: Children's understanding of division. *Cognitive Development, 20,* 388–406. https://doi.org/10.1016/j.cogdev.2005.05.004.

Kouba, V. L. (1989). Children's solution strategies for equivalent set multiplication and division word problems. *Journal for Research in Mathematics Education, 20,* 147–158.

Lautert, S. L. & Spinillo, A. G. (2004). Inverse relations between division terms: A difficulty children are able to overcome. In M. J. Hoines & A. B. Fuglestad (Eds.). *Proceedings of the 28th Conference of the International Group for the Psychology of Mathematics Education,* (Vol. 1, p. 397). Bergen, Norway: PME.

Lautert, S. L. & Spinillo, A. G. (2005). What's wrong with this solution procedure? Asking children to identify incorrect solutions in division-with-remainder (DWR) problems. In. H. L. Chick & J. L. Vincent (Eds.). *Proceedings of the 29th Conference of the International Group for the Psychology of Mathematics Education,* (Vol. 1, pp. 320). Melbourne; PME.

Lautert, S. L., Spinillo, A. G., & Correa, J. (2012). Children's difficulties with division: An intervention study. *Educational Research, 3,* 447–456.

Lesh, R., Post, T., & Behr, M. (1988). Proportional reasoning. In J. Hiebert & M. Behr (Eds.), *Number concepts and operations in the middle grades* (pp. 93–118). Reston, VA: National Council of teachers of Mathematics.

Lobato, J., Hawley Orrill, C., Druken, B., & Jacobson, E. (2011, April). *Middle school teachers' knowledge of proportional reasoning for teaching.* Paper presented at the Annual Meeting of the American Educational Research Association. New Orleans, US.

Mack, N. (1995). Confounding whole-number and fraction concepts when building on informal knowledge. *Journal for Research in Mathematics Education, 26,* 422–441.

Mamede, E., Nuñes, T., & Bryant, P. (2005). The equivalence and ordering of fractions in part-whole and quotient situations. In H. L. Chick & J. L. Vincent (Eds.). *Proceedings of the 29th Conference of the International Group for the Psychology of Mathematics Education* (Vol. 3, pp. 281–288). Melbourne; PME.

Matalliotaki, E. (2012). Resolution of division problems by young children: What are children capable of and under what conditions? *European Early Childhood Education Research Journal, 20,* 283 299. https://doi.org/10.1080/1350293x.2012.681132.

McCrink, K., & Wynn, K. (2007). Ratio abstraction by 6-month-old infants. *Psychological Science, 18,* 740–745.

Mix, K., Huttenlocher, J., & Levine, S. (2002). Multiple cues for quantification in infancy: Is number one of them? *Psychological Bulletin, 128,* 278–294. https://doi.org/10.1037//0033-2909.128.2.278.

Mix, K. S., Levine, S. C., & Huttenlocher, J. (1999). Early fraction calculation ability. *Developmental Psychology, 35,* 164–174.

Möhring, W., Newcombe, N. S., & Frick, A. (2015). The relation between spatial thinking and proportional reasoning in preschoolers. *Journal of Experimental Child Psychology, 132,* 213–220. https://doi.org/10.1016/j.jecp.2015.01.005.

Moore, C. F., Dixon, J. A., & Haines, B. A. (1991). Components of understanding in proportional reasoning: A fuzzy set representation of developmental progressions. *Child Development, 62,* 441–459.

Mulligan, J. T., & Mitchelmore, M. C. (1997). Young children's intuitive models of multiplication and division. *Journal for Research in Mathematics Education, 28,* 309–330.

Mulligan, J., & Watson, J. (1998). A developmental multimodal model for multiplication and division. *Mathematics Education Research Journal, 10,* 61–86.

National Governors Association Center for Best Practices, & Council of Chief State School Officers. (2010). *Common core state standards.* Washington, DC: Authors. www.corestandards.org.

National Mathematics Advisory Panel. (2008). *Foundations for Success: The Final Report of the National Mathematics Advisory Panel,* U.S. Department of Education: Washington.

Ng, R., Heyman, G. D., & Barner, D. (2011). Collaboration promotes proportional reasoning about resource distribution in young children. *Developmental Psychology, 47,* 1230–1238. 10.103/a0024923.

Ni, Y., & Zhou, Y.-D. (2005). Teaching and learning fraction and rational numbers: The origins and implications of whole number bias. *Educational Psychologist, 40,* 27–52.

Noelting, G. (1980a). The development of proportional reasoning and the ratio concept: Part I—Differentiation of stages. *Educational Studies in Mathematics, 11,* 217–253.

Noelting, G. (1980b). The development of proportional reasoning and the ratio concept: Part II Problem-structure at successive stage; problem-solving strategies and the mechanism of adaptive restructuring. *Educational Studies in Mathematics, 11,* 331–363.

Nuñes, T., Desli, D., & Bell, D. (2003). The development of children's understanding of intensive quantities. *International Journal of Educational Research, 39,* 651–675. https://doi.org/10.1016/j.ijer.2004.10.002.

Offenback, S. I., Gruen, G. E., & Caskey, B. J. (1984). Development of proportional response strategies. *Child Development, 55,* 963–972.

Piaget, J., & Inhelder, B. (1975). The origin of the idea of chance in children. New York: Norton [Original work published 1951].

Siegler, R. S., Fazio, L. K., Bailey, D. H., & Zhou, X. (2013). Fractions: The new frontier for theories of numerical development. *Trends in Cognitive Science, 17,* 13–19. https://doi.org/10.1016/j.tics.2012.11.004.

Siegler, R. S., & Vago, S. (1978). The development of proportionality concept: Judging relative fullness. *Journal of Experimental Child Psychology, 25,* 371–395.

Singer-Freeman, K. E., & Goswami, U. (2001). Does half a pizza equal half a box of chocolates? Proportional matching in an analogy task. *Cognitive Development, 16,* 811–829.

Sophian, C., Garyantes, D., & Chang, C. (1997). When three is less than two: Early developments in children's understanding of fractional quantities. *Developmental Psychology, 33*, 731–744. https://doi.org/10.1037//0012-1649.33.5.731.

Sophian, C., & Wood, A. (1997). Proportional reasoning in young children: The parts and the whole of it. *Journal of Educational Psychology, 89,* 309–317.

Spinillo, A. G., & Bryant, P. (1991). Children's proportional judgements: The importance of "half". *Child Development, 62,* 427–440.

Spinillo, A. G., & Bryant, P. E. (1999). Proportional reasoning in young children: Part-part comparisons about continuous and discrete quantities. *Mathematical Cognition, 5,* 181–197.

Spinillo, A. G., & Lautert, S. L. (2006). Exploring the role played by the remainder in the solution of division problems. In J. Novotna, H. Moravova, M. Kratka, & N. Stehlikova (Eds.), *Proceedings 30th Conference of the International Group for the Psychology of Mathematics Education* (Vol. 5, pp. 153–160). Prague: PME.

Squire, S., & Bryant, P. (2002a). From sharing to dividing: Young children's understanding of division. *Developmental Science, 5,* 452–466.

Squire, S., & Bryant, P. (2002b). The influence of sharing on children's initial concept of division. *Journal of Experimental Child Psychology, 81,* 1–43. https://doi.org/10.1006/jcep.2001.2640.

Squire, S., & Bryant, P. (2003a). Children's understanding and misunderstanding of the inverse relation in division. *British Journal of Developmental Psychology, 21,* 507–526.

Squire, S., & Bryant, P. (2003b). Children's models of division. *Cognitive Development, 18,* 355–376. https://doi.org/10.1016/s0885-2014(03)00039-x.

Streefland, L. (1991). *Fractions in realistic mathematics education: A paradigm of developmental research*. Dordrecht, The Netherlands: Kluwer Academic Publishers.

Streefland, L. (1997). Charming fractions or fractions being charmed? In T. Nunes & P. Bryant (Eds.), *Learning and teaching mathematics: An international perspective* (pp. 347–371). East Sussex: Psychology Press Ltd.

Wilkerson, T. L., Cooper, S., Gupta, D., Montgomery, M., Mechell, S., Arterbury, K., Moore, S., ... Sharp, P. T. (2015). An investigation of fraction models in early elementary grades: A mixed-methods approach. *Journal of Research in Childhood Education, 29*, 1–25. https://doi.org/10.1080/02568543.2014.945020.

Xu, F., & Denison, S. (2009). Statistical inference and sensitivity to sampling in 11-month-old infants. *Cognition, 112,* 97–104. https://doi.org/10.1016/j.cognition.2009.04.006.

Xu, F., & Garcia, V. (2008). Intuitive statistics by 8-month-old infants. *Proceedings of the National Academy of Sciences of the United States of America, 105,* 5012–5015. https://doi.org/10.1073/pnas.0704450105.

Yost, P. A., Siegel, A. E., & Andrews, J. M. (1962). Non-verbal probability judgments by young children. *Child Development, 33,* 769–780.

Index

K. M. Robinson et al. (eds.), *Mathematical Learning and Cognition in Early Childhood*, https://doi.org/10.1007/978-3-030-12895-1

Printed by Printforce, the Netherlands